Self-Incompatibility in Flowering Plants

Evolution, Diversity, and Mechanisms

Vernonica E. Franklin-Tong

Self-Incompatibility in Flowering Plants

Evolution, Diversity, and Mechanisms

Prof. Dr. Vernonica (Noni) E. Franklin-Tong
School of Biosciences
University of Birmingham
Edgbaston
Birmingham, B15 2TT, UK
v.e.franklin-tong@bham.ac.uk

Cover illustration: Primula which Charles Darwin recognized had dimorphic flowers (pin and thrum, as shown; photo Noni Franklin-Tong) and carried out crossing experiments, showing that they were self-incompatible.

ISBN 978-3-540-68485-5   e-ISBN 978-3-540-68486-2

DOI 10.1007/978-3-540-68486-2

Library of Congress Control Number: 2008927185

© 2008 Springer-Verlag Berlin Heidelberg

This work is subject to copyright. All rights are reserved, whether the whole or part of the material is concerned, specifically the rights of translation, reprinting, reuse of illustrations, recitation, broadcasting, reproduction on microfilm or in any other way, and storage in data banks. Duplication of this publication or parts thereof is permitted only under the provisions of the German Copyright Law of September 9, 1965, in its current version, and permission for use must always be obtained from Springer. Violations are liable to prosecution under the German Copyright Law.

The use of general descriptive names, registered names, trademarks, etc. in this publication does not imply, even in the absence of a specific statement, that such names are exempt from the relevant protective laws and regulations and therefore free for general use.

Cover design: WMX Design GmbH, Heidelberg

Printed on acid-free paper

9 8 7 6 5 4 3 2 1

springer.com

I dedicate this book to my parents, Vernon and Renata, who encouraged me in the quest for knowledge.

Preface

When I started work on poppy self-incompatibility (SI) in 1982 as a PhD student, de Nettancourt's "Incompatibility in Angiosperms" (1977) was the classic textbook for those in the field of pollen–pistil interactions and SI. It has remained so, with a second, revised edition published in 2001, providing all the basic background to the topic of SI, including a huge resource of the early literature, including the fundamentals of the basic features and genetical studies on SI carried out during the nineteenth century, which remain a valid and strong basis for understanding these systems, as well as recent advances. When I was approached regarding this project, "de Nettancourt" as everyone in the field refers to this book, was still on the lips of everyone, as there has really been no other monograph on SI since then. However, there was unanimous opinion that researchers needed another book on this topic and that the time was right. Thus, this book was born.

I have not attempted to reproduce the vast wealth of knowledge covered in "de Nettancourt," as it is still in print and the page limits imposed preclude this. Moreover, this book, being an edited volume, cannot deal with the topic in a similar manner, and because of the sheer volume of research, it has to be selective rather than all-inclusive. Here, I have attempted to cover the breadth of the field, giving readers an idea of the diversity of SI species and to cover our current understanding of how SI systems evolved. The book is written by internationally renowned scientists with a wide range of expertise. This multi-authorship allows the book to cover a huge breadth of topics with in-depth knowledge and great authority, which would be virtually impossible for a single-author book. This gives the book its unique breadth and flavour, encompassing many different, contrasting ways of examining this fascinating topic.

This book focuses on two major aspects where progress has been made in the last 25 years or so, since the advent of molecular biology as a tool. This period has seen an explosion of work in areas such as exploring evolutionary aspects using sequence information, identification of genes and the proteins encoding them, which are involved in specifying and mediating SI in a number of systems, and studies attempting to uncover mechanisms involved in these SI responses. Thus, here we focus on ecology, evolution, phylogeny, molecular genetics and cell biology, investigating the proposed mechanisms involved in regulating self-incompatibility. In doing so, I hope this book provides the reader with a good understanding of the diversity and complexity of these cell–cell recognition and rejection systems, as well as providing them with information on the great progress that has been made in our understanding of SI in flowering plants in the last few decades.

This book attempts to cover a broad spectrum of research into SI. For completeness, we have included both heteromorphic and homomorphic SI systems as they are not often discussed side-by-side. The book is divided into two sections. The first deals with "The Evolution and Population Genetics of SI." It begins with a chapter on comparative biology, ecology and the genetics of heterostyly, a topic which deserves a volume in its own right and which has not been reviewed for some time. (Later, in Part II, we also have a chapter on heteromorphic SI in Primula, which has fascinated many biologists, including Darwin). We then have a chapter that deals with interesting and important questions that are not often considered, relating to the complexity of the expression of SI in some species, permitting sophisticated regulation of SI during a plant's lifetime or in extreme environmental conditions. The evolution of SI is a long-standing question which has been much debated (and continues to be so), so several chapters are devoted to this. A theoretical consideration of how clonal reproduction, which occurs in many species, may play an important role in affecting the evolutionary dynamics of SI systems, provides possible explanations and insights into unexplained features of allele genealogies. One theme that is highlighted in this section is the switch from self-incompatibility to self-fertility. Chapter 4 considers the diversity of SI systems in conjunction with recent advances in understanding angiosperm phylogenies enabled by molecular studies. This provides an illuminating in-depth picture of the phylogenetic distribution of SI and evolution of SI in the lineages, suggesting that SI was probably the ancestral angiosperm mating system. Continuing the theme of evolution of SI from another perspective, molecular sequence data has provided much information about S-locus determinants. Chapter 5 considers S-locus genealogies using molecular sequence information from the S-RNase system. It examines the S-locus polymorphism, tracing the history of mating-system transitions across entire families and examining the demographic history of lineages. It suggests some possible reasons why there are problems in reconstructing the history of SI and draws attention to surprising data that suggests that the challenge of understanding how SI operates and evolved is far from over. The final chapter in Part I also deals with the evolution of SI, focusing on the Brassicaceae, discussing how new SI specificities may be generated and switches from self-incompatibility to self-fertility in this family, which includes Arabidopsis. Together these chapters provide an overview of the considerable progress that has been made in recent years. However, the overwhelming message is that despite this, our understanding of the evolution of SI is far from complete, and some major questions remain unresolved.

The second part of this volume is concerned with what we know about the genes and proteins encoding the pistil and pollen components of the S-locus, together with other components that are not located at the S-locus but are involved in SI. Together, they emphasize the great progress made, particularly in the last decade, in our understanding of some of the possible regulatory mechanisms involved in pollen inhibition by SI. Chapters in this section deal with various SI systems where S-locus determinants have been identified. Although investigations of the cell biology and understanding of the mechanisms involved lag behind genetical analyses, what is striking is the apparently wide diversity of cellular mechanisms recruited to ensure inhibition of "self" pollen. Two chapters deal with SI in Brassica. The first provides a historical background, introducing key breakthroughs in establishing the molecular basis of SI in this species, with the second describing the first insights into cell biology and possible mechanisms involved. The next two chapters cover what we know about SI in the families that use S-RNases as SI determinants. These chapters highlight the fact that we still know relatively little about how "self" pollen is inhibited in these systems. A chapter on SI in Papaver deals with both what we know about the components involved in SI and describes progress in understanding the network of mechanisms involved in self-pollen inhibition. These chapters reveal the huge progress made in recent years, but emphasize that there is much still to be understood about the mechanisms involved in regulating incompatible pollen tube growth.

The next group of chapters deals with those SI systems where the determinants of SI have not yet been established. This is an extremely important area, as it is already apparent that since all three of the SI systems that have been investigated in detail so far have different SI mechanisms, it is clear that there are at least several other SI systems which will have S-determinants that are different from those already identified. Thus, encouragement in pursuing such investigations is required, as without such studies we will not begin to get an idea of the real number, breadth and diversity of SI components and mechanisms. Chapters on molecular analysis of SI in Ipomoea, which has a sporophytic SI system, but does not use the Brassica S-determinants; the grasses, which have a multi-locus (S- and Z-) SI system; and of the heteromorphic SI system of Primula; are included here, as although identification of the molecular determinants of SI are yet to be made, considerable effort has been invested in them (as with several other species which we have not been able to include). In the case of Ipomoea considerable progress has been made, and the S-locus is defined; tantalizingly, potential candidates for pistil and pollen S-determinants have been identified. In the grasses, the genetics of SI has been well analyzed and molecular markers have confirmed synteny for S- and Z-loci in several species, but identification of the determinants of SI has been fraught with problems. This chapter discusses the various approaches (both forward and reverse genetic approaches have been taken) and provides a perspective of where research is at this point. In the final chapter, historical data regarding the classic heteromorphic SI system in Primula is reassessed and re-interpreted in the light of new developments, and progress in beginning to understand the molecular basis of SI in this species is discussed.

This book attempts to cover an enormous breadth of studies into various aspects of SI. I hope this book will be a valuable resource for those working in the field, and provide a good introduction to those to whom it serves as an introduction to this fascinating subject, which has interested an extremely broad group of biologists, from theoretical and population geneticists, to ecologists and evolutionary biologists, through to molecular geneticists and cell biologists. It is my hope that this book will be accessible to a broad readership, whatever their specialist field or understanding of self-incompatibility systems. For this reason, I have included an abbreviations list for each chapter, together with a substantial glossary to make it easier for the reader to follow the terminology and jargon associated with particular areas of biology, with clear explanations of the sometimes obscure terminologies used, compiled by experts in the various fields covered by this topic.

Finally, I would like to thank all the authors who have contributed to this volume, for their overwhelming enthusiasm for this project, which meant it was initiated, and for their cooperation in providing their chapters in good time and for dealing with queries and requests with good humour. Without them, this book would not have happened.

Birmingham, *Noni Franklin-Tong*
May 2008

Contents

Part I Evolution and Population Genetics of Self-Incompatibility

1 **New Insights on Heterostyly: Comparative Biology, Ecology and Genetics** 3
 S.C.H. Barrett and J.S. Shore
 1.1 Introduction 4
 1.2 Comparative Biology and Evolutionary History of Heterostyly ... 6
  1.2.1 Phylogeny Reconstruction and Character Evolution 7
 1.3 Function and Reproductive Ecology of Heterostyly 10
  1.3.1 Function of Heterostyly 10
  1.3.2 Floral Morph Ratios and Reproductive Success 12
 1.4 Inheritance of Heterostyly and the Supergene Model 14
  1.4.1 Inheritance of Distyly and Tristyly 14
  1.4.2 Supergene Model.................................... 15
  1.4.3 Mutational Analyses and the Study of Genetic Variants .. 19
 1.5 Molecular Genetics.. 19
  1.5.1 Theoretical Models and Predictions 20
  1.5.2 Protein Profiles 22
  1.5.3 mRNA Expression.................................. 22
  1.5.4 Genetic Localization 23
 1.6 Concluding Remarks 25
 References 26

2 **Genetic and Environmental Causes and Evolutionary Consequences of Variations in Self-Fertility in Self Incompatible Species** 33
 S.V. Good-Avila, J.I. Mena-Alí, and A.G. Stephenson
 2.1 Introduction 34
 2.2 Genetics of Self-Fertility 35
  2.2.1 Mutations Affecting the *S*-locus 36
  2.2.2 Unlinked Modifiers of SI 37

		2.2.3	Plasticity in Self-Fertility	40
		2.2.4	Summary of Genetics of Self-Fertility	41
	2.3	Fate of Self-Fertility Genes		42
		2.3.1	Conditions for Stable Polymorphisms	43
		2.3.2	Summary and Conclusions Concerning Stable Polymorphisms	45
	References			47
3	**On the Evolutionary Modification of Self-Incompatibility: Implications of Partial Clonality for Allelic Diversity and Genealogical Structure**			**53**
	M. Vallejo-Marín and M.K. Uyenoyama			
	3.1	Introduction		54
	3.2	Mating System Dynamics		55
		3.2.1	Relative Transition Rates	55
		3.2.2	Multiple Origins of SC in *Arabidopsis*	55
		3.2.3	Modified Forms of SI	56
	3.3	*S*-Locus Evolution Under Partial Clonality		57
		3.3.1	Diffusion Approximation	57
		3.3.2	*S*-Allele Number and Frequency	60
		3.3.3	Age of the Root	62
	3.4	Discussion		65
		3.4.1	Clonality in the Solanaceae	65
		3.4.2	Evolutionary Stability of Partial SI	65
		3.4.3	Paradoxical Effects on Mating Systems	66
	3.5	Conclusions		67
	Appendix 1: Diffusion Equation Approximation			68
	Appendix 2: Simulations			69
	References			69
4	**Evolution and Phylogeny of Self-Incompatibility Systems in Angiosperms**			**73**
	A.M. Allen and S.J. Hiscock			
	4.1	Introduction		74
		4.1.1	Diversity of SI Systems in Angiosperms	75
		4.1.2	Evolutionary Origin(s) of SI Systems	76
	4.2	Was Self-Incompatibility Present in the First Angiosperms?		77
		4.2.1	Self-Incompatibility in Basal Angiosperms	78
		4.2.2	Self-Incompatibility in the Monocots	80
	4.3	Phylogenetic Distribution of SI Systems		81
		4.3.1	Late-Acting Ovarian Self Incompatibility (OSI)	84
		4.3.2	Gametophytic Self-Incompatibility (GSI)	86
		4.3.3	Sporophytic Self-Incompatibility (SSI)	88
	4.4	The Relationship Between GSI and SSI		89
	4.5	Discussion		91
	References			95

5 What Genealogies of S-alleles Tell Us 103
J.R. Kohn
- 5.1 Introduction 104
- 5.2 Long-Term Demographic Information from the *S*-locus 105
- 5.3 Implications of Shared Ancestral Polymorphism 108
  - 5.3.1 Tracing the History of Mating System Change 108
  - 5.3.2 Diversification Rate Differences and Character State Reconstruction 110
- 5.4 The Pace of New Allele Formation 111
- 5.5 Remaining Issues of S-RNase Evolution 112
- 5.6 Pollen Specificity Genes 114
- 5.7 Conclusions 117
- References 117

6 Self-Incompatibility and Evolution of Mating Systems in the Brassicaceae 123
S. Sherman-Broyles and J.B. Nasrallah
- 6.1 Introduction 124
- 6.2 Structural and Sequence Diversification of *S*-locus Haplotypes and Their Recognition Genes 125
  - 6.2.1 Conserved and Diverged Features of the *S*-locus 127
  - 6.2.2 Intra-Specific Structural Heteromorphism and Sequence Polymorphism: Suppressed Recombination and Maintenance of SRK-SCR Linkage 128
  - 6.2.3 Diversification of the *S*-locus Genes and the SI Recognition Repertoire 130
- 6.3 Evolutionary Switches from Self-Incompatibility to Self-Fertility 135
  - 6.3.1 Molecular Genetics of Switches to Self-Fertility 136
  - 6.3.2 Breakdown of SI by Disruption of *S*-locus Gene Expression in Inter-Specific Hybrids 137
  - 6.3.3 The Case of Self-Fertility in *A. thaliana* 138
- 6.4 Future Prospects 142
- References 142

Part II Molecular and Cell Biology of Self-Incompatibility Systems

7 Milestones Identifying Self-Incompatibility Genes in Brassica Species: From Old Stories to New Findings 151
M. Watanabe, G. Suzuki, and S. Takayama
- 7.1 Self-Incompatibility as an Agriculturally Important Trait 152
- 7.2 The First Milestone: Bateman's Idea for Sporophytic Control of the *S*-locus 154
- 7.3 The Second Milestone: Identification of SLG by Using IEF 155
- 7.4 The Third Milestone: Identification of *SRK*, the Female *S* Determinant Gene 156

	7.5	The Fourth Milestone: Functional Evidences of *SRK* in SI 157
	7.6	The Fifth Milestone: Establishment of Bioassay System 158
	7.7	The Sixth Milestone: Identification of *SP11/SCR*, the Male *S* Determinant Gene 159
	7.8	After Identifying the SI Genes and Future Milestones 160
		7.8.1 Demonstrating Physical Interaction Between SRK and SP11/SCR 160
		7.8.2 Downstream of the Interaction: Identifying Components and Mechanisms Involved in Mediating the Rejection of Self Pollen 162
		7.8.3 Molecular Mechanisms of Dominance Relationships 163
		7.8.4 Evolution of SI Genes 164
	7.9	Prospects 165
	References 166

8 'Self' Pollen Rejection Through the Intersection of Two Cellular Pathways in the Brassicaceae: Self-Incompatibility and the Compatible Pollen Response 173

M.A. Samuel, D. Yee, K.E. Haasen, and D.R. Goring

	8.1	Introduction 174
	8.2	The Early Stages of Compatible Pollen–Stigma Interactions in the Brassicaceae 175
		8.2.1 Pollen Capture and Adhesion 175
		8.2.2 Pollen Hydration 176
		8.2.3 Pollen Germination and Pollen Tube Penetration 178
	8.3	The SI Response Causes Pollen Arrest at the Stigmatic Surface .. 179
		8.3.1 The S Receptor Kinase Activates a Cellular Signalling Pathway in the Stigmatic Papilla to Trigger Self Pollen Rejection 179
		8.3.2 The M Locus Protein Kinase acts Together with the S Receptor Kinase to Promote SI 180
		8.3.3 The SRK Kinase Domain can Interact with a Range of Intracellular Proteins 181
		8.3.4 Thioredoxin h Inhibits SRK Activity in the Absence of Self Pollen 182
		8.3.5 ARC1 Functions Downstream of SRK to Promote SI ... 182
		8.3.6 BnExo70A1 is a Potential Substrate for ARC1 and is Required for Compatible Pollen–Stigma Interactions 184
		8.3.7 Endomembrane Changes in the Stigmatic Papillae Following Compatible and Self-Incompatible Pollinations in the Brassicaceae 185
	8.4	Conclusions and Prospects 186
	References 187

9 Molecular Biology of S-RNase-Based Self-Incompatibility 193
Y. Zhang and Y. Xue

- 9.1 Introduction 194
- 9.2 S-RNase Determines *S*-Specificity in Pistil 196
  - 9.2.1 Isolation and Identification of S-RNase as the Pistil S ... 196
  - 9.2.2 S-RNase Sequence Features and the Specificity Determinant 196
  - 9.2.3 The Role of S-RNase: A Cytotoxin Specifically Inhibits Self Pollen 198
- 9.3 F-Box Proteins Determine *S*-Specificity in Pollen 199
  - 9.3.1 Clues from Pollen-Part Self Compatible Mutants 199
  - 9.3.2 Isolation of the Pollen SI Determinant, SLF/SFB 200
  - 9.3.3 Sequence Analysis of SLFs and SFBs 201
  - 9.3.4 Identification of SLF as the Pollen S 201
  - 9.3.5 SFBs from Rosaceae Likely Represent Another Class of F-Box Genes 202
- 9.4 Other Genes That Modulate the SI Response 203
  - 9.4.1 The Pistil Modifier Factors 203
  - 9.4.2 The Pollen Modifier Factors 204
- 9.5 Molecular Mechanisms for S-RNase-based SI 205
  - 9.5.1 Pollen S, the Positive or Negative Regulator of S-RNase? Clues from Genetic Evidence 205
  - 9.5.2 The Fate of S-RNases: S-RNase Restriction is Likely to Involve Ubiquitination 206
  - 9.5.3 Future Perspectives 208
- References 210

10 Comparing Models for S-RNase-Based Self-Incompatibility 217
B. McClure

- 10.1 The Biology of S-RNase-Based SI 218
  - 10.1.1 Genetic Breakdown 218
- 10.2 S-RNase and *S*-locus F-box Proteins 219
  - 10.2.1 S-RNase Structure and Specificity 219
  - 10.2.2 *S*-locus F-Box Genes 221
- 10.3 Non-*S*-Specific Genes 222
  - 10.3.1 HT Genes 223
  - 10.3.2 S-RNase Binding Proteins in the Transmitting Tract Extracellular Matrix 224
  - 10.3.3 Non-S-Specific Factors in Pollen..................... 226
- 10.4 Comparing Models for S-RNase-Based SI 227
  - 10.4.1 How do Compatible Pollen Tubes Resist S-RNase Cytotoxicity? 227
  - 10.4.2 Is There a Separate Mechanism in the Rosaceae? 227
  - 10.4.3 Inhibiting S-RNase Enzyme Activity 228

	10.4.4	S-RNase Degradation 228
	10.4.5	S-RNase Compartmentalisation 231
References 233		

11 Self-Incompatibility in *Papaver Rhoeas*: Progress in Understanding Mechanisms Involved in Regulating Self-Incompatibility in *Papaver* 237
V.E. Franklin-Tong

- 11.1 Introduction 238
  - 11.1.1 Genetics and Cell Biology of Self-Incompatibility in *Papaver* 238
  - 11.1.2 How Studies on Self-Incompatibility in *Papaver* Started 239
  - 11.1.3 The *Papaver in Vitro* SI System 241
- 11.2 *S* Proteins Determine *S*-Specificity in the Pistil 242
  - 11.2.1 Identification of Pistil *S*-locus Components 242
  - 11.2.2 Pistil *S*-Protein Sequence Information and Residues Required for Function 243
- 11.3 Identification of the Pollen *S*-Determinant 243
- 11.4 Mechanisms Involved in SI in the *Papaver* System 244
  - 11.4.1 Calcium Signalling Mediates *Papaver* SI 244
  - 11.4.2 A Role for Soluble Inorganic Pyrophosphatases (sPPases) in *Papaver* SI 245
  - 11.4.3 Alterations to the Actin Cytoskeleton are Triggered by *Papaver* SI 246
  - 11.4.4 SI Triggers Programmed Cell Death 247
- 11.5 An Overall Model for Mechanisms Involved in Regulating SI in *Papaver* 251
  - 11.5.1 A Contrast to the S-RNase System and *Brassica* SI Systems 253
- 11.6 Future Perspectives 254
- References 255

12 Molecular Genetics of Sporophytic Self-Incompatibility in *Ipomoea*, a Member of the Convolvulaceae 259
Y. Kowyama, T. Tsuchiya, and K. Kakeda

- 12.1 Introduction 260
- 12.2 Sexual Reproduction in the Genus *Ipomoea* 261
- 12.3 Genetics of Self-Incompatibility in *Ipomoea* 261
- 12.4 Stigma-Specific Proteins 263
- 12.5 Physical Size of the *S*-locus 264
- 12.6 Genomic Organisation of the *S*-locus 266
- 12.7 *S*-locus Genes in *Ipomoea* 268
- 12.8 Diversity of the SI Systems 270
- References 271

13 Self-Incompatibility in the Grasses ... 275
P. Langridge and U. Baumann
- 13.1 Introduction ... 276
- 13.2 Genetic Control of SI in the Grasses ... 277
  - 13.2.1 Features of the *S-Z* System ... 278
  - 13.2.2 The Third Locus 'T' ... 279
  - 13.2.3 Mechanism of Action of *S*- and *Z*-Gene Products ... 280
- 13.3 Approaches and Progress in Cloning *S* and *Z* ... 280
  - 13.3.1 Reverse Genetics ... 281
  - 13.3.2 Forward Genetics ... 281
- 13.4 Conclusions ... 284
- References ... 285

14 Heteromorphic Self-Incompatibility in *Primula*: Twenty-First Century Tools Promise to Unravel a Classic Nineteenth Century Model System ... 289
A. McCubbin
- 14.1 Introduction ... 290
- 14.2 Floral Characteristics of the Mating Types of *Primula* ... 291
  - 14.2.1 Style ... 292
  - 14.2.2 Stigma ... 292
  - 14.2.3 Corolla Mouth Size ... 292
  - 14.2.4 Anthers ... 293
  - 14.2.5 Pollen ... 293
  - 14.2.6 Self-Incompatibility Specificity ... 293
- 14.3 Functions of Heteromorphic Characters ... 293
- 14.4 Physiological Nature of SI ... 295
  - 14.4.1 Site of Operation ... 295
  - 14.4.2 Candidate Molecules in the Operation of Heteromorphic SI ... 297
- 14.5 The *Primula S*-locus ... 297
  - 14.5.1 Genetic Structure ... 297
  - 14.5.2 Location and Size of the *S*-locus ... 301
  - 14.5.3 Allelic Dominance ... 301
- 14.6 Floral Development ... 302
- 14.7 Molecular Genetic Characterisation of the *Primula S-locus*: Current Status ... 303
- 14.8 Conclusions and Future Prospects ... 305
- References ... 306

Index ... 309

Contributors

A. Allen
School of Biological Sciences, University of Bristol, Woodland Road, Bristol, BS8 1UG, UK

S.C.H. Barrett
Department of Ecology and Evolutionary Biology, University of Toronto, 25 Willcocks Street, Toronto, ON, Canada M5S 3B2, e-mail: barrett@eeb.utoronto.ca

U. Baumann
Australian Centre for Plant Functional Genomics, School of Agriculture Food and Wine, University of Adelaide, Urrbrae, SA 5064, Australia, e-mail: ute.baumann@acpfq.com.au

V.E. Franklin-Tong
School of Biosciences, University of Birmingham, Edgbaston, Birmingham B15 2TT, U.K., e-mail: v.e.franklin-tong@bham.ac.uk

S.V. Good-Avila
Acadia University, Department of Biology, 24 University Avenue, Wolfville, NS, Canada, B4P 2R6, e-mail: sara.good-avila@acadiau.ca

D.R. Goring
Department of Cell and Systems Biology, University of Toronto, Toronto, ON, Canada M5S 3B2, e-mail: d.goring@utoronto.ca

K.E. Haasen
Department of Cell and Systems Biology, University of Toronto, Toronto, ON, Canada M5S 3B2

S.J. Hiscock
School of Biological Sciences, University of Bristol, Woodland Road, Bristol, BS8 1UG, UK, e-mail: simon.hiscock@bristol.ac.uk

K. Kakeda
Graduate School of Bioresources, Mie University, Tsu 514-8507, Japan, e-mail: kakeda@bio.mie-u.ac.jp

J.R. Kohn
Section of Ecology, Behavior and Evolution, Division of Biological Sciences, University of California San Diego, 9500 Gilman Drive, La Jolla, CA 92093–0116 USA e-mail: jkohn@ucsd.edu

Y. Kowyama
Graduate School of Bioresources, Mie University, Tsu 514-8507, Japan, e-mail: kouyama@bio.mie-u.ac.jp

P. Langridge
Australian Centre for Plant Functional Genomics, School of Agriculture Food and Wine, University of Adelaide, Urrbrae, SA 5064, Australia, e-mail: Peter.Langridge@adelaide.edu.au

B. McClure
Department of Biochemistry, Interdisciplinary Plant Group, Christopher S. Bond Life Sciences Center, 240a Christopher S. Bond Life Sciences Center, 1201 East Rollins Street, Columbia, MO 65211, USA, e-mail: mcclureb@missouri.edu

A. McCubbin
School of Biological Sciences and Center for Reproductive Biology, Washington State University, Pullman, WA 99164, USA, e-mail: amccubbin@wsu.edu

J.I. Mena-Alí
Department of Biology, Amherst College, 311 McGuire LSB, Amherst, MA 01002, USA, e-mail: jmenaali@amherst.edu

J.B. Nasrallah
Department of Plant Biology, Cornell University, Ithaca, NY 14853, USA, e-mail: jbn2@cornell.edu

M.A. Samuel
Department of Cell and Systems Biology, University of Toronto, Toronto, ON, Canada M5S 3B2

S. Sherman-Broyles
Department of Plant Biology, Cornell University, Ithaca, NY 14853, USA, e-mail: sls98@cornell.edu

J.S. Shore
Department of Biology, York University, 4700 Keele Street, Toronto, ON, Canada M3J 1P3, e-mail: shore@yorku.ca

A.G. Stephenson
Department of Biology, 208 Mueller Lab, University Park, PA 16802, USA, e-mail: as4@psu.edu

G. Suzuki
Division of Natural Science, Osaka Kyoiku University, Kashiwara 582-8582, Japan, e-mail: gsuzuki@cc.osaka-kyoiku.ac.jp

S. Takayama
Graduate School of Biological Sciences, Nara Institute of Science and Technology, Ikoma 630-0192, Japan, e-mail: takayama@bs.naist.jp

T. Tsuchiya
Life Science Research Center, Mie University, Tsu 514-8507, Japan
e-mail: tsuchiya@gene.mie-u.ac.jp

M.K. Uyenoyama
Department of Biology Box 90338, Duke University, Durham, NC 27708–0338, USA, e-mail: marcy@duke.edu

M. Vallejo-Marín
Department of Ecology and Evolutionary Biology, 25 Willcocks Street, University of Toronto, Toronto, ON, Canada, M5S 3B2, e-mail: mvallejo@eeb.utoronto.ca

M. Watanabe
Graduate School of Life Sciences, Tohoku University, Sendai 980-8577, Japan, e-mail: nabe@ige.tohoku.ac.jp

and

The 21st Century Center of Excellence Program, Iwate University, Morioka 020-8550, Japan

and

Laboratory of Plant Reproductive Genetics, Graduate School of Life Sciences, Tohoku University, 2-1-1, Katahira, Aoba-ku, Sendai 980-8577, Japan

Y. Xue
Institute of Genetics and Developmental Biology, Chinese Academy of Sciences, and National Centre for Plant Gene Research, Beijing 100101, China, e-mail: ybxue@genetics.ac.cn

D. Yee
Department of Cell and Systems Biology, University of Toronto, Toronto, ON, Canada M5S 3B2

Y. Zhang
Institute of Genetics and Developmental Biology, Chinese Academy of Sciences, and National Centre for Plant Gene Research, Beijing 100101, China

Glossary

Additive genetic variation Genetic variation associated with the average effects of substituting one allele for another.

Allele extinction Disappearance of an allele from the gene pool.

Allelic diversification The evolution of multiple alleles from a common ancestor.

Allelic turnover The replacement in a population of old alleles by new ones over time.

Androecium The male reproductive organs of a flower.

Anti-parallel β-sheet A protein secondary structure motif including extended polypeptide conformations running in opposite orientation and with a characteristic hydrogen-bonding pattern.

Antisense A method for gene silencing, i.e. reducing expression of a specific gene. A gene construct is produced such that transcription produces an RNA with the opposite orientation, or sense, of the target gene. The sense and antisense transcripts hybridise and are processed to form small RNAs that stimulate degradation of the target gene transcript, thus inhibiting its expression. See RNAi.

Antisense oligonucleotides A short segment of DNA or RNA complementary to a chosen sequence, which is thought to act by preventing translation by binding to the mRNA.

Arabinogalactan A carbohydrate moiety consisting of a branched galactose backbone decorated with arabinosyl residues. Arabinogalactan proteins (AGPs) consist of a protein backbone decorated with arabinogalactan.

Assortative mating Non-random mating based on the characteristics of partners. Usually used for *positive assortative mating*, whereby the traits of mating partners are more similar than random pairing. *Negative assortative mating* (or *disassortative mating*) occurs when the traits of mating partners are less similar than expected from random pairing.

Autogamy Transfer of pollen from male to female parts within the same flower.

Autotetraploid A tetraploid formed from the doubling of a single genome. In an autotetraploid all the chromosomes come from the same species.

Balancing selection A form of selection that maintains genetic diversity or multiple alleles within a population. A genetic polymorphism (such as polymorphism at the *S*-locus) is maintained and is said to be under balancing selection. *S*-haplotypes rarely go extinct in a population because mating success is inversely proportional to haplotype abundance.

Barstar A small protein inhibitor that binds and inactivates a ribonuclease produced in *Bacillus amyloliquefaciens*, barnase.

Basal angiosperm A member of one of the earliest diverging lineages of angiosperms, e.g., Amborellaceae, Nymphaeales, and Austrobaileyales.

Bi-molecular Fluorescence Complementation (BiFC) A technique to detect interactions between two proteins in vivo, as well as their subcellular localization. The putative partners are fused to complementary parts of a fluorescent protein; fluorescence occurs only when the putative partners interact, bringing the two parts of the fluorescent protein together.

Bottleneck A reduction in population size that may be followed by a subsequent increase.

Ca^{2+} influx Movement of extracellular Ca^{2+} into a cell via a Ca^{2+}-permeable channel.

Ca^{2+} signalling Calcium ions (Ca^{2+}) act as second messengers, which are involved in key cellular signalling processes by affecting the activity of key enzymes that are sensitive to calcium, in order to ultimately alter numerous cellular physiological processes. Ca^{2+} increases are achieved by either release from intracellular stores or via channel activity, from extracellular sources.

Calcium signature A highly specific temporal and spatial Ca^{2+} pattern (analogous to a handwritten signature that is specific and identifies the writer), that contributes to the specificity of the cellular response.

Callose A plant polysaccharide deposited by pollen tubes during pollen tube growth.

Canonical SI A form of self-incompatibility characterized by full expression in every reproductive episode.

Carpel The female ovule-bearing organ of a flower, usually consisting of a swollen basal region (ovary), an elongated portion of the ovary (style), and a receptive tip (stigma). One, several, or many carpels may occur in a flower and they may be separate from each other or fused together. Collectively carpels are called the gynoecium (syn. pistil).

Caspases Cysteine aspartate-specific proteases, which are key mediators in the initiation and execution of apoptosis/programmed cell death. They cleave their substrates after an aspartic acid residue.

Catalytically inactive The state for an enzyme that has lost catalytic activity.

cDNA Complementary DNA (usually generated from mRNA reverse transcriptase), which contains coding sequences of DNA.

Centromere A region, often found in the middle of the chromosome, involved in cell division and the control of gene expression.

Chimera An experimental gene where DNA sequences from different alleles are joined to determine the functionally or selectively important portions of the sequence.

Chromosome walking A technique used to clone a gene that is known to be genetically linked to a gene that has already been cloned. Thus one would "walk" and sequence from a starting point towards another gene in this region that may be many kb away.

Clade A taxonomic group comprising a single common ancestor and all the descendants of that ancestor.

Clonal reproduction Propagation through vegetative structures such as root fragments, stolons, bulbils, tubers, etc. The products of clonal reproduction are plants that are potentially or effectively physiologically independent units.

Clonality Clonal reproduction.

Coalescent approach Retrospective approach using gene genealogies to infer demographic and selective history.

Co-dominance The independent expression of both alleles in a heterozygote, thereby resulting in manifestation of both phenotypes.

Co-evolution Evolutionary process whereby change in one player selects for change in another and vice-versa.

Colinearity The one to one linear correspondence between the order of genes along a chromosome or chromosome section.

Colonizing Organisms that become established in environments previously unoccupied by conspecifics.

Comb-like Another representation of a star-like phylogeny. A phylogeny that looks like a comb, where all extant forms appear to have separated from all others at the base of the phylogenetic tree.

Common frequencies In SI systems, allele frequencies close to the inverse of allele number.

Comparative mapping Alignment of chromosomes of related species based on genetic mapping of common DNA markers.

Competitive interaction A condition that arises in heteroallelic pollen expressing two different pollen S-alleles. The pollen S-alleles effectively compete with one another such that pollen is not inhibited in pistils expressing either of the matching S-haplotypes.

Complementary interaction The different components of the system must work together to exert an effect.

Conspecifics Individuals of the same species.

Convergent evolution Evolution of the same character state in different lineages, such the presence of wings in birds and bats.

Core eudicots Members of a large monophyletic eudicot group that includes most angiosperms.

Corolla The name for the petals of a flower, considered as a unit.

Co-sexual Having male organs (stamens) and female organs (carpels/pistils) present in the same flower (synonym: 'bisexual').

Co-suppression The effect of introducing a gene transgenically results in silencing or reduced expression of the endogenous gene.

Cross-hybridize A condition where distinct gene sequences are sufficiently similar that they can form duplex DNA. For example, DNA from two allelic *S-RNase* genes will cross-hybridize if their sequences are very similar but more distantly related sequences will not cross-hybridize.

Cross-incompatibility The failure of fertilization in crosses between pairs of individuals despite their formation of normal gametes.

Cryptic self-incompatibility A form of physiological SI found in self-compatible plants in which out-crossing is promoted through differential pollen-tube growth between outcross- and self-pollen tubes. It is usually detected using genetic markers. It is also used as a term for a situation where there is a GSI system operating alongside SSI.

Cytosol The soluble part of the cell, which together with the organelles (which form the microsomal fraction), forms the cytoplasm.

Cytotoxin A toxin whose action is in the cytoplasm. For example, ribonucleases are cytotoxins because they attack RNA located in the cytoplasm and, thus, inhibit growth. They thus have cytotoxic activity.

Deterministic equilibrium frequency Equilibrium frequency of alleles expected in the absence of stochasticity.

Diallel A crossing scheme used by plant breeders and geneticists. In a full diallel, both parents are crossed in all possible combinations; in a half diallel just one half of the diagonal is completed (i.e. not duplicating reciprocal combinations).

Dichogamy Maturation of male and female organs in hermaphrodite flowers at different times. Dichogamy is subdivided into two forms: (1) Protandry - Dehiscence of mature pollen before the stigma is receptive and (2) Protogyny - Functional stigma receptivity before pollen dehiscence.

Dimorphic Possessing two morphologically distinct forms.

Dioecious With unisexual male and female flowers borne on separate plants.

Disassortative pollination A breeding system in which one mating type preferentially fertilizes another; see assortative mating.

Dicotyledons ('dicots') Plants that produce two embryonic leaves or cotyledons. There are around 200,000 species within this group. The other group is the monocots.

Differential AFLP DNA fragments that were amplified by using the polymerase chain reaction method and that differ in length between organisms of the same species due to allelic variation. AFLP = amplified fragment length polymorphism.

Diffusion approximation In population genetics, an approximation of a stochastic process that is discrete in state (number of genes) and time (generations) by a stochastic process that is continuous in state and time.

Divergence time Time since the most recent common ancestor.

Diversification rate The rate of production of new forms such as species or alleles.

DNA methylation Modification of DNA involving addition of a methyl group(s); this can be inherited without changing the original DNA sequence, and is the best characterized mechanism involved in epigenetics.

Dominance A situation when in a heterozygous individual, the only one of the two alleles present control the phenotype. This allele is dominant (the other being said to be recessive).

Dominant allele An allele that expresses its phenotypic effect even when heterozygous with a recessive allele; usually designated by a capital letter (eg. T).

Dominant-recessive Interaction between two different alleles of a gene: the phenotype determined by the dominant allele is manifested, whereas that of the recessive allele is masked in the heterozygote.

"Dry" stigma In dry stigmas, stigma exudates are absent on the stigmatic papillae; i.e. they lack free flowing secretions. Typically they have intact papillar cells covered by primary cell wall, a waxy cuticle and an extracellular protein layer.

Dual-specificity chimeric An artificial gene made by combining parts of two genes to achieve two specificities within one gene.

Duplication A genetic change in which a chromosomal segment is present in two copies. Duplicated segments can be large or small and may or may not be linked to the original segment.

E3 ubiquitin ligase complex The enzyme complex that transfers ubiquitin from the E2-ubiquitin carrier to a target protein. Ubiquitylation is a three step process; each step is catalyzed by a different enzyme, E1, E2, and E3. The final step, catalyzed by the E3 ubiquitin ligase complex determines which cellular proteins are ubiquitylated, usually, as a prelude to their degradation. See also ubiqutination.

Early Cretaceous The beginning of the Cretaceous period, approximately between 140 and 120 million years ago.

Effective population size Size of an idealized unstructured population in which stochastic changes would occur at the same rate as in a more complex population.

Emasculated Usually referring to flowers in which the anthers have been removed using fine forceps for experimental purposes.

Endocytosis Any of several cellular processes that allow material from outside a cell to be internalized in membrane-bound vesicles.

Endomembrane system The internal membrane system in eukaryotic cells; it includes the endoplasmic reticulum, Golgi, vacuole, or prevacuolar compartment.

Epigenetic Processes that lead to inheritance that is not reflected in the sequences of the nuclear DNA; often DNA and/or chromatin modifications that alter gene function without any change in DNA sequence.

Epigenetic gene silencing The "switching off" of a gene by a mechanism than does not involve changes in the underlying DNA sequence of the organism.

ERAD pathway Endoplasmic reticulum associated degradation pathway. This pathway is usually regarded as a system for disposal of mis-folded or incorrectly glycosylated proteins. It involves exit from the endoplasmic reticulum and degradation in the cytoplasm.

Eudicot A eudicotyledon ('true' dicotyledon), plant with two cotyledons and pollen grains with predominantly three apertures (tricolpate).

Evolutionary conflict Discordance of evolutionary interests.

Exine The exine is a multi-layered and highly-sculptured outer wall of the pollen grain made up of the polymer, sporopollenin. It is involved in the initial adhesion of pollen grains to the stigma.

Exon Portion of a gene transcribed into the mRNA which codes for a polypeptide, multiple exons may be separated by non-coding regions (i.e., introns) that are spliced together forming the final mRNA.

Extinction rate The rate of species extinction in a lineage.

F_1-hybrid An agricultural terminology to describe a cultivar generated by crossing two distinct inbred parents to create a new, uniform variety with desirable characteristics derived from the parents.

F-box genes A class of genes that has a conserved F-box motif at the N-terminus which has the potential to bind SKP1 to form SCF (Skp1/Cullin/F-box) complex (a type of E3 ubiquitin ligase) and usually carries one of a variety of typical protein–protein interaction domains that confers substrate specificity to an SCF complex.

Female S-determinant/Pistil S The protein encoded by the *S*-locus to determine the SI specificity in the pistil.

Fine-scale mapping Construction of a high resolution genetic linkage map of numerous DNA markers and genes at a level of less than 0.1 cM using thousands of progeny from a single cross to enable detection of infrequent recombination events.

Forward genetics A genetic method of gene identification, starting with a genetically distinct phenotype and then working towards the identification of the gene underlying the phenotype.

Founder event A severe demographic bottleneck caused when a few individuals found a new population; founders of new population carry only a small fraction of the total genetic variation in the source population.

Frameshift mutation A mutation that causes a change in reading frame in a gene and, therefore, a substantial change in amino acid sequence.

Frequency-dependent selection A form of selection in which the selection coefficients associated with genes or genotypes depend on the frequencies of those genes or genotypes. It favors low frequency alleles; common alleles are selected against.

Gametes Products of meiosis.

Gametic selection The forces acting to cause differential reproductive success of gametes. Examples are pollen viability and sperm cell competition.

Gametophytic SI (GSI) A form of SI genetically distinct from the sporophytic SI (SSI). In the GSI response, the recognition specificity is determined by the pollen (gametophyte) haploid *S*-haplotype of the polymorphic *S*-locus: pollen tube growth is arrested when its *S*-haplotype is the same as either of the *S*-haplotypes of the diploid pistil; conversely, pollen with any *S*-haplotype not present in the pistil is compatible. Representative families include the Solanaceae, Plantaginaceae and Papaveraceae.

Gene genealogy Relationship among genes through direct descent.

Genealogical depth Age of the most recent common ancestor of a sample of alleles.

Genealogy Phylogenetic history of genes, populations, or species.

Genetic drift Changes in the frequencies of alleles or phenotypes in a population. Often results from periods of small population size or because the genes appearing in offspring do not represent a perfectly representative sampling of the parental genes.

Genetic maps A genetic map describes the order of genes or loci along chromosomes but in contrast to physical maps uses indirect measurements, based on recombination frequency in centiMorgan (cM), as the distance measure.

Genetic modifiers See modifier gene.

Genetic polymorphism A difference in DNA sequence at a specific locus among individuals, groups, or populations.

Gene silencing Gene regulation by epigenetic processes, "turning off" a gene by a mechanism other than genetic modification.

Genets Plants derived by sexual reproduction.

Genomic heteromorphism Variations in genomic nucleotide sequences in different genotypes due to insertion or deletion of some nucleotide sequences, or accumulation of base-pair changes.

Genotype by environment interactions The condition in which the advantage of a particular genotype depends upon the environment in which it is expressed.

Genetic load Deleterious mutations that are not exposed to selection in heterozygotes, which accumulate in outcrossing individuals as masked recessive alleles. If the breeding system changes to selfing, this deleterious load is exposed to selection as recessive homozygotes and may lead to inbreeding depression.

Glycoprotein A post-translational modification to a protein, comprising addition of sugars.

Gynoecium Collective term for the female organs (carpels) of a flower (synonym: pistil).

Heritability In the narrow sense, it is the proportion of phenotypic variance that can be attributed to additive genetic variance.

Herkogamy The spatial separation of dehiscing anthers and receptive stigmas within flowers. Common in flowering plants and reduces intra-flower self-pollination.

Hermaphrodite Possessing both male and female reproductive organs.

Heteroallelic pollen Pollen with two different S-alleles. Normal pollen is haploid, having only one allele at each locus, including the genes at the S-locus. Duplication of a locus, as in a tetraploid, allows the possibility of two different alleles being present and expressed in haploid pollen.

Heteroallelic multimers Multimers formed from products of different allelic genes. For example, in heteroallelic S_1S_2 pollen, multimers of a pollen S protein could be composed of products of both the pollen S_1-gene and the pollen S_2-gene (heteroallelic multimer), or products of just the pollen S_1-gene, or products of just the pollen S_2-gene (homoallelic multimers).

Heterosis Often called hybrid vigour, which describes the improved characteristics of hybrids over their parents.

Heterostyly A sexual polymorphism in which populations are composed of two (distyly) or three (tristyly) floral morphs with reciprocal arrangements of anthers and stigmas (reciprocal herkogamy). Usually associated with *heteromorphic incompatibility* and ancillary pollen and stigma polymorphisms. Maintained in populations by disassortative mating between the floral morphs.

Heterotetramers A tetramer composed of different subunits. See Heteroallelic multimer.

Hexaploid Cells or organisms having six full sets (6x) of the basic chromosome number (x) in somatic cells. Hexaploidy is cytogenetically classified into two types: autohexaploidy, consisting of six copies of the same genome (e.g., sweet potato); and allohexaploidy, consisting of three different genomes (e.g., bread wheat).

High-dimensional equation Equation in terms of multiple variables.

High resolution mapping The production of a detailed genetic map with recombination frequencies of less than 1cM either for the full genome or for a defined region of interest on a chromosome.

Homology Similarity attributed to descent from a common ancestor.

Homostyle A variant of a distylous species in which flowers have anthers and stigmas at equivalent levels and plants are usually self-compatible. Long homostyles have long styles and long-level anthers and short homostyles have short styles and short-level anthers.

Homoallelic pollen Pollen grains containing two copies of the same pollen S-allele.

Homozygosity This occurs when the alleles at a particular locus are identical on both chromosomes; the frequency with which a randomly-sampled pair of genes are identical.

Hybridization The crossing of two different species, races or varieties to form a hybrid.

Hydropathy plot A measure of the hydrophobicity of a protein region, predicting the likelihood of signal peptides and transmembrane domains.

Hypomorphic allele An allele that has reduced levels of gene activity.

Hypervariable regions Regions of a protein that are especially variable and, thus, could be involved in recognition.

Hypervariability High degree of variation among individuals at a given genetic marker.

Immunolocalization A technique involving the use of an antibody to establish where a particular protein is distributed in a cell.

Immunoprecipitation Use of an antibody to "pull-down" the antigen that cross-reacts with it, to enrich for a specific protein.

Inbred lines Inbred lines are produced by successive cycles of self-pollination. They are characterised by increased levels of homozygosity as they move through each cycle of selfing.

Inbreeding depression A reduction in fitness and vigour of inbred progeny (resulting from self-fertilization or sib-matings), relative to outbred progeny.

Intergenic recombination Breakage and reunion event (crossover) between non-sister chromatids of homologous chromosomes to produce new combinations of genes along a chromosomal arm.

Interspecific incompatibility An active pre-zygotic mechanism preventing fertilization (hybridization) between two different species.

Invasion probability Probability that an allele introduced in very low frequencies becomes relatively common before becoming extinct.

Kinase An enzyme that modifies a target protein by phosphorylation during a signalling cascade, leading to a variety of cellular responses.

Linear dominance hierarchy Interactions between multiple alleles that, in the different heterozygous combinations, show a range of dominant-recessive relationships.

Linkage A situation when genetic loci or alleles are inherited together due to their close physical location on a chromosome.

Linkage disequilibrium The non-random association of alleles at two or more loci.

Macroevolutionary Evolution at and above the species level.

Male S-determinant/Pollen S The protein encoded by the S-locus to determine the SI specificity in the pollen.

MAPK Mitogen activated protein kinase. A specific type of protein kinase activated by phosphorylation, and ultimately in a variety of cellular responses.

Mating-system transitions Changes in the outcrossing rate of populations, as may occur, for instance, after the transition from self-incompatibility to self-compatibility.

Maximum-likelihood analysis A statistical approach to estimation based on maximizing the probability of the observations.

Metapopulation dynamics Migration of genes (via pollen, seed or individuals) among assemblages of spatially delimited local populations.

Microsomal Generally the fraction of a cell extract that comprises the organelles (as opposed to the cytosol).

Microspores Immature pollen grains.

Microtubule A major cytoskeletal component, comprising alpha and beta tubulin polymers. They are very dynamic and require GTP for polymerization. Their dynamics are regulated by microtubule-associated proteins (MAPs).

Modifier gene A gene that affects the phenotypic expression of another gene or that modifies the action of another gene.

Modifier alleles Alleles that change the conditions under which other genes evolve.

Molecular marker Polymorphic DNA sequence that can be used for distinguishing different genotypes, constructing linkage maps, and for selection and identification of genes.

Monocotyledons Monocotyledons or 'monocots' produce only a single embryonic leaf. They comprise many agricultural plants including the grasses and cereals, orchids, palms and lilies. The other group of plants is the dicots.

Monoecy Unisexual flowers with male and female flowers separately located on the same plant.

Monomorphic A population or species with only one allele or phenotype where polymorphism is usually evident in other populations or related species.

Monophyletic clade A group with a single common ancestor that contains all of the descendants of that ancestor.

Multiallelic loci A locus where multiple variants or alleles may be present within a population.

Multigene complex A group of linked genes.

Multi-locus Controlled by more than one genetic locus.

Mutagenesis The process of inducing mutations. For example, chemical mutagenesis causes changes in the base sequence of a gene. In molecular biology, the creation of a mutation at a defined site in a DNA molecule.

Natural populations Populations unaffected human disturbance or selection.

Negative-frequency-dependent selection Natural selection in which the fitness of individual genotypes or phenotypes depend on their frequency in a population. In negative-frequency-dependent selection rare genotypes or phenotypes have higher fitness than those that are common; i.e. selection that always favours rare forms. Commonly results in polymorphic equilibria.

Noncanonical Outside the accepted dogma. With reference to noncanonical protein complexes, the term refers to complexes with an unusual composition or unusual subunits.

Non-conservative replacement A situation where an amino acid is changed for one which has a different charge, which may impact on conformation.

Non-contiguous Not connected.

Non-synonymous nucleotide polymorphism Nucleotide substitutions in coding regions that alter amino acid composition.

Non-synonymous substitutions Nucleotide substitution that results in a different amino acid change.

Normed variability index A window-averaged measure of variability across a protein sequence that is calculated from a protein sequence alignment. A plot of the normed variability index is used to identify protein regions that are especially variable.

Nucleotide variation Variation in the base pair constitution at a given nucleotide site.

Nucleotide substitution Replacement of one nucleotide by another.

Obligate outcrossers Plants that are not capable of self-fertilization and must mate with a different genetic individual to produce viable offspring, generally due to a genetically determined self-incompatibility mechanism, dioecy, or herkogamy.

Obligately heterozygous A locus for which individuals are always heterozygous, as in the gametophytic S-locus.

Oligomers A polymer comprising a short chain of monomers.

One-dimensional diffusion equation Diffusion equation describing change in a single state variable.

Orthologues A gene/protein in two or more species that has evolved from a common ancestor and so have similar structure and function across species.

Papilla Hair-like cells found on the stigma surface.

Parthenocarpy The development of fruit without sexual fertilization.

Pellicle The proteinaceous outermost layer of the stigmatic papilla that is involved in the selective adhesion of compatible pollen grains.

Phenotypic plasticity Ability of a single genotype to express different phenotypes across environments.

Phosphorylation Addition of one or more phosphate (PO_4) groups, usually to a protein, in order to modify its activity. Reversible phosphorylation is an important signalling and regulatory mechanism.

Phylogenetic reconstructions Reconstruction of relationships among genes or species using phylogenetic methods.

Phylogenetic A branching diagram representing the evolutionary history of a group of organisms.

Physical distance The distance, expressed in base pairs (bp), between two genes or loci, generally obtained by direct sequencing of genomic clones. In contrast to genetically estimated distances, expressed in centimorgans (cM), derived from linkage analysis. See also 'physical map.'

Physical linkage When genetic loci or alleles are inherited together, occurring on the same chromosome.

Physical map A physical map describes the orders of genes along chromosomes of a genome using base pairs, i.e. the number of DNA nucleotides, as the distance measurement.

Pin morph A flower morph with short stamens and long pistils (their positioning being reciprocal to the thrum morph).

Pollen coat Secreted material present in the exine cavities of the pollen grain. It is composed of lipids, proteins, pigments, and aromatic compounds.

Pollen discounting Reduction in the amount of pollen available for export due to self-pollination.

Pollen-part mutations Mutations affecting only pollen functions in SI.

Polygenic A quantitative trait whose phenotypic variation is due to the interaction of a large number of genes, each with a small additive effect on the trait.

Polymorphism The existence of a gene or an individual in several forms in a single species or population.

Polymorphic equilibria Equilibrium points where two or more alleles or discrete phenotypes are maintained at particular frequencies in a population as a result of selection.

Polyphyletic origin A character that has evolved on more than one occasion in two or more different lineages.

Polyploidy The possession of more than two complete sets of homologous chromosomes.

Population bottleneck A severe reduction in the size of a population usually as a result of ecological factors.

Population genetics The study of genetic variation in populations.

Population size The number of individuals in a population.

Poricidal anther dehiscence Anthers in which pollen is dispersed through a small pore at the apical end. Associated with buzz pollination by bees.

Positional cloning A method of obtaining an unknown gene of interest that uses information based upon its position on a chromosome, obtained by gene mapping techniques (see high resolution mapping).

Posterior probability The probability of a random variable given the value of another variable, using Bayes' theorem, multiplying the prior probability distribution by the likelihood function and then dividing by the normalizing constant.

Post-zygotic self-sterility SI manifesting itself after fertilization, e.g., embryo abortion.

Proteasome A large protein complex, comprising several subunits that function to selectively degrade poly-ubiquitylated proteins in an ATP-dependent manner.

Proteasomal activity Proteolytic activity presented by the proteasome. The proteasome protease relies on a threonine residue in the active site. Proteins processed by the proteasome are reduced to small polypeptides.

Pseudo-gene A gene that has lost its protein coding ability.

Pseudogenization Loss of protein-coding function (making a pseudo-gene).

Pseudo-self-fertility/pseudo-compatibility Condition in which genes can modify or inhibit S gene activity in plants with functional self-incompatibility systems. Situation where compatibility can be induced between normally incompatible pollen and stigma by environmental circumstances, such as elevated temperatures.

Purifying selection A type of natural selection in which genetic diversity decreases because of selection against extreme values of the character.

Radiations Rapid and dramatic increases in diversity. Usually where a taxonomic group occupied a fairly small ecological niche but rapidly expanded out into new environments and regions with an associated increase in diversity.

Ramets A collection of plants derived from the same genetic individual via clonal reproduction.

Ribosome A complex of proteins and RNAs that functions in protein synthesis.

Realized heritabilities The ratio of the single-generation progress of selection to the selection differential of the parents.

Receptor Usually a transmembrane protein with an extracellular domain that binds to a ligand (such as a hormone); this binding stimulates a signalling cascade resulting in a biological response by the cell.

Receptor-like kinases (RLKs) Transmembrane proteins which are important in cell signalling. They have amino-terminal extracellular domains, transmembrane domains and carboxy-terminal intracellular kinase domains. The kinase domain is responsible for transferring phosphate from ATP to a second substrate, often forming the first step in a signalling cascade.

Receptor kinase Receptor kinases are single pass transmembrane proteins located at the plasma membrane, and composed of an extracellular receptor domain, a transmembrane domain, and an intracellular kinase domain. The receptor binds a specific extracellular signal (ligand) which leads to activation of the intracellular kinase domain.

Recessive allele An allele whose phenotypic effect is not expressed in the heterozygous condition; usually designated by a small letter (e.g., t).

Recombinant protein An artificially produced protein, often made by inserting DNA sequence into a plasmid, allowing the rapid production of large amounts of protein.

Recombination The process of breaking a DNA molecule, and repairing the broken DNA using the other homologous chromosome.

Recombination breakpoint The position of a reciprocal exchange (crossover) in meiosis between homologous chromosomes.

Recombination value Frequency of recombination events between two loci expressed in centiMorgans (cM).

Reproductive assurance Mechanism that permits self-seed production when cross pollen limits seed production.

Reproductive resources Resources invested in reproductive structures or functions.

Retrograde transport Movement of material 'backwards' through the endomembrane system. Secretion is regarded as anterograde transport, movement from the endoplasmic reticulum through the Golgi and hence to the plasma membrane. Material captured by endocytosis moves in the opposite, retrograde, direction from the plasma membrane to a variety of compartments including the vacuole, Golgi, and endoplasmic reticulum.

Retrotransposon Mobile genetic elements that can amplify themselves in a genome and are ubiquitous components of the DNA of many eukaryotic organisms.

RNAi Abbreviation for RNA interference. A method for reducing expression of a specific gene. Often, a gene construct is produced such that transcription produces a hairpin RNA that is processed to form small RNAs that stimulate degradation of the target gene transcript, thus, inhibiting its expression. See also, 'Antisense.'

RNase (ribonuclease) An enzyme that catalyzes the hydrolysis of RNA.

***S*-locus** The self-incompatibility *S*-locus. It comprises a single polymorphic locus that controls SI in most species. A physical locus containing at least two genes, responsible for pollen and pistil recognition specificity, thereby determining the SI specificity in pollen and pistil.

S-allele As the *S*-genes are polymorphic, *S*-alleles are variants of these genes (e.g., the pollen *S*-allele and the pistil *S*-allele). While classical genetic studies predicted that the self-incompatibility trait was determined by a single polymorphic *S*-locus, molecular studies have found that there are generally at least two tightly-linked polymorphic *S*-genes involved (e.g., the pollen *S*-gene and the pistil *S*-gene). The linked combination of alleles from these genes is usually referred to as *S-haplotype* in current literature.

S-haplotype The *S*-haplotype is determined by the specific allele combinations inherited from at least of two physically-linked, multi-allelic genes (e.g., the pollen *S*-gene plus the pistil *S*-gene). For example, a specific combination of *Brassica* SRK and SP11/SCR alleles that are inherited together represent an *S*-haplotype and encode a specific ligand–receptor interaction. Because of this, "*S*-haplotype," rather than *S*-allele, is used for the SRK + SP11/SCR allele combinations that determine the self-incompatibility trait.

S protein Self-incompatibility protein (the pistil *S*-determinant in *Papaver*).

Scaled mutation rate Product of the rate of mutation and population size.

Secondary structure prediction Computer-assisted prediction of the two-dimensional structure of a protein based on the arrangement of the polypeptide chain in alpha helices and beta sheets.

Second messenger A component of a signal transduction system used to transduce and amplify a signal into a cellular change. Ca^{2+} is a second messenger.

Seed bank Collection of viable seeds in the soil, accumulated over multiple reproductive seasons.

Seed discounting Reduction in the number of ovules available for outbreeding due to self-pollination.

Self-incompatibility (SI) Prevention of self-fertilization by otherwise normal and viable gametes, due to their genetic similarity.

Self sterility An older term for self-incompatibility (SI).

Sequence identity A measure of how similar two proteins are. This refers to the number of aligned positions where the amino acids are identical, expressed as a percentage.

Sequence polymorphism Variations in nucleotide sequences among different genotypes, which can be single base pair differences, identified by direct sequencing of the genome, or by RFLP (restriction fragment length polymorphism) or AFLP (amplified fragment length polymorphism) analyses.

Selective sweep Fixation of a selectively advantageous allele.

Sheltered load Genetic load that is closely linked to the *S*-locus of self-incompatible plants.

Signalling ligand Generally a small molecule or protein that binds to a target protein (often a receptor) to elicit a signalling response.

Speciation rate The rate of formation of new species in a lineage.

Signal peptide Short peptide that directs protein transport from the cytoplasm to membrane bound organelles or cell membrane.

Soluble inorganic pyrophosphatase (sPPase) An essential enzyme responsible for catalyzing the hydrolysis of inorganic pyrophosphate (PPi) to phosphate (Pi). This drives many biosynthetic reactions, which otherwise would not go to completion.

Sporophytic SI (SSI) A form of SI genetically distinct from gametophytic SI (GSI). In the SSI response, the recognition specificity of the pollen is determined by the diploid genotype of the parent (the sporophyte). Pollen tube growth is arrested when one of its *S*-haplotype matches either of the *S*-haplotypes of the diploid pistil; when pollen carries an *S*-haplotype not present in the pistil, it is compatible. This form of SI is best characterized in Brassica, though it has been identified in other families.

Staminode A sterile stamen.

Star-like A phylogeny in which all diversification appears to occur at the base.

Stationary distribution Distribution of the frequency of an allele at stochastic steady-state.

Stigmatic papillae Stigmatic papillae are elongated epidermal cells that protrude from the surface of the plant stigma and function to receive pollen grains.

Supergene complex A group of two or more tightly linked genes that function together to produce adaptive characters.

Synonymous nucleotide polymorphism Nucleotide substitutions in coding regions that do not alter amino acid composition.

Synonymous substitution A nucleotide substitution that does not result in an amino acid change.

Synteny The physical co-localization of genetic loci on the same chromosome within an individual or species.

Tapetum In the anther, the nutritive tissue of the inner-most layer of cells of the pollen sac wall.

Tertiary conformation Three-dimensional structure of a protein.

Tetraploid An organism with four copies of each chromosome (two pairs of chromosomes to yield four of each chromosome). For example, *Nicotiana alata* has a base chromosome number of ten. Therefore, a normal diploid sporophyte has 20 chromosomes and a tetraploid has 40 chromosomes.

Thrum morph A flower morph with long stamens and short pistils (their positioning being reciprocal to the pin morph).

Transcriptome The set of all mRNAs produced by an organism, tissue, or cell type.

Transient expression A relatively rapid way to assess the function of a gene of interest and/or to assess the localization of the protein encoded using GFP-constructs, in live cells.

Transitions Changes among states.

Transition rate The rate of transitions between character states in a lineage. For instance, the rate at which species switch from self-incompatibility to self-compatibility.

Translocation A genetic change where a chromosomal segment is moved to a new location, thus, changing the linkage relationships in the genome.

Transmembrane protein A protein with amino acids that form a domain that is predicted to span (usually several times) a plasma membrane. Many transmembrane proteins are receptors.

Transmitting tract The central tissue of styles through which pollen tubes grow on their pathway to the ovary.

Transmitting tract ECM (extracellular matrix) Material secreted by cells of the pollen tube transmitting tract in the pistil.

Trans-specific polymorphism Greater similarity between nucleotide sequences sampled from different species than from the same species; the sharing of polymorphisms among closely related species.

Transposon/transposable element A heterogeneous class of genetic "elements" (DNA sequences) that are able to move and insert at numerous sites within the genome and, occasionally, induce an insertion mutation or genome rearrangement.

Transformation A process by which the genetic material carried by an individual cell is altered by the incorporation of foreign (exogenous) DNA into its genome.

Triplex A tetraploid plant in which two alleles are present in a 3:1 ratio.

Two-Locus Two genetically independent loci or genes are involved in controlling the process.

Ubiquitination (Also referred to, more correctly, as ubiquitylation) The process of adding ubiquitin to a protein through an iso-peptide linkage. In poly-ubiquitination, this is often refereed to as the "kiss of death" process for a protein, as several ubiquitin molecules attach to a specific protein and acts as a tag that signals the protein-transport machinery to ferry the protein to the proteasome for degradation. See also 'E3 ligase complex.'

Unisexual flowers Flowers with either male reproductive organs (stamens) or female reproductive organs (carpels/pistils).

"Wet" stigma A stigma with a surface secretion, either lipidic or aqueous.

Yeast two-hybrid This is an interaction-based screening technique carried out in yeast where the interaction between two proteins (called a prey and bait) leads to activation of selectable markers and/or reporter genes.

Zygomorphic Flowers with bilateral symmetry and therefore divisible into two equal halves in one plane.

Zygotes The product of gamete fusion.

Part I
Evolution and Population Genetics of Self-Incompatibility

Chapter 1
New Insights on Heterostyly: Comparative Biology, Ecology and Genetics

S.C.H. Barrett and J.S. Shore

Abstract Here, we review recent progress on the evolutionary history, functional ecology, genetics and molecular biology of heterostyly using a variety of taxa to illustrate advances in understanding. Distyly and tristyly represent remarkable examples of convergent evolution and are represented in at least 28 flowering plant families. The floral polymorphisms promote disassortative mating and are maintained in populations by negative frequency-dependent selection. Comparative analyses using phylogenies and character reconstruction demonstrate multiple independent origins of heterostyly and the pathways of evolution in several groups. Field studies of pollen transport support the Darwinian hypothesis that the reciprocal style–stamen polymorphism functions to increase the proficiency of animal-mediated cross-pollination. Although the patterns of inheritance of the style morphs are well established in diverse taxa, the identity, number and organization of genes controlling the heterostylous syndrome are unknown, despite recent progress. In future, it will be particularly important to establish the contribution of 'supergenes' vs. regulatory loci that cause morph-limited expression of genes.

Abbreviations

2D PAGE	Two-dimensional polyacrylamide gel electrophoresis
GPA locus	Gynoecium, Pollen size, Anther height
IEF	Isoelectric focussing
L-morph	Long-styled morph

S.C.H. Barrett
Department of Ecology and Evolutionary Biology, University of Toronto, 25 Willcocks Street, Toronto, Ontario, Canada M5S 3B2, e-mail: barrett@eeb.utoronto.ca

J.S. Shore
Department of Biology, York University, 4700 Keele Street, Toronto, Ontario, Canada M3J 1P3

M-morph	Mid-styled morph
mRNA	messenger RNA
P*	Abnormal pin morph
PvSLL1	cDNA from *Primula* linked to the *S-locus*; encodes a small putative transmembrane protein of unknown function
PvSLL2	cDNA from *Primula* closely linked to the *S-locus*; has homology to the *CONSTANS-LIKE* gene
SDS-PAGE	Sodium dodecyl sulfate polyacrylamide gel electrophoresis
SI	Self-incompatibility
S-locus	Self-incompatibility locus
S-morph	Short-styled morph

1.1 Introduction

The sexual organs of most flowering plants exhibit a small degree of continuous variation resulting from quantitative inheritance and environmental influences. A strikingly different pattern of variation is evident in populations of some species, where hermaphroditic individuals fall into two or three morphologically distinct mating groups, which differ in style length, anther height and a suite of ancillary pollen and stigma polymorphisms (Darwin 1877; Vuilleumier 1967; Ganders 1979; Barrett 1992). Populations with this type of polymorphic sexual variation are distylous or tristylous, respectively, and the general condition is referred to as heterostyly.

The defining feature of heterostylous populations is a reciprocal arrangement of sex-organ heights in the floral morphs (Fig. 1.1), also known as reciprocal herkogamy (Webb and Lloyd 1986). By convention, the morphs are referred to as long- and short-styled (hereafter L- and S-morphs) in distylous populations, and long-, mid- and short-styled (hereafter L-, M- and S-morphs) in tristylous populations. In most heterostylous species, reciprocal herkogamy is associated with a hetermorphic self-incompatibility (SI) system that limits or prevents self- and intra-morph mating. Therefore, compatible pollinations occur only between anthers and stigmas of equivalent height, termed 'legitimate pollinations' (Fig. 1.1; Darwin 1877). Understanding the evolution, function and genetic basis of heterostyly has attracted considerable attention since Darwin's classic book on polymorphic sexual systems in plants (Darwin 1877). The sustained fascination with heterostyly occurs because the sexual polymorphisms are a remarkable example of convergent evolution. In addition, they represent one of the classic research paradigms for the study of evolution and adaptation in plants (Barrett 1992).

Darwin (1877) provided the earliest adaptive explanation for the function of heterostyly. He proposed that the reciprocal placement of anthers and stigmas was a mechanism to promote pollinator-mediated cross-pollination between floral morphs.

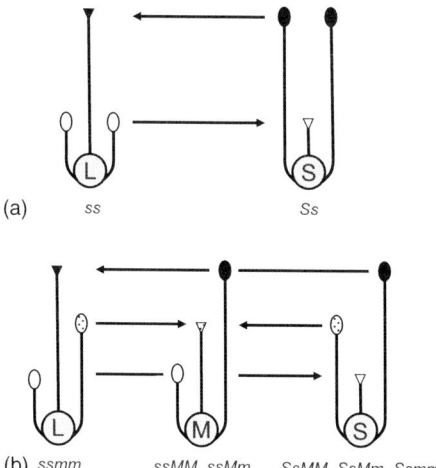

Fig. 1.1 The heterostylous floral polymorphisms: (**a**) distyly and (**b**) tristyly. L, M and S refer to the long-, mid- and short-styled morphs, respectively. The arrows indicate cross-pollinations between anthers and stigmas of equivalent height. In the majority of heterostylous species these are the only compatible pollinations. Genotypes for the floral morphs with the most common patterns of inheritance are indicated (see Sect. 1.4.1. for further details)

According to Darwin's hypothesis, pollinators visiting heterostylous flowers pick up pollen on different parts of their bodies during nectar feeding, and cross-pollen transfer between floral morphs is promoted by this segregated pollen deposition. Several lines of evidence support Darwin's cross-pollination hypothesis (Kohn and Barrett 1992; Lloyd and Webb 1992b) and heterostyly is generally described as an 'outcrossing mechanism'. However, this interpretation is insufficient for two reasons. First, self-incompatibility (SI) in heterostylous populations guarantees outcrossing, and, second, by preventing intra-morph mating heteromorphic SI restricts mating opportunities with one half (distyly), or one third (tristyly), of the plants in a population. A more complete interpretation of the adaptive significance of heterostyly recognizes different functional roles for the morphological and physiological components of the polymorphism in promoting fitness through male and female function, respectively.

Reciprocal herkogamy, as Darwin (1877) proposed, functions to promote proficient cross-pollination by reducing male gamete wastage on incompatible stigmas and increasing fitness through male function (Lloyd and Webb, 1992a, b). In contrast, SI safeguards against self-fertilization and inbreeding depression, thereby promoting the maternal component of fitness. Recognition of these different paternal and maternal functions resolves the apparent redundancy of two mechanisms with essentially the same role. The evolution of heterostyly reduces the conflict that can occur in sexually monomorphic animal-pollinated species—achieving efficient cross-pollination but simultaneously avoiding self-interference between female and male sexual organs (Barrett 2002).

Studies of heterostyly have largely concentrated on a few well-characterized taxa (e.g. *Primula, Linum, Lythrum*) originally studied by Darwin (1877). Of these, *Primula* has attracted most attention and is often represented in the literature as the model system for heterostyly (Mast and Conti 2006; see also Chap. 14, this volume). However, heterostyly is now reported from at least 28 angiosperm families and the polymorphism has evolved on numerous occasions. In addition, other stylar polymorphisms have also been recognized with their own distinctive features [e.g. stigma-height dimorphism (Baker et al. 2000a, b; Barrett et al. 2000a); enantiostyly (Barrett 2002; Jesson and Barrett 2003); flexistyly (Li et al. 2001; Renner 2001; Sun et al. 2007); inversostyly (Pauw 2005)] raising new questions about their evolution, function and relation to heterostyly.

Early research on heterostyly largely focused on genetical aspects of the polymorphism. Indeed, fundamental concepts in Mendelian and population genetics, including patterns of inheritance, linkage, supergenes, epistasis and polymorphic equilibria, were initially studied in *Primula* and *Lythrum* by leading geneticists, including W. Bateson, R.A. Fisher, J.B.S. Haldane, A. Ernst. A.B. Stout, K. Mather and D. Lewis. Today, a much broader range of questions are being addressed in heterostylous species employing diverse approaches. The objective of this chapter is to review recent advances on the study of heterostyly by examining progress made since the last general treatment (Barrett 1992). We review work on the comparative biology, ecology and genetics of heterostyly and conclude by briefly outlining future research for solving outstanding problems remaining in the study of heterostylous plants. A goal of our review is to demonstrate that although heterostyly is probably the most well studied plant sexual polymorphism, there still remain many unanswered questions that require future investigation.

1.2 Comparative Biology and Evolutionary History of Heterostyly

Heterostyly has a scattered distribution among at least 28 angiosperm families with new heterostylous taxa continuely reported [e.g. distyly in *Aliciella* formerly *Gilia* (Polemoniaceae; Tommerup 2001); *Salvia* (Lamiaceae; Barrett et al. 2000b); and *Tylosema* (Caesalpinioideae; Hartley et al. 2002); tristyly in *Hugonia* (Linaceae; Thompson et al. 1996)]. Lloyd and Webb (1992a) surveyed the character states of 25 families with heterostylous species to identify why the polymorphism may have evolved in some families and not others. Their analysis indicated that there are constraints on the types of flowers in which reciprocal herkogamy is likely to evolve. Heterostylous flowers are usually actinomorphic with a simple open corolla and a floral tube with nectar concealed at the base. These flowers are described as stereomorphic or 'depth-probed' (Lloyd and Webb 1992a), with sexual organs contacted by long-tongued pollinators in succession during nectar feeding. Heterostyly is rarely associated with strongly zygomorphic flowers, probably because in such groups effective cross-pollen transfer is achieved through pollinator positioning.

Families in which flowers possess numerous stamens, free carpels, open-dished shaped corollas and exposed nectar usually lack the precision in pollen transfer required for the evolution of reciprocal herkogamy. However, exceptions to these patterns exist in heterostylous taxa [e.g. zygomorphy in *Salvia* (Barrett et al. 2000b); open dish- or bowl-shaped flowers in *Fagopyrum* (Bjorkman 1995) and *Turnera* (Rama Swamy and Bahadur 1984); numerous stamens and nectar-less flowers in *Hypericum* (Ornduff 1975)], raising the question of how reciprocal herkogamy evolved and is maintained in these taxa.

1.2.1 Phylogeny Reconstruction and Character Evolution

The advent of molecular systematics, phylogeny reconstruction and character mapping has led to interest in the evolutionary history of SI. See Chap. 4 for consideration of the evolutionary history of homomorphic SI. Here, we consider studies of heterostylous taxa, which include Pontederiaceae (Kohn et al. 1996), *Amsinckia* (Schoen et al. 1997), *Houstonia* (Church 2003); *Primula* (Mast et al. 2004), *Narcissus* (Graham and Barrett 2004); *Linum* (Armbruster et al. 2006), *Turnera* (Truyens et al. 2005) and Lythraceae (Morris 2007). Phylogenetic analyses of these groups have been conducted to address questions concerning the origin and evolutionary history of heterostyly and related sexual systems, and these have included the following: (1) Has heterostyly evolved more than once in a particular lineage? (2) What are the ancestral states and intermediate stages involved in the evolution of heterostyly? (3) What is the order of establishment of morphological and physiological traits in the heterostyous syndrome? (4) What are the evolutionary relationships between heterostyly and related stylar conditions, including homostyly? Part of this work has been motivated by efforts to distinguish between the predictions of competing theoretical models on the evolution of heterostyly (Charlesworth and Charlesworth 1979a; Lloyd and Webb 1992a, b).

Lloyd and Webb (1992a) estimated that heterostyly originated on at least 23 separate occasions, based on its distribution among 19 orders of flowering plants, but conceded that many more origins may be involved if multiple origins have occurred within heterostylous taxa. Although several studies have assumed that heterostyly is the basal condition in lineages (Schoen et al. 1997; Truyens et al. 2005), other evidence points to multiple origins of heterostyly within some genera (Graham and Barrett 2004). Inferences on the number of origins of heterostyly within particular groups may be particularly sensitive to taxon sampling and the weighting schemes employed for the gain and loss of heterostyly [e.g. Pontederiaceae (Kohn et al. 1996); *Linum* (Armbruster et al. 2006); *Primula* (Mast et al. 2006)]. In large geographically widespread families with numerous heterostylous species, such as Rubiaceae and Oxalidaceae, the polymorphism may have had multiple origins. If this turns out to be the case, it will be interesting to investigate the details of each transition, including their ecological and developmental basis and if the patterns of inheritance are similar.

Determining the ancestral states of heterostylous lineages is of importance for distinguishing models of the evolution of heterostyly. Lloyd and Webb (1992a, b) proposed that the immediate ancestors of distylous species were monomorphic for style length and possessed approach herkogamy, with long styles and stamens positioned below the stigma. In contrast, in the model of Charlesworth and Charlesworth (1979a), the ancestral condition involves a population with long styles and long-level stamens. Both models assume that distyly evolves via an intermediate stage of stigma-height dimorphism with short-styled variants invading long-styled monomorphic populations.

Narcissus offers an opportunity to evaluate these models because the genus contains species with stylar monomorphism, stigma-height dimorphism and distyly (Barrett et al. 1996). Reconstruction of the evolutionary history of stylar conditions provides evidence to support the stages envisioned in the Lloyd and Webb model (Graham and Barrett 2004). As shown in Fig. 1.2, the immediate ancestors of

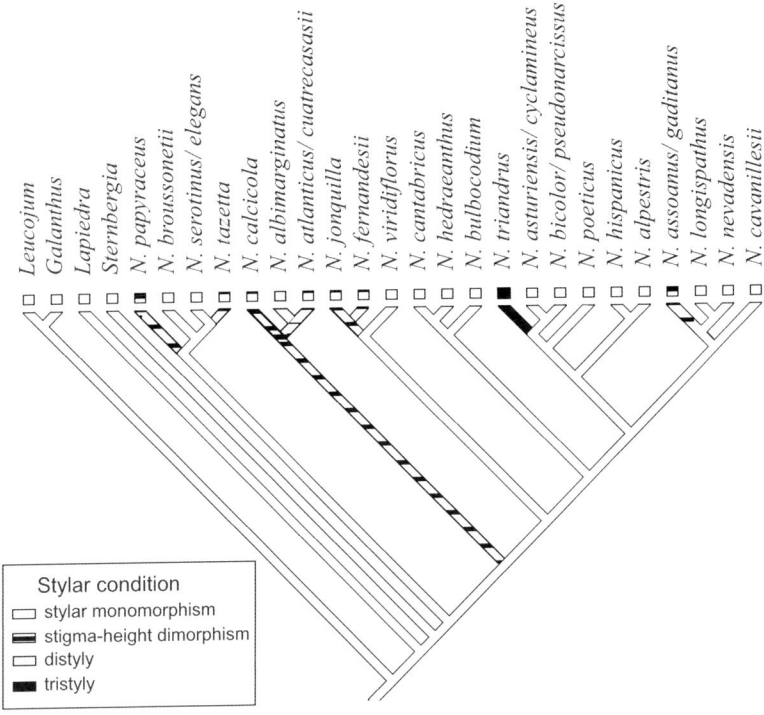

Fig. 1.2 Reconstruction of the evolutionary history of stylar polymorphisms in *Narcissus* based on parsimony analysis of combined *ndh*F and *trn*L–*trn*F DNA sequences. Some taxa with identical stylar conditions are reduced to single termini. Six origins of stylar polymorphism are inferred, with independent origins of distyly and tristyly and multiple origins of stigma-height dimorphism. The single distylous species, *N. albimarginatus*, is descended from ancestors with stigma-height dimorphism supporting theoretical models that propose stigma-height dimorphism as an intermediate stage in the evolution of distyly, see Sect. 1.2.1 for further details. After Graham and Barrett (2004), figure published with permission from *New Phytologist*

the sole distylous member of the genus (*N. albimarginatus*) possess stigma-height dimorphism, and the basal condition in the genus is stylar monomorphism and approach herkogamy. The rarity of stigma-height dimorphism in flowering plants (Barrett et al. 2000a) limits opportunities to examine these models more generally. However, reconstruction of the evolutionary history of stylar conditions in Boraginaceae would be of value because this family includes genera (e.g. *Lithodora*) with both distyly and stigma-height dimorphism.

Models of the evolution of heterostyly also differ in the order in which the morphological and physiological components of heterostyly are established. In the Charlesworth and Charlesworth (1979a) model, the origin of diallelic SI is a prerequisite for the subsequent evolution of reciprocal herkogamy, whereas in Lloyd and Webb's (1992a,b) model, the style–stamen polymorphism establishes first and diallelic incompatibility may, or may not, follow depending on the importance of selection against the genetic costs of self-fertilization. It has been difficult to obtain comparative evidence to support either sequence, in part, because SI systems in heterostylous groups commonly exhibit considerable variation in expression. This ranges from self-compatibility (SC), through cryptic self-incompatibility, to differences in the strength of incompatibility among the floral morphs, to rigid SI in all morphs (reviewed in Barrett and Cruzan 1994). This variation complicates character reconstructions, as does the problem of determining whether the self-compatible status of some heterostylous species represents an ancestral or derived condition. These issues were encountered in the only explicit attempt to determine the order of establishment of reciprocal herkogamy and incompatibility, in tristylous Pontederiaceae (Kohn et al. 1996). The isolated case of distyly in self-compatible *Salvia brandegeei* of the Lamiaceae is noteworthy because there are no reliable reports of SI in this family (Barrett et al. 2000b). This observation supports the Lloyd and Webb view that reciprocal herkogamy can establish independently of diallelic incompatibility.

The status of stylar monomorphism is contentious in heterostylous groups as it could represent either an ancestral or derived condition. In *Primula*, 'secondary homostyly' through recombination of the linkage group controlling distyly is well known (Charlesworth and Charlesworth 1979b; and see Sect. 1.4.2.). However, several authors have interpreted monomorphism in some *Primula* species as representing the ancestral condition in the genus prior to the origin of distyly (Ernst 1955; Al Wadi and Richards 1993). Recent phylogenetic analyses of *Primula* cast doubt on this interpretation, suggesting instead that species with stylar monomorphism are more likely to be derived from distylous ancestors (Mast et al. 2006). Among the derived monomorphic taxa are members of *Dodecatheon*, a group of buzz-pollinated species (Mast et al. 2004, 2006). This indicates an intriguing shift in pollination system from distyly, in which pollen dispersal is based on nectar feeding and the geometry of pollinator contacts, to a strikingly different strategy based on nectarless flowers and metered pollen dispensing from anthers with poridical dehiscence. Mast et al. (2004) detail the character-state changes associated with this transition, but nothing is known about the ecological mechanisms responsible for this transition in pollination system.

Self-pollinating homostyles commonly occur in heterostylous groups. In *Eichhornia* (Kohn et al. 1996), *Amsinckia* (Schoen et al. 1997) and *Turnera* (Truyens et al. 2005), phylogenetic reconstructions have provided good evidence that homostyles are derived from heterostylous ancestors. The shift to predominant selfing has occurred on multiple occasions within each genus, probably as a response to selection for reproductive assurance under conditions of unreliable pollinator service. Although the acquisition of homostyly is commonly associated with the evolution of small flowers and the loss of floral traits promoting outcrossing, as in *Amsinckia* and *Eichhornia*, this is not the only evolutionary outcome. In *Turnera*, some homostylous populations have large flowers, display a moderate degree of herkogamy and exhibit mixed mating systems (Belaoussoff and Shore 1995). Homostylous populations retain residual diallelic incompatibility reactions, suggesting that they originated as long-homostyled recombinants with stigmas and anthers at equivalent heights (Tamari et al. 2001 and see Sect. 4.2.). Selection on quantitative genetic variation governing the degree of herkogamy has led to increased outcrossing rates (Barrett and Shore 1987; Shore and Barrett 1990), illustrating the potential evolutionary lability of mating patterns in some homostylous populations. The transition from heterostyly to homostyly is the most frequent evolutionary change in floral biology and mating system that has occured among heterostylous groups. Future work on the phylogeny of heterostylous taxa is critical for providing a comparative context for investigating the ecological, genetic and developmental basis of this shift from outcrossing to selfing.

1.3 Function and Reproductive Ecology of Heterostyly

Heterostyly is a conspicuous floral polymorphism that can be readily identified under field conditions. Consequently, there is growing interest in using the polymorphism to address diverse questions in functional ecology, ecological genetics and conservation biology. Here we review recent studies on these topics.

1.3.1 Function of Heterostyly

The occurrence of pollen-size heteromorphism in heterostylous species enables measurement of disassortative (inter-morph) pollen transfer based on the analysis of stigmatic pollen loads. Data obtained from 'pollen flow' studies can be used to evaluate Darwin's cross-promotion hypothesis for the adaptive significance of heterostyly. Ganders (1979) and Lloyd and Webb (1992b) summarized experimental evidence supporting the Darwinian hypothesis, based largely on studies of bee-pollinated species from temperate ecosystems. Since then additional studies, particularly on tropical members of Rubiaceae with diverse pollinators (e.g. bird, butterfly), have provided further support that reciprocal herkogamy promotes inter-morph pollen transfer (Stone 1995; Ree 1997; Pailler et al. 2002; Lau and Bosque 2003; Massinga et al. 2005; Hernandez and Ornelas 2007). In common with earlier work, marked asymmetries between the floral morphs in total pollen transfer and the capture of compatible pollen were evident. Stigmas of the L-morph capture more total pollen, but the proportion of compatible pollen on stigmas of the S-morph is generally higher.

A survey of pollen transfer efficiency in distylous species found that in 13 of 17 studies pollen grains of the S-morph were more likely to be transferred to compatible stigmas than pollen grains of the L-morph (Stone and Thomson 1994). This probably occurs because long-level organs are more accessible to pollinators. However, the direction of asymmetry reported can depend on the particular estimates of efficiency used, whether morph ratios and pollen production are taken into account, and if flowers are left intact or are emasculated (Ganders 1979; Stone 1995; Pailler et al. 2002). The cause of pollen transfer asymmetries lie in the interaction between floral morphology and the entry and exit paths of pollinators during nectar feeding (Ganders 1979; Lloyd and Webb 1992b; Stone 1995). Although the specific mechanical details need to be determined for individual cases, the overall patterns appear to be quite general, given the consistent results reported across diverse floral morphologies and pollinator groups.

Darwin's cross-promotion hypothesis has recently been examined in *Narcissus* species with stylar polymorphisms. This group presents a particular challenge because the floral morphs produce pollen of uniform size and it is not possible to determine directly the source of pollen deposited on stigmas. Instead, field manipulations have been employed to investigate pollen transfer. By altering the morph composition of local patches of *N. assoanus*, a species with stigma-height dimorphism, Cesaro and Thompson (2004) inferred higher rates of inter-than intra-morph pollination based on measurements of female fertility. However, this effect was evident only in pollen transfer from the L-morph to the S-morph, not in the reverse direction. Despite asymmetrical disassortative pollination, the levels reported were sufficient to maintain style-length dimorphism, and also satisfy theoretical conditions necessary for the evolution of the polymorphism under pollen-limited conditions (Lloyd and Webb 1992b). Other manipulative field experiments with *N. assoanus*, demonstrating negative frequency-dependent reproductive success also provide evidence for inter-morph pollen transfer in maintaining stylar dimorphism (Thompson et al. 2003).

In most heterostylous populations, regardless of the patterns of pollen transfer, diallelic incompatibility governs realized mating patterns, resulting in symmetrical disassortative mating and equal morph ratios, at least in equilibrium populations. However in *Narcissus* species, late-acting ovarian SI limits selfing but permits both intra- and inter-morph mating (Sage et al. 1999; see also Chap. 4, this volume). Therefore, differences in the floral morphology of the morphs influence mating patterns more strongly than those in typical heterostylous species. Specifically, opportunities for intra-morph mating in the L-morph occur because this morph has anthers positioned close to stigmas of long styles. Theoretical models indicate that asymmetrical mating resulting from incomplete sex-organ reciprocity could explain the prevalence of L-morph biased ratios in *Narcissus* species with stylar polymorphisms (Barrett et al. 1996; Baker et al. 2000b; Hodgins and Barrett 2006). Recent investigations provide empirical support for these models by showing a functional link between floral morphology, asymmetrical mating and biased morph ratios.

In tristylous *N. triandrus*, the L-morph possesses long-level rather than mid-level stamens (Barrett and Hodgins 2006). Populations exhibit strongly L-morph biased ratios throughout the range and pollen transfer models implicate assortative mating in the L-morph as the cause of biased ratios (Barrett et al. 2004; Hodgins and Barrett 2006). Paternity analysis of offspring from open-pollinated progenies was recently used to estimate rates of intra- and inter-morph mating (Hodgins and Barrett 2008). Patterns of outcrossed siring success indicated levels of assortative mating in the L-morph sufficient to cause L-morph bias. However, despite this result, overall levels of inter-morph transfer exceeded intra-morph transfer, supporting predictions of the Darwinian hypothesis that reciprocity of sex-organ position promotes pollinator-mediated cross-pollination.

1.3.2 Floral Morph Ratios and Reproductive Success

Reproductive success in species with stylar polymorphisms should be sensitive to plant density and the morph composition of local neighbourhoods. To investigate the spatial ecology of mating, Stehlik et al. (2006) mapped the location of floral morphs in *N. assoanus* populations and determined their female fertility. They found that pollen transfer and mating was context dependent, with the floral morphs responding differently to the density and morph identity of plants in local patches. Studies of *Narcissus* demonstrate that biased style morph ratios can be an equilibrium expectation if the strength of negative frequency-dependent selection varies among the morphs because of asymmetrical mating. However, because most heterostylous species possess heteromorphic incompatibility, biased morph ratios usually result from non-equilibrium conditions associated with the ecology of populations.

Biased morph ratios are particularly common in species with prolific clonal propagation (Barrett and Forno 1982; Castro et al. 2007). The signature of founder effects is often preserved over long periods, and progress towards morph-frequency equilibrium depends on the regularity of sexual reproduction and the demographic characteristics of populations (Eckert and Barrett 1995). In species with strong heteromorphic incompatibility, extensive clonal spread can interfere with sexual reproduction, resulting in a deficit of inter-morph cross-pollination and reduced seed set (Thompson et al. 1998; Ishihama et al. 2003; Brys et al. 2007; see also Chap. 3, this volume, with respect to the evolutionary implications of clonal propagation on the evolutionary dynamics of homomorphic SI). Species of *Nymphoides*, a genus of clonal aquatics, commonly exhibit biased morph ratios or stylar monomorphism, resulting in pollen limitation of fruit set and restricted sexual recruitment (Ornduff 1966; Shibayama and Kadono 2003; Wang et al. 2005a). By manipulating the frequency of flowers of the L- and S-morphs in an experimental population

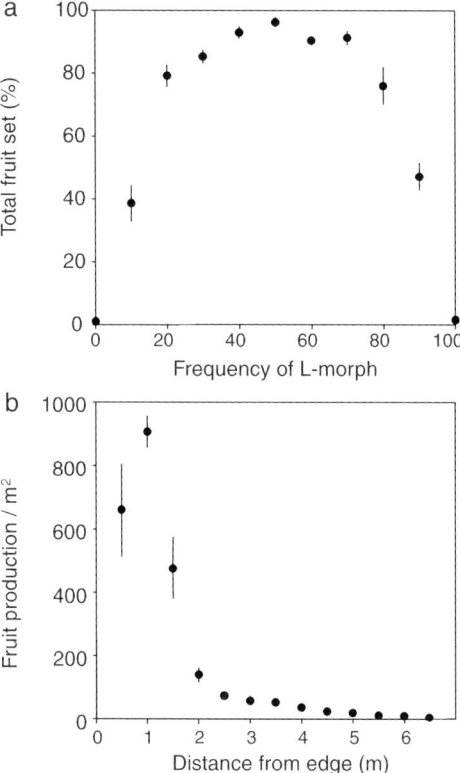

Fig. 1.3 Frequency-dependent (**a**) and proximity-dependent (**b**) reproductive success in distylous *Nymphoides peltata* (Menyanthaceae), an aquatic species with prolific clonal growth and strong dimorphic incompatibility. In (**a**) the ratio of long- and short-styled flowers in an experimental population influence total fruit set and biased ratios reduce maternal fertility because of a deficit of compatible pollen. In (**b**) fruiting density per square meter declines steeply with distance from the nearest compatible mating partner (S-morph) in a large clonal patch of the L-morph. After Wang et al. (2005a), figure published with permission from *New Phytologist*

of *Nymphoides peltata*, Wang et al. (2005a) demonstrated the frequency-dependent nature of reproductive success in distylous populations and also how extensive clone size can limit fruit set (Fig. 1.3).

Founder events, population bottlenecks and genetic drift can have significant ecological and evolutionary consequences, particularly through the loss of morphs from heterostylous populations (Barrett et al. 1989; Eckert and Barrett 1992). Also, small populations are often less attractive to pollinators, resulting in pollinator limitation of seed set in self-incompatible species (Ågren 1996). The growth of conservation biology has led to interest in the extent to which small population size resulting from habitat fragmentation influences the reproductive success of heterostylous populations. This work has largely focused on declining populations of distylous species of *Primula* occurring in human dominated landscapes of Japan [e.g. *P. sieboldii*

(Washitani et al. 2005)] and Europe [e.g. *P. vulgaris* (Jacquemyn et al. 2003; Kéry et al. 2003; Brys et al. 2004)]. These studies demonstrate that in small fragmented populations, the combined effects of pollinator losses and the scarcity of compatible morphs can limit reproductive success.

1.4 Inheritance of Heterostyly and the Supergene Model

The genetics of heterostyly was last reviewed by Lewis and Jones (1992); here we briefly outline salient work on the inheritance of distyly and tristyly to provide a background for our review of molecular investigations aimed at identifying the 'heterostyly genes'. We follow the convention of referring to the locus (or loci) determining distyly and tristyly as the *S*-locus, and the *S*- and *M*-loci, respectively. However, it is important to recognize that the '*S*-locus' in species with heteromorphic self-incompatibility is unlikely to be homologous with the *S*-locus of species with homomorphic SI (Gibbs 1986), described elsewhere in this volume; see also Chap. 14. The heterostyly loci are diallelic and appear to determine the entire syndrome of morphological and physiological traits that distinguish the floral morphs. A few cases are known in which self-incompatible heterostylous species do not possess diallelic incompatibility and instead may possess a multi-allelic incompatibility system (e.g. *Anchusa*, *Narcissus*, reviewed in Barrett and Cruzan 1994). In these genera, SI is not associated with the morphological traits that distinguish the morphs. However, the details of the type of SI involved (gametophytic vs. sporophytic) and its genetic basis have not been established, although at least two loci appear to be responsible in *A. officinalis* (Schou and Philipp 1984).

1.4.1 Inheritance of Distyly and Tristyly

Prior to the rediscovery of Mendelian inheritance, Darwin (1877) had already carried out the crosses necessary to partly, or wholly, deduce the patterns of inheritance of heterostyly in several taxa. Subsequently, Bateson and Gregory (1905) elucidated the inheritance of distyly in *Primula sinensis*, demonstrating the segregation of a dominant (*S*) and a recessive (*s*) allele governing the L-morph (*ss*) and S-morph (*Ss*) at what was considered to be a single locus. Today, the inheritance of distyly has been investigated in species from ~10 heterostylous families (Lewis and Jones 1992). All species exhibit the *Primula* inheritance pattern, although the dominance relations are reversed in *Armeria*, possibly *Limonium* (Baker 1966), and in *Hypericum* (Orndurff 1979). Lloyd and Webb (1992a) provide a compelling evolutionary explanation for the similar pattern of inheritance of distyly among heterostylous families.

The genetics of tristyly has also had a long history of investigation through studies of Lythraceae and Oxalidaceae (Barlow 1923; Fisher and Mather 1943;

Weller 1976; Eckert and Barrett 1993). Inheritance involves two diallelic loci with the *S*-locus epistatic to the *M*-locus. Plants carrying at least one copy of the *S*-allele are of the S-morph, and those that are homozygous recessive at the *S*-locus are either of the M-morph (*ssMM* and *ssMm*) or the L-morph (*ssmm*). The two loci segregate independently in *Decodon verticillatus* (Eckert and Barrett 1993) and *Lythrum salicaria*, which exhibits tetrasomic inheritance (Fisher 1941; Fisher and Mather 1943). In contrast, the loci are linked in most *Oxalis* species (Fisher and Martin 1948; Fyfe 1950; Weller 1976), and in *Eichhornia paniculata* of Pontederiaceae (Barrett unpublished data), but not in *Pontederia cordata* (Gettys and Wofford, 2008). More complex patterns of inheritance, including recessive epistasis and an additional locus, are reported from other *Oxalis* species (Fyfe 1956; Bennett et al. 1986; Trognitz and Hermann 2001). As yet there is no obvious explanation as to why similar patterns of inheritance involving two loci and epistasis have originated independently in three unrelated tristylous families.

1.4.2 Supergene Model

Largely as a result of work by Alfred Ernst on *Primula*, details of the genetic architecture of distyly emerged from studies of the inheritance of novel floral phenotypes obtained from intra- and inter-specific crosses (Ernst 1955). Ernst demonstrated that self-compatible long homostyles were inherited as if determined by additional alleles at the *S*-locus, and proposed that two (Ernst 1928), and later showed that three, tightly linked loci were responsible for the inheritance of distyly (Ernst 1955). This early work essentially established the 'supergene model' for *Primula*. A supergene is a series of two or more tightly linked genes that function together to produce adaptive characteristics (Lewis and Jones 1992). Later, Dowrick (1956) extended the supergene model for *Primula* by revising the number and order of loci. The inference that a series of tightly linked diallelic loci determines distyly was made, in part, because the rate of appearance of novel floral phenotypes by recombination was greater than could be accounted for by mutation to new alleles.

Today, the number of loci comprising the supergene in *Primula* is usually assumed to be three linked diallelic loci (Charlesworth and Charlesworth 1979a; Lewis and Jones 1992; but see below and Chap. 14, this volume). The *G* locus determines characteristics of the gynoecium, including style length and its incompatibility response, *P* determines pollen size and its incompatibility response and *A* determines anther height. The three dominant alleles at each locus are linked in coupling and comprise the apparent '*S*-allele' (*GPA*) with the three recessive alleles comprising the '*s*-allele' (*gpa*). Lewis and Jones (1992) re-analyzed the data of Ernst and determined the order and map distance for the three loci. However, additional loci have been proposed to account for other features of dimorphism. For example, Dowrick (1956) argued that there are an additional four loci for the incompatibility specificities of pollen and styles, area of style transmitting tissue and stigmatic papillae length, thus increasing the count to seven loci with all alleles linked in coupling.

This inference was not based on the analysis of recombinants, which in some cases would be very hard to detect. Alternatively, it has been suggested that some of the ancillary dimorphisms (e.g. papillae length) result from the pleiotropic effects of other gene(s) at the *S*-locus (Dulberger 1992; Al Wadi and Richards 1993).

More recently, Kurian and Richards (1997) proposed that there are at least seven loci at the distyly supergene in *Primula*. This suggestion was, in part, based on their studies of a curious novel phenotype (P*) discovered in a hybrid cultivar (*P.* × *tommasinii*). This plant resembles a self-compatible L-morph with a shortened style that produces dimorphic pollen, both sizes of which exhibit the incompatibility phenotype of large-sized pollen of the S-morph. Following a study of the inheritance of the P* variant, Kurian and Richards (1997) proposed a gene for pollen incompatibility separate from the gene determining pollen size. Richards (2003) subsequently proposed as many as nine linked genes comprising the *Primula* supergene. However, there are no convincing genetic data to support this claim.

Evidence that distyly is determined by a supergene in other heterostylous families is limited and no study has unambiguously documented recombinants arising from controlled crosses. The inheritance and compatibility behaviour of homostyles have been investigated in *Fagopyrum* (Woo et al. 1999; Wang et al. 2005b; Fesenko et al. 2006) and *Turnera* (Shore and Barrett 1985; Barrett and Shore 1987; Tamari et al. 2001) using inter-specific crosses between distylous and self-compatible long homostylous species. In both taxa, the dominance hierarchy among alleles at the *S*-locus is consistent with the *Primula* inheritance pattern (i.e. $S > S^h > s$), where the S^h-allele confers the long homostyle phenotype in $S^h S^h$ and $S^h s$ genotypes, although a second unlinked locus may be involved in *Fagopyrum* (Wang et al. 2005b; Fesenko et al. 2006). The compatibility relations between distylous and homostylous species exhibit a pattern comparable to that observed for *Primula*, with the style of the homostyle retaining the incompatibility phenotype of long styles, and pollen of the homostyle exhibiting the large size and incompatibility phenotype of pollen of the S-morph (Fig. 1.4). The inheritance and compatibility data have been interpreted as evidence that distyly is controlled by a *Primula* type supergene. The apparent S^h allele is inferred to result from recombination within the supergene, giving rise to long homostyles (*gPA/gpa* or *gPA/gPA*) as observed in *Primula* spp. (Figs. 1.4 and 1.5; see also Chap. 14, this volume). However, in *Fagopyrum* and *Turnera*, evidence supports the occurrence of only two loci at the supergene because plants have not been recovered in which pollen size and stamen height are dissociated (Woo et al. 1999; Matsui et al. 2003; Tamari et al. 2005).

There are few cases outside of *Primula* where the inheritance of mutants or putative recombinants has been studied to explore the genetic architecture of distyly. A homostylous somatic mutant arose as a shoot on an otherwise S-morph F_1 plant from the cross between distylous *Turnera subulata* and distylous *T. krapovickasii*. The mutant behaved as if determined by an allele, S^h, at the distyly locus, with the dominance and compatibility relations as observed for long homostyles of *Primula* (Tamari et al. 2005; Fig. 1.4). While clearly not the result of meiotic recombination, its mutational origin and inheritance are consistent with the possibility that a separate mutable gene for style length and associated style incompatibility might

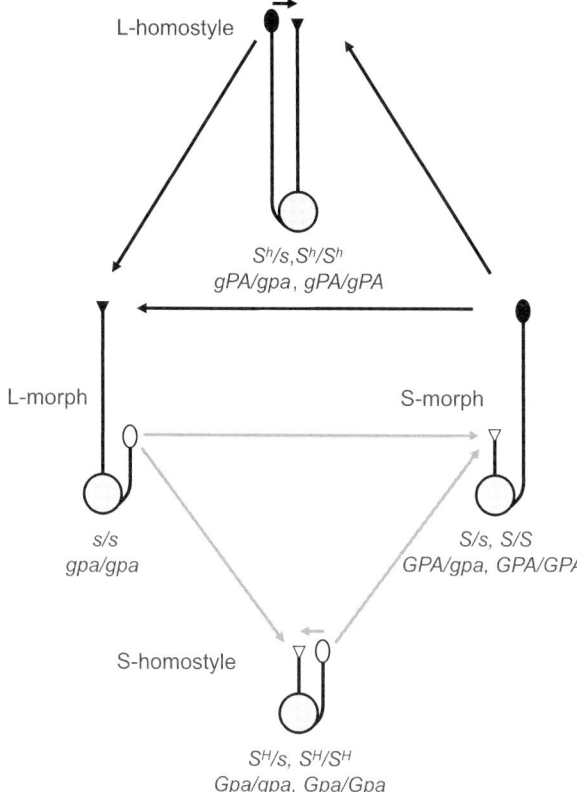

Fig. 1.4 Compatible pollinations between distylous and homostylous morphs. The L- and S-morphs are self-incompatible, whereas the long and short homostyle morphs (L- and S-homostyle, respectively) are self-compatible. Arrows indicate compatible pollinations. Genotypes of the morphs are indicated as alleles at the S-locus with the dominance hierarchy $S > S^h$, $S^H > s$, and as a series of three linked alleles comprising a supergene. For the genotypes illustrated, the S^h allele determines the L-homostyle, while S^H determines the S-homostyle (the superscripts for the alleles conferring homostyly do not imply that one allele is dominant to the other). Both homostyles are incompatible with one another; however, if the homozygous genotypes were able to be crossed, and assuming the supergene model, only S-morph progeny would be produced because of complementation

reside adjacent to a gene(s) determining stamen characteristics. A long homostylous 'mutant' of distylous autotetraploid *T. scabra* exhibited comparable behaviour coupled with tetrasomic inheritance (Tamari et al. 2005). Therefore, outside of *Primula* conclusive evidence of a supergene based on clear documentation of recombination yielding novel floral morphs is lacking. The available genetic data are consistent with the occurrence of two linked loci in *Turnera* and *Fagopyrum*, but are not unequivocal demonstrations of the existence of supergenes.

Supergene control of tristyly is often implied in the literature, although there is actually no direct genetic evidence (Charlesworth 1979; Lewis and Jones 1992;

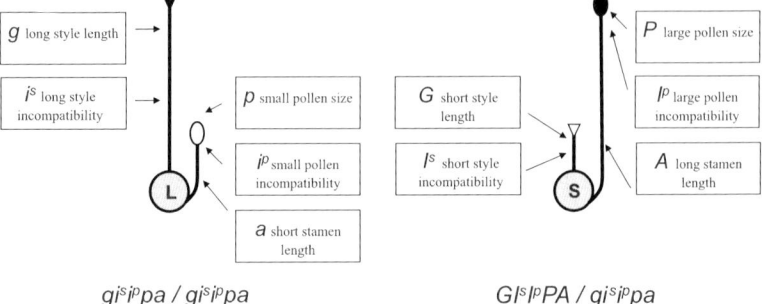

Fig. 1.5 A supergene model for the inheritance of distyly involving five tightly linked genes. In addition to the G/g, P/p and A/a genes, two further genes determining pollen, I^p/i^p, and style, I^s/i^s, incompatibility are included, as often inferred in the literature (see Sect. 1.4). Each gene acts independently to determine the morphological and incompatibility phenotypes of morphs, as indicated by the arrows

Barrett 1993). Because of the occurrence of two stamen levels within tristylous flowers, homostylous variants (referred to as semi-homostyles) are usually characterized by a change in the position of a single stamen level only. However, subsequent selection associated with the evolution of small flowers and self-fertilization can result in both stamen levels in close proximity to stigmas (Ornduff 1972; Barrett 1988). For reasons that still need to be determined, semi-homostylous variants most often involve the M-morph. Under the supergene model, semi-homostyles should be inherited as if determined by a new allele at the S or M locus (Charlesworth 1979). Genetic analysis of semi-homostyly is limited; however, what data are available [e.g. *Lythrum salicaria* Stout (1925); *Eichhornia paniculata* Fenster and Barrett (1994)] do not indicate recombinational origins, but rather mutations non-allelic to the S and M loci.

A supergene model for tristyly requires the occurrence of two supergenes, one residing at each of the S and M loci. Genes within each supergene complex would each exhibit the usual pattern of dominance and epistatic interactions. The difficulty in imagining how supergenes may have evolved in tristylous species led Charlesworth (1979) to invoke pleiotropy and 'incompatibility gradients' in her model of the evolution of tristyly. Several lines of evidence support elements of her model. For example, changes in incompatibility response and pollen size are associated with alterations in the position of stamens in tristylous species (Barlow 1913; Weller 1976; Manicacci and Barrett 1995). Regulatory genes may play a role in altering the incompatibility phenotype of pollen within individual flowers and perhaps also in controlling other features of tristyly. If this is true, we see no reason why comparable regulatory genes may not also be involved in controlling elements of the distylous syndrome, especially in cases where distyly has evolved from tristyly, as in *Lythrum* and *Oxalis* (reviewed in Weller 1992).

1.4.3 Mutational Analyses and the Study of Genetic Variants

A useful approach for probing the genetic architecture of heterostyly is to investigate mutants. Under the supergene model, mutant alleles determining novel floral morphs (e.g. homostyles) should map to the *S*-locus and involve predictable phenotypes (Mather 1950; Matsui et al. 2007). In an ongoing experiment, pollen carrying the dominant *S*-allele from homozygous (*SS*) S-morph plants of *Turnera subulata* was X-ray irradiated and used to pollinate L-morph (*ss*) plants (Shore et al. 2006; Labonne and Shore, unpublished data). The cross produced three L-morph mutants and one long homostyle mutant. Although the inheritance of the mutants has yet to be determined, a molecular marker tightly linked to the *S*-locus was deleted in each of the four mutants, suggesting that the mutations occur at the *S*-locus. The occurrence of a long homostyle mutant supports earlier work on *Turnera*, suggesting that distyly is controlled by at least two linked genes.

It is unlikely that a single gene determines all of the heteromorphic features of distylous plants unless it is a complex regulatory gene that resides at the *S*-locus. Alternatively, it seems unreasonable to us that each character of the suite of dimorphic traits results from a separate linked gene. Elements of one possible supergene model are illustrated in Fig. 1.5 (Dowrick 1956; Ganders 1979, Lewis and Jones 1992; Matsui et al. 2003 and Chap. 14, this volume). Could a single regulatory gene at the *S*-locus be responsible for all of the trait differences among the morphs? In *Primula*, *Turnera* and *Fagopyrum*, inheritance data for homostyles appear to discount this possibility but may indicate that at least two (*Fagopyrum* and *Turnera*) and possibly three (*Primula*) linked regulatory loci occur. One of the regulatory loci could determine gynoecial characteristics (style length and incompatibility) and the other androecial characteristics (pollen size, production and incompatibility). These loci could conceivably regulate genes that might be distributed elsewhere in the genome (Fig. 1.6). The capacity of the dominant *S*-allele to directly or indirectly up-regulate genes located outside the *S*-locus has been demonstrated in *Turnera* (Athanasiou et al. 2003; Khosravi et al. 2004; Tamari and Shore 2006) and *Primula vulgaris* (McCubbin et al. 2006). This raises the possibility that 'incompatibility genes' may not reside at the *S*-locus, but instead are turned on by signals generated by the *S*-locus (Fig. 1.6). If correct, then it would be difficult to recover mutant alleles at the *S*-locus in which style length and pollen size are separated from their respective incompatibility responses, although non-allelic mutations could be found, as may be the case for a self-compatible plant of the S-morph of *T. subulata* (Shore and Barrett 1986; Athanasiou and Shore 1997).

1.5 Molecular Genetics

Despite the long interest in heterostyly as a plant model system for linking genetics and adaptation, the molecular basis of the polymorphism has yet to be determined. Attempts to find the genes and proteins responsible for heteromorphic SI

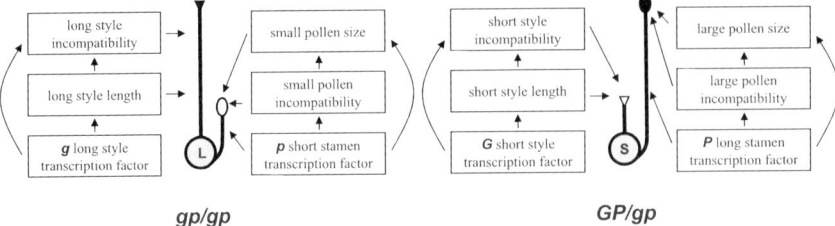

Fig. 1.6 A model of the genetic architecture of distyly that involves morph-limited gene expression. In the model, two closely linked genes (G/g and P/p) occur at the S-locus. G/g encodes a transcription factor responsible for the direct or indirect regulation of genes for gynoecial characters (e.g. style length, style incompatibility), whereas P/p encodes a transcription factor responsible for the direct or indirect regulation of genes for androecial characters (e.g. stamen length, pollen size, pollen incompatibility). The two alleles of each transcription factor may regulate different suites of genes distributed in the genome. The arrows indicate possible patterns of regulation in which the initial transcription factor directly regulates one or more genes, and/or the regulation of one of these genes causes a cascade of regulatory effects on other genes

in distylous species were initiated during the past decade. These included efforts to discover proteins or RNA transcripts unique to pollen and/or styles of the morphs, and genetic mapping and positional cloning of the putative genes determining distyly. Discovery of the genes governing distyly may depend on specific details of how the polymorphism evolved, because the assembly of the polymorphism seems likely to have influenced the underlying molecular mechanism(s). Theoretical models for the evolution of heterostyly may be useful for informing future molecular investigations. As mentioned previously (see Sect. 1.2.1.), there are two widely discussed models for the evolution of distyly (Charlesworth and Charlesworth 1979a; Lloyd and Webb 1992a, b). Although other models have been proposed (Richards 2003; Sakai and Toquenaga 2004), they have attracted little attention, probably because they contain several unlikely assumptions, and we therefore do not consider them further. Below we briefly review relevant features of the Charlesworth and Charlesworth (1979a) and Lloyd and Webb, (1992a,b) models and consider the molecular methods best suited to discover the genes determining distyly, based on predictions of these models.

1.5.1 Theoretical Models and Predictions

Charlesworth and Charlesworth (1979a) used a population-genetic approach to model the evolution of distyly. In their model, the SI system evolves first, followed by reciprocal herkogamy. There is a linkage constraint in which a recessive mutation to a new pollen type arises first, and the population is then invaded by a dominant mutation to a new style type, provided that the mutation occurs at a linked locus. Subsequently, morphological differences evolve enhancing compatible inter-morph pollinations, and there are likely to be linkage constraints involved in their evolution as well. A number of linked genes determine distyly according to this model. Stylar incompatibility is governed by allelic variation at one of the supergene loci, whereas pollen incompatibility results from variation at a linked locus. Likewise, genes for style and stamen length result from allelic variation linked to the incompatibility loci. Fine-scale genetic mapping and positional cloning would appear to be the best approach to test predictions of this model. Under this model all genes that determine the polymorphism reside at the S-locus supergene (e.g. Fig. 1.5).

Lloyd and Webb (1992a, b) provide an alternative phenotypic selection model for the evolution of distyly based on the efficacy of cross-pollen transfer. In their model reciprocal herkogamy evolves prior to the evolution of diallelic incompatibility. A variant with reduced style length determined by a dominant mutation invades a population composed of plants exhibiting approach herkogamy, resulting in polymorphism for stigma height. According to the model, higher inter- than intra-morph pollen transfer maintains stylar dimorphism, although the mechanism responsible for promoting disassortative pollination is unclear (Stone and Thomson 1994). There is then selection for anther-height dimorphism and associated ancillary features, which improves the efficacy of inter-morph pollen transfer. Finally, SI (if it develops) evolves gradually as a result of co-adaptation between pollen-tube growth and the styles in which they most commonly grow. Under this scenario, SI could result from genes with morph-limited expression, or the genes involved may be linked to genes for style length and/or anther position. Morph-limited expression of the SI response may also be increased by subsequent selection. That is, mutations arising anywhere in the genome could cause pollen tubes to grow less well in one or the other stylar environment.

In contrast to the Charlesworth and Charlesworth (1979a) model, the genes responsible for distyly in the Lloyd and Webb model could be distributed throughout the genome, although some could also be linked (Lloyd and Webb 1992b, p 198). Lloyd and Webb (1992a, b) predict that SI is unlikely to involve the type of S-gene specificities found in homomorphic systems. They also propose that the SI mechanism may be different between the morphs. Genes exhibiting differential expression between the morphs could also be responsible for the regulation of various other features of distyly (Fig. 1.6). Comparing protein or mRNA profiles of the morphs may be the best approach for testing the predictions of the Lloyd and Webb model since this would allow the discovery of expression differences, and possibly allelic variation manifested in proteins or mRNA. Fine-scale genetic mapping and positional cloning of gene(s) at the S-locus should lead to the discovery of the gene determining style-length variation, and possibly a regulatory gene(s) responsible for the expression of incompatibility and various ancillary features. Should such a regulatory gene(s) be discovered, it might be possible to find its interacting partners using, for example, chromatin immunoprecipitation assays or in vitro genomic selection (Wang 2005). Below we review progress made in attempts to discover the molecular basis of heterostyly using three main approaches: protein profiles, mRNA expression and genetic localization. See other chapters in this volume (Part B) for accounts of investigations into the molecular basis of homomorphic SI.

1.5.2 Protein Profiles

The first attempts to investigate the molecular basis of heterostyly used methods largely designed to detect protein differences among the morphs. The earliest experiments involved cytochemical staining of tissues (Dulberger 1974), as well as physiological approaches, exploring the effects of style and/or stigma extracts on pollen germination in vitro (reviewed in Gibbs 1986; Dulberger 1992; Chap. 14, this volume). More recent efforts have used protein separation techniques, including isoelectric focussing (IEF), one-dimensional SDS-polyacrylamide gel electrophoresis (SDS-PAGE) and 2D PAGE. Wong et al. (1994) used both SDS-PAGE and 2D PAGE to compare protein profiles of floral organs in distylous *Averrhoa carambola* (Oxalidaceae). They discovered an abundant 72 kDa protein unique to styles of the L-morph but the identity of the protein was not determined. Miljuš-Đukić et al. (2004) compared protein profiles of the L- and S-morphs of *Fagopyrum esculentum* using similar techniques and also found differences between the morphs. Once again the identity or function of proteins was not determined.

Studies of *Turnera subulata* and related species resulted in the discovery of two proteins: putative polygalacturonase and α-dioxygenase, unique to styles of the S-morph (Athanasiou and Shore 1997; Athanasiou et al. 2003; Khosravi et al. 2003, 2004). The proteins were localized to the transmitting tissue of styles and stigmas using immunocytochemistry. The gene encoding the polygalacturonase was linked but distal to the *S*-locus (Athanasiou et al. 2003; Tamari and Shore 2006), while the gene for the α-dioxygenase was unlinked (Khosravi et al. 2004). Using 2D PAGE and mass spectrometry, Khosravi et al. (2006) identified two additional proteins unique to the S-morph of *Turnera*: putative cysteine protease and β-expansin. No confirmation of morph- or style-specificity has yet been undertaken nor are the possible roles of the proteins known.

Recently, protein expression differences have been reported for the first time in pollen from different stamen levels within flowers of a tristylous species (Kalinowski et al. 2007). Using 2D PAGE of proteins extracted from pollen from each of the six style length-stamen level combinations in *Lythrum salicaria*, these workers report surprisingly large differences in the number of proteins present in different morphs. For example, in the most extreme case, they report 177 proteins present in pollen of mid-level anthers of the S-morph that are not represented in pollen of long-level anthers of the S-morph. Further work is needed to corroborate these findings and determine their relevance, if any, to the functioning of tristyly.

1.5.3 mRNA Expression

Two studies have attempted to find transcripts unique to the floral morphs of distylous *Primula vulgaris*. McCubbin et al. (2006) used suppressive subtractive hybridization to discover 11 cDNAs differentially expressed between the morphs. None of these were linked to the *S*-locus and the authors suggest that the genes may be downstream components of developmental pathways, leading to floral dimorphism and/or physiological differences between the morphs. By contrast, Li et al. (2007) used differential display to identify transcripts unique to the floral morphs undertaking an extensive sampling of the transcriptome of a single size class of flower bud. Only one amplification product proved to be differentially expressed and sequencing provided no information on its identity. However, significantly, two of the initial amplification products proved to be linked to the *S*-locus (see Sect. 1.5.4; see also Chap. 14, this volume). These results from *Primula vulgaris* differ with respect to the number of differentially expressed genes found. This may have occurred because McCubbin et al. (2006) used a range of bud sizes and mature tissues in their analysis.

1.5.4 Genetic Localization

Linkage maps were first constructed for distylous species early in the last century (De Winton and Haldane 1935). Today the advent of hypervariable molecular markers allows high-resolution linkage maps to be constructed and, because the markers are typically DNA fragments, they can be used to probe cDNA and/or genomic libraries to discover genes of interest.

Progress towards positional cloning of the *S*-locus has been made in *Fagopyrum*. Aii et al. (1998) constructed a linkage map targeting the *S*-locus. They used 181 F_2 progeny from an inter-specific cross of distylous *Fagopyrum esculentum* with homostylous *F. homotropicum*. Three RAPD markers were linked in coupling with the S^h-allele, the closest at a distance of 0.6 cM. Both Nagano et al. (2001) and Yasui et al. (2004) extended this analysis and discovered AFLP markers closely linked to the *S*-locus. The linked markers should provide useful starting points for high-resolution mapping and positional cloning of *S*-locus genes. A bacterial artificial chromosome (BAC) library has been constructed from *F. homotropicum* that should enable positional cloning (Nagano et al. 2005).

Manfield et al. (2005) discovered a 323 bp RAPD marker tightly linked (~0.5 cM) in coupling to the *S*-allele in a diploid hybrid horticultural variety of *Primula vulgaris*. They probed a phage λ genomic DNA library and obtained a clone of 8.8 kb, approximately 1 kb of which showed homology to regions of Ty3/gypsy-like retrotransposons. The remainder of the sequence consisted of multiple short repetitive elements. To determine on which side of the *S*-locus the linked marker resides, these authors attempted to amplify the 323 bp fragment from genomic DNA of a long homostyle, with predicted genotype *gPA/gPA*, and a short homostyle, predicted genotype *Gpa/gpa*. The homostyles were developed in a wild *P. vulgaris* genetic background. The 323 bp fragment amplified in the long homostyle but not in the short homostyle. Assuming the homostyles arose by recombination within the supergene, and their genetic backgrounds and the cultivar are similar in the region around the *S*-locus, these results indicate that the 323 bp fragment lies on the *A*-locus side of the supergene, or possibly within the supergene but adjacent to the *P*- or *A*-locus.

Li et al. (2007) used Southern analyses to demonstrate that two differential display products were closely linked to the *S*-locus. One gene, *PvSLL1*, co-segregated with the *S*-locus, while the second, *PvSLL2*, was closely linked to it. *PvSLL1* encodes a putative small transmembrane protein with similarity to an *Arabidopsis* gene of unknown function. These authors used long and short homostyles of *P. vulgaris* to map the location of both genes (as above). The *S*-coupled allele of *PvSLL1* occurred in the homozygous (*gPA/gPA*) long homostyle, while one of the two *s*-coupled alleles was found in the short homostyle. These results suggest that *PvSLL1* lies on the *A*-locus side of the supergene within 0.57 cM of the *S*-locus, or even possibly within the supergene, since no recombinants were found. Further analysis revealed that *PvSLL2* lies on the *G*-locus side of the supergene at a map distance of approximately 1.4 cM. The identification of these closely linked genes and the marker identified by Manfield et al. (2005) will provide valuable probes for screening a BAC library of *Primula* and should enable positional cloning of *S*-locus genes.

In distylous *Turnera subulata*, isozyme loci (*Aco-1* and *Pgd-c*) map 3.3 cM and 1.8 cM, respectively, on either side of the *S*-locus (Athanasiou and Shore 1997; Athanasiou et al. 2003). Labonne et al. (2007) demonstrated significantly greater female meiotic recombination rates in *T. subulata*, expanding the map distance between the two isozyme loci approximately 4–6-fold. Further mapping efforts involved greater precision because of higher recombination rates in female meiosis. Labonne et al. (2008) obtained a fine-scale map of the *S*-locus region based upon 736 backcross progeny. Two closely linked molecular markers on opposite sides of the *S*-locus have been mapped, as well as two co-segregating markers. Chromosome walking has been initiated to identify gene(s) that reside at the *S*-locus in *T. subulata*.

As with the homomorphic SI systems (Charlesworth 2006), balancing selection has likely maintained the distylous polymorphism over long evolutionary time scales. As such, the application of molecular population genetic approaches should aid in identifying the target(s) of balancing selection and testing whether there is recombination suppression in the *S*-locus region. Genes comprising, or tightly linked to, the *S*-locus identified through mapping studies should show elevated nucleotide diversity (i.e. the footprint of balancing selection), as has been demonstrated for homomorphic SI systems (Kamau and Charlesworth 2005; Charlesworth 2006). For genes less tightly linked, infrequent recombination over evolutionary time will erode this diversity (Charlesworth 2006). While conclusive proof of the function of genes at the *S*-locus can best be demonstrated using transformation and knockout strategies, molecular population genetics will be important in identifying putative *S*-locus genes.

1.6 Concluding Remarks

Heterostyly continues to fascinate evolutionary biologists interested in the genetics, evolution and function of adaptations. The polymorphisms illustrate how simply inherited developmental changes in the positions of plant sexual organs have important consequences for the pollination and mating biology of populations. Unlike homomorphic SI, only a small number of mating phenotypes can be maintained in populations because of the constraints in achieving multiply-segregated sites for pollen deposition on animal pollinators. The three floral morphs in tristylous populations probably represent the upper limit for mating groups that can be achieved for sexual polymorphisms that function to promote cross-pollination by the geometry of pollinator contacts.

Our review has highlighted recent work on the comparative biology, ecology and genetics of heterostyly. Over the past few decades, studies on a wider range of heterostylous species have broadened our understanding of the polymorphisms and demonstrated that considerable variation in expression occurs among taxonomic groups. This variation should caution against overgeneralization of the *typical features* of heterostyly, based on studies of models systems such as *Primula*. Comparisons of variation in heterostylous traits, particularly the occurrence of diallelic incompatibility and ancillary pollen and stigma polymorphisms, in lineages that differ in the number of heterostylous species may be instructive. The expression of the heterostylous syndrome in genera with single isolated heterostylous species could provide clues on the early stages in the assembly of the polymorphism and shed light on models of the evolution of heterostyly.

A major theme in the literature concerns the evolutionary pathways responsible for the buildup and breakdown of heterostyly (Ganders 1979; Weller 1992). Evidence for character state transitions has come largely from comparative arguments (Baker 1966; Ornduff 1966) or population-level studies of intra-specific variation (Piper et al., 1986; Barrett et al. 1989; Weller et al. 2007). Future work using phylogenies and character reconstructions are needed to provide convincing evidence linking observed microevolutionary changes to putative macroevolutionary patterns. Boraginaceae, Menyanthaceae, Oxalidaceae, Plumbaginaceae and Rubiaceae are groups that contain particularly high sexual-system diversity and warrant serious attention by comparative biologists interested in the evolution of reproductive systems.

As evident from other chapters in this volume, progress on determining the genes controlling heterostyly has lagged behind research on homomorphic incompatibility. Lewis and Jones (1992) suggested that the incompatibility genes in heterostylous species should be of first priority for molecular dissection, predicting that they would have considerable output of transcribed mRNA and translated protein. However, subsequent work on *Primula* did not find abundant and/or differentially expressed mRNA encoded by genes linked to the *S*-locus (McCubbin et al. 2006 and this volume; Li et al. 2007). Similarly, identifying *S*-locus proteins distinguishing the morphs in distylous *Fagopyrum* (Milјuš-Đukić et al. 2004) and *Turnera subulata* (Athanasiou et al., 2003; Khosravi et al., 2004) has also proven elusive. Hence, collectively there is no compelling evidence for highly abundant transcripts or proteins governing diallelic incompatibility.

Genes regulating floral development may exhibit differential expression, obscuring the signal of incompatibility genes and making their identification more difficult (McCubbin et al. 2006). Alternatively, there may be no abundantly expressed incompatibility genes and/or proteins, because, as Lloyd and Webb (1992a, b) suggested, incompatibility may largely result from genes elsewhere in the genome affecting pollen-tube growth. The up- vs. down-regulation of genes reported by McCubbin et al. (2006) and the observed protein differences between styles of the floral morphs may be key elements in the expression of diallelic incompatibility. Identification of the genes controlling heterostyly will help to illuminate theoretical models for the evolution of the polymorphism. When this is achieved, the long quest for determining the causal relations between genes, form and adaptive function in heterostylous plants will have finally been achieved.

Acknowledgements We thank Deborah Charlesworth, Ivan Fesenko, Peter Gibbs, Jorg Grigull, Kathi Hudak, David Lloyd and Stephen Wright for stimulating discussions on heterostyly, Andy McCubbin for providing us with a copy of his chapter in this volume, graduate students who have worked with us on heterostyly for their insights and hardwork, and the Natural Sciences and Engineering Research Council of Canada for Discovery Grants that have funded our research on heterostyly.

References

Ågren J (1996) Population size, pollinator limitation, and seed set in the self-incompatible herb *Lythrum salicaria*. Ecology 77:1779–1790
Aii J, Nagano M, Penner GA, Campbell CG, Adachi T (1998) Identification of RAPD markers linked to the homostylar (*Ho*) gene in buckwheat. Breed Sci 48:59–62
Al Wadi H, Richards AJ (1993) Primary homostyly in *Primula* L. subgenus *Sphondylia* (Duby) Rupr. and the evolution of distyly in *Primula*. New Phytol 124:329–338
Armbruster WS, Peréz-Barrales R, Arroyo J, Edwards ME, Vargas P (2006) Three-dimensional reciprocity of floral morphs in wild flax (*Linum suffructicosum*): A new twist on heterostyly. New Phytol 171:581–590
Athanasiou A, Shore JS (1997) Morph-specific proteins in pollen and styles of distylous *Turnera* (Turneraceae). Genetics 146:669–679
Athanasiou A, Khosravi D, Tamari F, Shore JS (2003) Characterization and localization of short-specific polygalacturonase in distylous *Turnera subulata* (Turneraceae). Am J Bot 90:675–682
Baker AM, Thomson JD, Barrett SCH (2000a) Evolution and maintenance of stigma-height dimorphism in *Narcissus* I. Floral variation and style-morph ratios. Heredity 84:502–513
Baker AM, Thomson JD, Barrett SCH (2000b) Evolution and maintenance of stigma-height dimorphism in *Narcissus* II. Fitness comparisons between style morphs. Heredity 84:514–524
Baker HG (1966) The evolution, functioning and breakdown of heteromorphic incompatibility systems. I. The Plumbaginaceae. Evolution 20:349–368
Barlow N (1913) Preliminary note on heterostylism in *Oxalis* and *Lythrum*. J Genet 3:53–65
Barlow N (1923) Inheritance of the three forms in trimorphic species. J Genet 13:133–146
Barrett SCH (1988) Evolution of breeding systems in *Eichhornia*: A review. Ann Mo Bot Gard 75:741–760
Barrett SCH (1992) Evolution and function of heterostyly. Springer, Berlin Heidelberg New York

Barrett SCH (1993) The evolutionary biology of tristyly. In: Futuyma D, Antonovics J (eds) Oxford surveys in evolutionary biology, vol 9. Oxford University Press, Oxford, UK, pp 283–326

Barrett SCH (2002) The evolution of plant sexual diversity. Nat Rev Genet 3:274–284

Barrett SCH, Cruzan MB (1994) Incompatibility in heterostylous plants. In: Williams EG, Clarke AE, Knox RB (eds) Genetic control of self-incompatibility and reproductive development in flowering plants. Kluwer, Dortrecht, pp 189–219

Barrett SCH, Forno IW (1982) Style morph distribution in New World populations of *Eichhornia crassipes* (Mart.) Solms-Laubach (Water Hyacinth). Aquat Bot 13:299–306

Barrett SCH, Hodgins KA (2006) Floral design and the evolution of asymmetrical mating. In: Harder LD, Barrett SCH (eds) Ecology and evolution of flowers. Oxford University Press, Oxford, UK, pp 239–255

Barrett SCH, Shore JS (1987) Variation and evolution of breeding systems in the *Turnera ulmifolia* complex (Turneraceae). Evolution 41:340–354

Barrett SCH, Morgan MT, Husband BC (1989) The dissolution of a complex genetic polymorphism: The evolution of self-fertilization in tristylous *Eichhornia paniculata* (Pontederiaceae). Evolution 43:1398–1416

Barrett SCH, Lloyd DG, Arroyo J (1996) Stylar polymorphisms and the evolution of heterostyly in *Narcissus* (Amaryllidaceae). In: Lloyd DG, Barrett SCH (eds) Floral biology: Studies on floral evolution in animal-pollinated plants. Chapman and Hall. New York, USA, pp 339–376

Barrett SCH, Jesson LK, Baker AM (2000a) The evolution of stylar polymorphisms in plants. Ann Bot 85(suppl A):253–265

Barrett SCH, Wilken DH, Cole WW (2000b) Heterostyly in the Lamiaceae: The case of *Salvia brandegeei*. Plant Syst Evol 223:211–219

Barrett SCH, Harder LD, Cole WW (2004) Correlated evolution of floral morphology and mating-type frequencies in a sexually polymorphic plant. Evolution 58:964–975

Bateson W, Gregory RP (1905) On the inheritance of heterostylism in *Primula*. Proc R Soc Lond ser B 76:581–586

Belaoussoff S, Shore JS (1995) Floral correlates and fitness consequences of mating-system variation in *Turnera ulmifolia*. Evolution 49:545–556

Bennett JH, Leach CR, Goodwins IR (1986) The inheritance of style length in *Oxalis rosea*. Heredity 56:393–396

Bjorkman T (1995) The effectiveness of heterostyly in preventing illegitimate pollination in dish-shaped flowers. Sex Plant Reprod 8:143–146

Brys R, Jacquemyn H, Endels P, Van Rossum F, Hermy M, Triest L, De Bruyn L, Blust GDE (2004) Reduced reproductive success in small populations of the self-incompatible *Primula vulgaris*. J Ecol 92:5–14

Brys R, Jacquemyn H, Hermy M (2007) Impact of mate availability, population size, and spatial aggregation of morphs on sexual reproduction in a distylous, aquatic plant. Am J Bot 94:119–127

Castro S, Loureiro J, Santos C, Ater M, Ayensa1 G, Navarro1 L (2007) Distribution of flower morphs, ploidy level and sexual reproduction of the invasive weed *Oxalis pes-caprae* in the western area of the Mediterranean region. Ann Bot 99:507–517

Cesaro AC, Thompson JD (2004) Darwin's cross-promotion hypothesis and the evolution of stylar polymorphism. Ecol Lett 7:1209–1215

Charlesworth D (1979) The evolution and breakdown of tristyly. Evolution 33:489–498

Charlesworth D (2006) Balancing selection and its effects on sequences in nearby genome regions. PLoS Genet 2:0379–0384

Charlesworth B, Charlesworth D (1979b) The maintenance and breakdown of distyly. Am Nat 111:499–513

Charlesworth D, Charlesworth B (1979a) A model for the evolution of distyly. Am Nat 114: 467–498

Church SA (2003) Molecular phylogenetics of *Houstonia* (Rubiaceae): Descending aneuploidy and breeding system evolution in the radiation of the lineage across North America. Mol Phylogenet Evol 27:223–238

Darwin C (1877) The different forms of flowers on plants of the same species. Murray, London, UK

De Winton D, Haldane JBS (1935) The genetics of *Primula sinensis* III linkage in the diploid. J Genet 31:68–100

Dowrick VPJ (1956) Heterostyly and homostyly in *Primula obconica*. Heredity 10:219–236

Dulberger R (1974) Structural dimorphism of stigmatic papillae in distylous *Linum* species. Am J Bot 61:238–243

Dulberger R (1992) Floral polymorphisms and their functional significance in the heterostylous syndrome. In: Barrett SCH (ed) Evolution and function of heterostyly. Springer, Berlin Heidelberg New York, pp 41–84

Eckert CG, Barrett, SCH (1992) Stochastic loss of style morphs from populations of tristylous *Lythrum salicaria* and *Decodon verticillatus* (Lythraceae). Evolution 46:1014–1029

Eckert CG, Barrett SCH (1993) The inheritance of tristyly in *Decodon verticillatus* (Lythraceae) Heredity 71:473–480

Eckert CG, Barrett SCH (1995) Style morph ratios in tristylous *Decodon verticillatus* (Lythraceae): Selection vs. historical contingency. Ecology 76:1051–1066

Ernst A (1928) Zur verebung der morphologischen heterostylie-merkmale. Ber Dtsch Bot Ges 46:573–588

Ernst A (1955) Self-fertility in monomorphic primulas. Genetica 27:391–448

Fenster CB, Barrett SCH (1994) Inheritance of mating-system modifier genes in *Eichhornia paniculata* (Pontederiaceae). Heredity 72:433–445

Fesenko NN, Fesenko IN, Ohnishi O (2006) Homostyly of two morphologically different lineages of *Fagopyrum homotropicum* Ohnishi is determined by locus $S4$, which is an S-locus related gene in the linkage group #4. Fagopyrum 23:11–15

Fisher RA (1941) The theoretical consequences of polyploid inheritance for the Mid style form in *Lythrum salicaria*. Ann Eugen 11:31–38

Fisher RA, Martin VC (1948) Genetics of style length in *Oxalis*. Nature 162:533

Fisher RA, Mather K (1943) The inheritance of style length in *Lythrum salicaria*. Ann Eugen 12:1–23

Fyfe VC (1950) The genetics of tristyly in *Oxalis valdiviensis*. Heredity 4:365–371

Fyfe VC (1956) Two modes of inheritance of the short-styled form in the 'genus' *Oxalis*. Nature (London) 177:942–943

Ganders FR (1979) The biology of heterostyly. N Z J Bot 17:607–635

Gettys LA, Wofford DS (2008) Genetic control of floral morph in tristylous pickerelweed (*Pontederia cordata* L.). J. Heredity, Advanced Access, published May 21, 2008 doi: 10.1093/jhered/esn031

Gibbs PE (1986) Do homomorphic and heteromorphic self-incompatibility systems have the same sporophytic mechanism? Plant Syst Evol 154:285–323

Graham SW, Barrett SCH (2004) Phylogenetic reconstruction of the evolution of stylar polymorphisms in *Narcissus* (Amaryllidaceae). Am J Bot 91:1007–1021

Hartley ML, Tshamekeng E, Thomas SM. (2002) Functional heterostyly in *Tylosema esculentum* (Caesalpinioideae). Ann Bot 89:67–76

Hernandez A, Ornelas JF (2007) Disassortative pollen transfer in distylous *Palicourea padifolia* (Rubiaceae), a hummingbird-pollinated shrub. Ecoscience 14:8–16

Hodgins KA, Barrett SCH (2006) Female reproductive success and the evolution of mating-type frequencies in tristylous populations. New Phytol 171:569–580

Hodgins KA, Barrett SCH (2008) Asymmetrical mating patterns and the evolution of biased morph ratios in a tristylous daffodil. Genet Res 90:3–15

Ishihama F, Nakano C, Ueno S, Ajima M, Tsumura Y, Washitani I (2003) Seed set and gene flow patterns in an experimental population of an endangered heterostylous herb with controlled local opposite-morph density. Funct Ecol 17:680–689

Jacquemyn H, van Rossum F, Brys R, Endels P, Hermy M, Triest L, de Blust G (2003) Effects of agricultural land use and fragmentation on genetics, demography, and population persistence of the rare *Primula vulgaris*, and implications for conservation. Belg J Bot 136:5–22

Jesson LK, Barrett SCH (2003) The comparative biology of mirror-image flowers. Int J Plant Sci 164:S237–S249

Kalinowski A, Bocian A, Kosmala A, Winiarczyk K (2007) Two-dimensional patterns of soluble proteins including three hydrolytic enzymes of mature pollen of tristylous *Lythrum salicaria*. Sex Plant Reprod 20:51–62

Kamau E, Charlesworth D (2005) Balancing selection and low recombination affect diversity near the self-incompatibility loci of the plant *Arabidopsis lyrata*. Current Biol 15:1773–1778

Kéry M, Matthies D, Schmid B (2003) Demographic stochasticity in population fragments of the declining distylous perennial *Primula veris* (Primulaceae). Basic Appl Ecol 4:197–206

Khosravi D, Joulaie R, Shore JS (2003) Immunocytochemical distribution of polygalacturonase and pectins in styles and pollen of distylous and homostylous Turneraceae. Sex Plant Reprod 16:179–190

Khosravi D, Yang CC, Siu KWM, Shore JS (2004) High level of α-dioxygenase in short styles of distylous *Turnera* spp. Int J Plant Sci 165:995–1006

Khosravi D, Siu KWM, Shore JS (2006) A proteomics approach to the study of distyly in *Turnera* species. In: Teixeira da Silva JA (ed) Floriculture, ornamental and plant biotechnology: Advances and topical issues, vol 1. Global Science Books, London, UK, pp 51–60

Kohn JR, Barrett SCH (1992) Experimental studies on the functional significance of heterostyly. Evolution 46:43–55

Kohn JR, Graham SW, Morton B, Doyle JJ, Barrett SCH (1996) Reconstruction of the evolution of reproductive characters in Pontederiaceae using phylogenetic evidence from chloroplast DNA restriction-site variation. Evolution 50:1454–1469

Kurian V, Richards AJ (1997) A new recombinant in the heteromorphy '*S*' supergene in *Primula*. Heredity 78:383–390

Labonne JDJ, Hilliker AJ, Shore JS (2007) Meiotic recombination in *Turnera* (Turneraceae): Extreme sexual difference in rates, but no evidence for recombination suppression associated with the distyly *S*-locus. Heredity 98:411–418

Labonne JDJ, Vaisman A, Shore JS (2008) Construction of a first genetic map of distylous *Turnera* and a fine-scale map of the *S*-locus region. Genome 51:471–478

Lau P, Bosque C (2003) Pollen flow in the distylous *Palicourea fendleri* (Rubiaceae): An experimental test of the disassortative pollen flow hypothesis. Oecologia 135:593–600

Lewis D, Jones DA (1992) The genetics of heterostyly. In: Barrett SCH (ed) Evolution and function of heterostyly. Springer, Berlin Heidelberg New York, pp 129–150

Li J, Webster M, Furuya M, Gilmartin PM (2007) Identification and characterization of pin and thrum alleles of two genes that co-segregate with the *Primula S*-locus. Plant J 51:18–31

Li QJ, Xu ZF, Kress WJ, Xia YM, Zhang L, Deng XB, Gao JY, Bai ZL (2001) Flexible style that encourages outcrossing. Nature 410:432

Lloyd DG, Webb CJ (1992a) The evolution of heterostyly. In Barrett SCH (ed) Evolution and function of heterostyly, Springer, Berlin Heidelberg New York, pp 151–178

Lloyd DG, Webb CJ (1992b) The selection of heterostyly. Barrett SCH (ed) Evolution and function of heterostyly, Springer, Berlin Heidelberg New York, pp 179–208

Manfield IW, Pavlov VK, Li J, Cook HE, Hummel F, Gilmartin PM (2005) Molecular characterization of DNA sequences from the *Primula vulgaris S*-locus. J Exp Bot 56:1177–1188

Manicacci D, Barrett SCH (1995) Stamen elongation, pollen size and siring ability in tristylous *Eichhornia paniculata* (Pontederiaceae). Am J Bot 82:1381–1389

Massinga PH, Johnson SD, Harder LD (2005) Heteromorphic incompatibility and efficiency of pollination in two distylous *Pentanisia* species (Rubiaceae). Ann Bot 95:389–399

Mast AR, Conti E (2006) The primrose path to heterostyly. New Phytol 171:439–442

Mast AR, Feller MS, Kelso S, Conti E (2004) Buzz-pollinated *Dodecatheon* originated from within the heterostylous *Primula* subgenus *Auriculastrum* (Primulaceae): A seven-region cpDNA phylogeny and its implications for floral evolution. Am J Bot 91:926–942

Mast AR, Kelso S, Conti E (2006) Are any primroses (*Primula*) primitively monomorphic? New Phytol 171:605–616

Mather K (1950) The genetical architecture of heterostyly in *Primula sinensis*. Evolution 4:340–352

Matsui K, Tetsuka T, Nishio T, Hara T (2003) Heteromorphic incompatibility retained in self-compatible plants produced by a cross between common and wild buckwheat. New Phytol 159:701–708

Matsui K, Nishio T, Tetsuka T (2007) Review: Use of self-compatibility and modifier genes for breeding and genetic analysis in common buckwheat (*Fagopyrum esculentum*). Jpn Agric Res Q 41:1–5

McCubbin AG, Lee C, Hetrick A (2006) Identification of genes showing differential expression between morphs in developing flowers of *Primula vulgaris*. Sex Plant Reprod 19:63–72

Miljuš-Đukic J, Ninković S, Radović S, Maksimović V, Brkljačić J, Neškovic M (2004) Detection of proteins possibly involved in self-incompatibility response in distylous buckwheat. Biol Plant 48:293–296

Morris JA (2007) A molecular phylogeny of the *Lythraceae* and inference of the evolution of heterostyly. Dissertation, Kent State University, USA

Nagano M, Aii J, Kuroda M, Campbell C, Adachi T (2001) Conversion of AFLP markers linked to the S^h allele at the *S*-locus in buckwheat to a simple PCR based marker form. Plant Biotechnol 18:191–196

Nagano M, Aii J, Campbell C, Adachi T, Kawasaki S (2005) Construction of a BAC library for the investigation of the *S*-locus in Buckwheat. Fagopyrum 22:13–20

Ornduff R (1966). The origin of dioecism from heterostyly in *Nymphoides* (Menyanthaceae). Evolution 20:309–314

Ornduff R (1972) The breakdown of trimorphic incompatibility in *Oxalis* section Corniculatae. Evolution 26:52–65

Ornduff R (1975) Heterostyly and pollen flow in *Hypericum aegypticum* (Guttiferae). Bot J Linn Soc 71:51–57

Ornduff R (1979) The genetics of heterostyly in *Hypericum aegypticum*. Heredity 42:271–272

Pailler T, Maurice S, Thompson JD (2002) Pollen transfer patterns in a distylous plant with overlapping pollen-size distributions. Oikos 99:308–316

Pauw A (2005) Inversostyly: A new stylar polymorphism in an oil-secreting plant, *Hemimeris racemosa* (Scrophulariaceae). Am J Bot 92:1878–1886

Piper JG, Charlesworth B, Charlesworth D (1986) Breeding system and evolution in *Primula vulgaris* and the role of reproductive assurance. Heredity 56:207–217

Rama Swamy N, Bahadur B (1984) Pollen flow in dimorphic *Turnera subulata* (Turneraceae). New Phytol 98:205–209

Ree RH (1997) Pollen flow, fecundity, and the adaptive significance of heterostyly in *Palicourea padifolia* (Rubiaceae.) Biotropica 29:298–308

Renner SS (2001) How common is heterodichogamy? Trends Ecol Evol 16:595–597

Richards J (2003) Primula, 2nd edn. Timber Press, Portland, OR, USA

Sage TL, Strumas F, Cole WW, Barrett SCH (1999) Differential ovule development following self- and cross-pollination: The basis of self-sterility in *Narcissus triandrus* (Amaryllidaceae). Am J Bot 86:855–870

Sakai S, Toquenaga Y (2004) Heterostyly: Speciation within a species. Popul Ecol 46:253–262

Schoen DJ, Johnston MO, L'Heureux AM, Marsoais JV (1997) Evolutionary history of the mating system in *Amsinckia* (Boraginaceae). Evolution 51:1090–1099

Schou O, Philipp M (1984) An unusual heteromorphic incompatibility system 3. On the genetic control of distyly and self-incompatibility in *Anchusa officinalis* L. (Boraginaceae). Theor Appl Genet 68:139–144

Shibayama Y, Kadono Y (2003) Floral morph composition and pollen limitation in the seed set of *Nymphoides indica* populations. Ecol Res 18:725–737

Shore JS, Barrett SCH (1985) The genetics of distyly and homostyly in *Turnera ulmifolia* L. (Turneraceae). Heredity 55:167–174

Shore JS, Barrett SCH (1986) Genetic modifications of dimorphic incompatibility in the *Turnera ulmifolia* complex (Turneraceae). Can J Genet Cytol 28:796–807

Shore JS, Barrett SCH (1990) Quantitative genetics of floral characters in homostylous *Turnera ulmifolia* var. *angustifolia* (Turneraceae). Heredity 64:105–112

Shore JS, Arbo MM, Fernández A (2006) Breeding system variation, genetics and evolution in the Turneraceae. New Phytol 171:539–551

Stehlik I, Casperson JP, Barrett SCH (2006) Spatial ecology of mating success in a sexually polymorphic plant. Proc R Soc Lond ser B 273:387–394

Stone JL (1995) Pollen donation patterns in a tropical distylous shrub (*Psychotria suerrensis*; Rubiaceae). Am J Bot 82:1390–1398

Stone JL, Thomson JD (1994) The evolution of distyly: Pollen transfer in artificial flowers. Evolution 48:1595–1606

Stout AB (1925) Studies of *Lythrum Salicaria* II A new form of flower in this species. Bull Torrey Bot Club 52:81–85

Sun S, Gao JY, Liao WJ, Li QJ, Zhang DY (2007) Adaptive significance of flexistyly in *Alpinia blepharocalyx* (Zingiberaceae): A hand-pollination experiment. Ann Bot 99:661–666

Tamari F, Shore JS (2006) Allelic variation for a short-specific polygalacturonase in *Turnera subulata*: Is it associated with the degree of self-compatibility? Int J Plant Sci 167:125–133

Tamari F, Athanasiou A, Shore JS (2001) Pollen tube growth and inhibition in distylous and homostylous *Turnera* and *Piriqueta* (Turneraceae). Can J Bot 79:578–591

Tamari F, Khosravi D, Hilliker AJ, Shore JS (2005) Inheritance of spontaneous mutant homostyles in *Turnera subulata* x *krapovickasii* and in autotetraploid *T. scabra* (Turneraceae). Heredity 94:207–216

Thompson FL, Hermanutz LA, Innes DJ (1998) The reproductive ecology of island populations of distylous *Menyanthes trifoliata* (Menyanthaceae). Can J Bot 76:818–828

Thompson JD, Paillier T, Strasberg D, Manicacci D (1996) Tristyly in the endangered Mascarene island endemic *Hugonia serrata* (Linaceae). Am J Bot 83:1160–1167

Thompson JD, Barrett SCH, Baker AM (2003) Frequency-dependent variation in reproductive success in *Narcissus*: Implications for the maintenance of stigma-height dimorphism. Proc R Soc Lond ser B 270:949–953

Tommerup MM (2001) The evolution of heterostyly and allopolypoidy in *Aliciella* (Polemoniaceae). Dissertation, Claremont Graduate University, USA

Trognitz BR, Hermann M (2001) Inheritance of tristyly in *Oxalis tuberosa*. Heredity 86:564–573

Truyens S, Arbo MM, Shore JS (2005) Phylogenetic relationships, chromosome and breeding system evolution in *Turnera* (Turneraceae): Inferences from ITS sequence data. Am J Bot 92:1749–1758

Vuilleumier BS (1967) The origin and evolutionary development of heterostyly in the angiosperms. Evolution 21:210–226

Wang JC (2005) Finding primary targets of transcriptional regulators. Cell Cycle 4:356–357

Wang Y, Wang QF, Guo YH, Barrett SCH (2005a) Reproductive consequences of interactions between clonal growth and sexual reproduction in *Nymphoides peltata* (Menyanthaceae). New Phytol 165:329–336

Wang Y, Scarth R, Campbell C (2005b) S^h and S_c – Two complementary genes that control self-compatibility in Buckwheat. Crop Sci 45:1229–1234

Washitani I, Ishihama F, Matsumura C, Nagai M, Nishihiro J, Nishihiro MA (2005) Conservation ecology of *Primula sieboldii*: Synthesis of information toward the prediction of the genetic/demographic fate of a population. Plant Species Biol 20:3–15

Webb CJ, Lloyd DG (1986) The avoidance of interference between the presentation of pollen and stigmas in angiosperms 2 Herkogamy. N Z J Bot 24:163–178

Weller SG (1976) The genetic control of tristyly in *Oxalis* section *Ionoxalis*. Heredity 37:387–393

Weller SG (1992) Evolutionary modifications of tristylous breeding systems. In: Barrett SCH (ed) Evolution and function of heterostyly. Springer, Berlin Heidelberg New York, pp 247–272

Weller SG, Dominguez CA, Molina-Freaner FE, Fornoni J, LeBuhn G (2007) The evolution of distyly from tristyly in populations of *Oxalis alpina* (Oxalidaceae) in the Sky islands of the Sonoran desert. Am J Bot 94:972–985

Wong KC, Watanabe M, Hinata K (1994) Protein profiles in pin and thrum floral organs of distylous *Averrhoa carambola* L. Sex Plant Reprod 7:107–115

Woo SH, Adachi T, Jong SK, Campbell CG (1999) Inheritance of self-compatibility and flower morphology in an inter-specific buckwheat hybrid. Can J Plant Sci 79:483–490

Yasui Y, Wang Y, Ohnishi O, Campbell G (2004) Amplified fragment length polymorphism linkage analysis of common buckwheat (*Fagopyrum esculentum*) and its wild self-pollinated relative *Fagopyrum homotropicum*. Genome 47:341–351

Chapter 2
Genetic and Environmental Causes and Evolutionary Consequences of Variations in Self-Fertility in Self Incompatible Species

S.V. Good-Avila, J.I. Mena-Alí, and A.G. Stephenson

Abstract Within many self incompatible species there is variation among plants in self-fertility. Mutations conferring partial or complete self-fertility occur in the *S*-alleles themselves, in genes closely linked to the *S*-alleles or in unlinked genes that affect the downstream rejection pathway, in unlinked genes that modify the expression or turnover of the *S*-allele products, and in unlinked genes that modify the pistil environment. Populations with genetic polymorphisms for self-fertility can be used as natural laboratories to examine the dynamic interplay of the forces that shape the evolution of plant breeding systems. We find that all self-fertility mutations are beneficial in populations with few *S*-alleles and/or high rates of pollen limitation and they may result in stable polymorphisms when there are high levels of inbreeding depression and/or *S*-linked load as might occur in highly fragmented or low density populations and in those species prone to repeated bouts of colonisation and extinction.

Abbreviations

120 K	120 kDa glycoprotein
GSI	Gametophytic SI
HT-B/HT	H-Top Band; a small, novel asparagine-rich protein
mRNA	messenger RNA
PSF	Pseudo-self-fertility

S.V. Good-Avila
Acadia University, Department of Biology, 24 University Avenue, Wolfville, Nova Scotia, B4P 2R6, Canada, e-mail: sara.good-avila@acadiau.ca

J.I. Mena-Alí
Department of Biology, Amherst College, 311 McGuire LSB, Amherst, MA 01002, USA, e-mail: jmenaali@amherst.edu

A.G. Stephenson
Department of Biology, 208 Mueller Lab, University Park, PA 16802, USA, e-mail: as4@psu.edu

V.E. Franklin-Tong (ed.) *Self-Incompatibility in Flowering Plants – Evolution, Diversity, and Mechanisms.*
© Springer-Verlag Berlin Heidelberg 2008

rRNA   ribosomal RNA
S allele   Self-incompatibility allele
SI   Self-incompatibility
S-locus   Self-incompatibility locus
S-RNase   S-ribonuclease
TTS   Transmitting tract specific glycoprotein

2.1 Introduction

Over the past two and half decades there has been an explosion of progress in a growing number of model self incompatibility (SI) systems on our understanding of the molecular, biochemical and cellular processes underlying the recognition of self pollen and the initiation of a cascade of biochemical and cellular events that prevent self fertilisation (see Chaps. 8, 10, and 11, this volume). These studies are unravelling the complexity of a trait (SI) whose sole purpose, as far as we know, is to exert a strong influence on the breeding system of plants. Evolutionary interest in floral traits that influence the breeding system and in the forces that shape these traits began with Darwin who devoted one complete book to the subject (Darwin 1876) and significant portions of a second book (Darwin 1877). Although the presence of an SI system in a species is often tacitly assumed to be a binary regulator of the breeding system (i.e. species with SI are not self-fertile while those lacking SI are self-fertile), it is becoming increasingly clear that there is variation in self-fertility both within and between populations of species with SI, including species with well-characterised SI systems such as those found in the Brassicaceae and Solanaceae (Stone 2002; Nielsen et al. 2003; Tsukamoto et al. 2003a,b; Sato et al. 2004; Ushijima et al. 2004; Busch 2005; Newbigin and Uyenoyama 2005; Willi et al. 2005; Goodwillie and Knight 2006; Hegedus et al. 2006; Kruszewski and Galloway 2006; Schierup et al. 2006; Stone et al. 2006; Vilanova et al. 2006; Ferrer and Good-Avila 2007; Mena-Alí and Stephenson 2007; Sherman-Broyles et al. 2007; Tang et al. 2007; Tao et al. 2007). Recently, evolutionary biologists have begun to view these populations as natural laboratories to examine the dynamic interplay of the forces that shape the evolution of breeding systems (Lloyd and Schoen 1992; Levin 1996; Good-Avila and Stephenson 2002; Barrett, 2003; Stephenson et al. 2003 and references therein). These studies have been enriched by the recent progress on the molecular biology of SI, by the influx of molecular, biochemical and quantitative techniques, and by the continuing development of formal theory and models (see other chapters in this volume, including Chaps. 3–5, this volume).

The evolution of plant breeding systems is often viewed as the interplay between the advantages and disadvantages of selfing. Evolutionary biologists have long noted that there are three primary advantages to selfing. First, there is an inherent genetic transmission advantage to selfing because a plant donates two haploid sets of chromosomes to each selfed seed and can still donate pollen to conspecifics (e.g. Fisher 1941). Second, selfing can provide reproductive assurance when pollinators are scarce or unreliable (i.e. selfed progeny are better than no progeny) (e.g. Baker 1955; Stebbins 1957; Schoen et al. 1996) and third, it often costs less, in terms of energy and other resources, to produce selfed seed (e.g. fewer resources are expended to attract and reward pollinators) (Schemske 1978; Waller 1979; Schoen and Lloyd 1984). Consequently, a mutation that promotes self-fertilisation should increase in a population unless it is counterbalanced by opposing selective pressures. These genetic and environmental advantages of selfing are potentially counterbalanced by inbreeding depression (a reduction in the fitness of selfed progeny relative to outcrossed progeny as a result of the increase in homozygosity in selfed progeny that exposes deleterious recessive alleles to selection and decreases the contribution of overdominance to fitness) (Charlesworth and Charlesworth 1987; Uyenoyama et al. 1993; Byers and Waller 1999; Glemin et al. 2001). Recent advances have also shown that traits that promote selfing frequently also decrease pollen donation to conspecifics (pollen discounting), which effectively reduces the transmission advantage of selfing (e.g. Nagylaki 1976; Holsinger et al. 1984), and traits that provide reproductive assurance when pollinators are scarce or unreliable often decrease outcross seed production (self progeny are produced at the expense of outcrossed progeny: seed discounting, Lloyd 1992). In this chapter, we review the genetic basis of population level polymorphisms in self-fertility; we examine the consequences of the various genetic mechanisms that enhance self-fertility for inbreeding depression, pollen and seed discounting. We also examine the environmental and genetic conditions that determine the fate of genes that enhance self-fertility. Finally, we explore the possibility that genetic polymorphisms for self-fertility are part of a stable mixed mating system.

2.2 Genetics of Self-Fertility

The genetic mechanisms underlying complete or partial self-fertility of individuals in self-incompatible species are as complex and varied as the SI systems in which they appear. Mutations that promote self-fertility occur in both heteromorphic and homomorphic SI systems and in both sporophytic and gametophytic SI (GSI) systems. Moreover, these mutations can occur in the S-alleles themselves, in genes closely linked to the S-alleles that modify their expression, in genes involved in the downstream rejection pathways, in unlinked genes that modify the expression or turnover of the S-allele products and in unlinked genes that modify the pistil environment in which the SI system operates. Below is a non-exhaustive peek at the diversity of genetic mechanisms that underlie self-fertility.

2.2.1 Mutations Affecting the S-locus

2.2.1.1 Recessive Alleles in Sporophytic Systems

In the Brassicaceae, there are two classes of *S*-alleles: Class I alleles display co-dominance in expression while Class II alleles are recessive when found in association with Class I *S*-alleles (Hatakeyama et al. 1998; Billiard et al. 2007) thereby permitting self-fertilisation in some individuals in populations with both classes of alleles. The mechanisms underlying these dominance relationships, however, are not clearly understood and several models have been proposed to explain the evolutionary dynamics of both classes of *S*-alleles in natural populations (Schierup et al. 1997; Vekemans et al. 1998). Curiously, brassicaceous species differ in the degree to which they conform to these models and individuals with higher than expected self-fertility have been found in natural populations of *Arabidopsis lyrata* (Schierup et al. 2006), *Laevenworthia alabamica* (Busch 2005) and *Brassica* (Nasrallah 1989; Hatakeyama et al. 1998), suggesting that the genetics of self-fertility in the Brassicaceae may be even more complicated. Dominance relationships among *S*-alleles have also been reported in the Convolvulaceae: in *Ipomoea trifida* the dominance relationships appear to be linear (Kowyama et al. 1994; see also Chap. 12, this volume).

2.2.1.2 Self-Fertility *S*-Alleles (S_f-Alleles)

In some populations of many species (including species in the Asteraceae, Fabaceae, Poaceae, Rosaceae and Solanaceae), self-fertility is known to segregate with specific *S*-alleles (Levin 1996), suggesting that mutations at the *S*-locus (or in a gene tightly linked to the *S*-locus) have converted the *S*-allele into a null (S_f) allele. Recent advances in the molecular biology of RNase-based GSI have allowed investigators to explore the nature of S_f-alleles in Solanaceae and Rosaceae. In *Petunia axillaris*, some of the self-compatible plants found in a population from Uruguay carry a loss of function on the pollen component of allele S_{17} (Tsukamoto et al. 2003a). In peach (*Prunus persicae*), all three of the self compatible haplotypes that have been studied carry pollen-part mutations that produce truncated peptides; one of these haplotypes, S_{2m}, also carries a pistil-part mutation that reduces the stability of its encoded *S*-RNase (Tao et al. 2007). In a second population of *P. axillaris* from Uruguay, the presence of self-compatible plants has been shown to be due to a linked gene that suppresses the expression of a specific *S*-RNase allele (Tsukamoto et al. 2003b). Several studies have also shown that duplication of the *S*-locus can lead to a breakdown of the GSI response, which is ascribed to 'competitive interaction' (de Nettancourt 1977, 1997; Stone 2002), specifically the duplication of the pollen-expressed component of the GSI reaction of the Solanaceae (Tsukamoto et al. 1999). This finding, in turn, has aided in the development of models to explain the specificity of the interaction of the male and female determinants of RNase-based GSI (Kao and Tsukamoto 2004; see also Chap. 9, this volume). The mutations described here result in populations consisting of self-incompatible and self-compatible individuals.

In a population of *Solanum carolinense* consisting of self-incompatible and partially self-fertile individuals, Mena-Alí and Stephenson (2007) used sequence analysis to show that partial self-fertility segregated with the S_9 allele. The observed patterns of leakiness suggest that self-fertility is not due to the mutation that rendered S_9 functionless. All six alleles found in the population were successfully amplified from RNA extractions of fresh styles; therefore, at least at the transcription level, the S_9 allele is expressed and functional (although S_9 could still be a weak allele if the mature protein shows reduced specificity or activity). If this were the case, however, then plants of genotype S_9S_j should still be able to recognise and reject the other S-allele carried (S_j). The presence of S_jS_j progeny revealed that any allele found in association with S_9 is also able to escape rejection and achieve fertilisation.

A modifier gene linked to allele S_9, on the other hand, may explain the association between the presence of this allele and a plant's ability to self. If some of the molecular machinery involved in the SI reaction is defective and linked to S_9, then even strong alleles should be able to escape the SI recognition and achieve fertilisation when found in association with S_9. Several genes are known to be involved in the SI response (Cruz-Garcia et al. 2003; Kao and Tsukamoto 2004; Goldraij et al. 2006; see also Chaps. 9 and 10, this volume), but none of them are found in close enough proximity to the S-locus to account for this tight association between S_9 and the weakened SI response. One possibility may be that the weakened SI response is caused by mutations in the flanking regions surrounding different S-alleles. These sequences are known to be highly divergent (Coleman and Kao 1992). Thus, it is possible that mutations in the regulatory regions required for the transcription of different S-alleles have occurred and led to the down regulation of specific S-allele transcripts (Tsukamoto et al. 2003b). These findings suggest that a gene linked to the S-locus can produce quantitative (rather than qualitative) variation in the strength of SI.

2.2.2 Unlinked Modifiers of SI

2.2.2.1 Single Genes

Although S-alleles determine SI specificity, other genes are now known, at least for the model SI systems, to be required for the rejection mechanism to operate. Consequently, there are many downstream genes that may potentially affect the timing or magnitude of the S-locus expression, the turnover of the products of the S-locus or the normal function of the pollen/pollen tube rejection pathways (Tsukamoto et al. 1999). For example, the current model of GSI in the Solanaceae evokes a cytotoxic response in which S-RNases are incorporated into pollen tubes, where they have the potential to degrade mRNA and rRNA. Furthermore, it now seems that the products of several unlinked genes are involved in the formation of a multi-protein complex that represents the active form of *S*-RNase (Cruz-Garcia et al. 2003). Other peptides, such as HT-B and a 120 kDa glycoprotein, are also required for pollen rejection but in a non-haplotype-specific manner (Goldraij et al. 2006; Hua and Kao 2006). The role of these peptides in the SI response is just starting to emerge (see also Chaps. 9 and 10, this volume). Briefly, in *Nicotiana*, *S*-RNases are compartmentalised into an endomembrane system (perhaps the vacuole) upon uptake by the pollen tubes; the stability of these endomembranes is somehow associated with HT. In compatible crosses, HT is degraded and the endomembrane compartments remain intact, thus preventing the release of *S*-RNase into the pollen tube while HT degradation following self-incompatible reactions mediates the cytotoxic response that leads to pollen tube growth arrest (Goldraij et al. 2006). Mutations in HT have been associated with the emergence of self-compatibility in cultivated tomato (Kondo et al. 2002).

Because of the complexity of the SI response, there is ample opportunity for mutations in unlinked genes involved in the rejection process to have a qualitative or a quantitative effect on self-fertility (depending upon the nature of the mutation) even in plants with a fully functional *S*-locus. These unlinked modifiers of the *S*-locus would be expected to segregate in the progeny. The presence of such a modifier has been found in several self compatible plants in a population of *P. axillaris* (Tsukamoto et al. 1999), and several such modifiers have been described in *Brassica* that suppress the sporophytic self incompatibility reaction (e.g., Nasrallah 1974, 1989; Hinata and Okasaki 1986; Nasrallah et al. 1992). Although only a few modifier genes have been identified, any mutation that affects the production, turnover or function of the genes involved in pollen–pistil interactions could also potentially modulate the SI response, as could mutations that affect the stigmatic or stylar environment in which these genes operate.

2.2.2.2 Polygenic Unlinked Modifiers and Pseudo Self-Fertility (PSF)

From the perspective of the discussion in the previous section, it is not surprising that SI in natural populations of many species (including species in the Asteraceae, Brassicaceae, Fabaceae, Poaceae, Ranunculaceae, Solanaceae and many others) is a quantitative, rather than a qualitative, trait due to the segregation of unlinked genes that modify the strength of SI. In fact, these populations include species that have served as model organisms for studies of the molecular basis of GSI and SSI such as *Brassica oleracea*, *B. napus*, *Lycopersicon peruvianum*, *Petunia hybrida*, and various species of *Nicotiana* (Levin 1996). This phenomenon (the segregation of unlinked genetic modifiers in a population) is termed pseudo-self-fertility: PSF (de Nettancourt 1977; Levin 1996). When multiple unlinked modifier genes are involved, this results in populations in which the self-fertility of individual plants varies continuously from self sterile to nearly self-fertile (as opposed to S_f alleles that result in self fertile and self sterile individuals). In species with PSF, the phenotypic expression of variability in self-fertility is due to several phenomena, including variability in the speed of self pollen germination, the attrition of self pollen tubes in the style and/or the growth rate of self pollen tubes relative to that of cross pollen (Stephenson and Bertin 1983; Stephenson et al., 1992, 2003; Levin 1996).

Genetic insights into the nature of the unlinked genes that modify the strength of SI have been obtained using quantitative genetics. Studies using *Nemesia stumosa*, *Petunia integrifolia*, *Medicago sativa* and *Phlox drummondia* have shown that it is possible to select for greater self-fertility by crossing the more self-fertile individuals within a population (Levin 1996). These studies demonstrate that there is additive genetic variation for self-fertility in these populations. For example, in *P. drummondii* self pollen grains are normally prevented from germinating on the stigmatic surface. Bixby and Levin (1996) intercrossed the more self-fertile individuals in each of two natural populations for two generations, and were able to substantially increase self fruit production from 23 to 41% in one population and from 4 to 56% in the other population. These represent realised heritabilities of 0.47 and 0.90. At the same time, they concomitantly increased the percentage of self pollen grains with tubes from 0.55 to 2.2% in the first population and from 0.01 to 7.5% in the second. In a population of the dichogamous species *Campanula rapunculoides*, seed set following self pollination on young female phase flowers ranged from 0 to 60% of that obtained by cross pollination.

Other studies have examined the number of unlinked genes that modify the strength of SI (Lipow and Wyatt 2000; Good-Avila and Stephenson 2002). For example, studies had shown that cross and self pollen tubes grow at different rates in the styles of *C. rapunculoides* (Richardson et al. 1990; Vogler and Stephenson 2001). Thirty-one plants from this population were randomly chosen and crossed to generate 101 F1 families and the self-fertility of 2–7 progeny per family was assessed. The analyses of these families and their F2 progeny suggested the presence of 3–5 unlinked, mostly recessive genetic modifiers affecting the strength of the SI response and that the narrow sense heritability of self-fertility was 0.27 (Good-Avila and Stephenson 2002). Finally, investigations of individuals in some populations of *Leptosiphon jepsonii*, which are known to be partially self fertile, also revealed evidence for genetic modifiers. When the more self fertile individuals from different populations were crossed, the F1 had strong SI (i.e. low self-fertility), and self-fertility reappeared, segregating in the F2. This suggested that (i) genetic modifiers of SI differ among populations, (ii) the genes are recessive, and (iii) complementation occurs in the F1. This suggests that different genes are responsible for self-fertility in different populations (Weber and Goodwillie 2007). Together these studies reveal that natural selection acting on the breeding system of plants can potentially operate on polygenic unlinked modifiers of SI that confer self-fertility.

Curiously, several species are known to exhibit a cryptic form of SI in which seed set following cross and self pollinations is similar but when mixed (self and cross) pollen loads are deposited on a stigma, all or nearly all of the seeds are sired by cross pollen. This suggests some form of competition between self and cross pollen. In *Eichornia paniculata*, self and cross pollen does not differ in the speed of germination, but cross pollen sires most of the seeds, even when self pollen is deposited onto the stigmas several hours before the cross pollen (Cruzan and

Barrett 1993). Species with cryptic SI are scattered throughout the angiosperms (e.g. Boraginaceae, Brassicaceae, Campanulaceae, Caryophylaceae, Liliaceae, Lythraceae, Onagraceae, Pontederiaceae) but seem to be especially prevalent in heteromorphic species whose close relatives have fully functional heteromorphic SI systems (Bateman 1956; Weller and Ornduff 1977; Ockendon and Currah 1978; Bowman 1987; Casper et al. 1988; Aizen et al. 1990; Cruzan and Barrett 1993; Rigney et al. 1993; Jones 1994; Eckert and Allen 1997; Kruszewski and Galloway 2006; see Chap. 1 for an account of heteromorphic SI). In *Amsinckia douglasiana*, the long and short styled morphs differ in their self-fertility (Casper et al. 1988), but to our knowledge there have been no comprehensive studies of either the variability among individuals in the relative growth rates of self pollen tubes or of the genetics underlying cryptic SI. It is tempting, however, to view cryptic SI as merely one end of the SI spectrum, resulting from the action of genetic modifiers (linked and unlinked) on functional SI systems. At this end of the spectrum, high self seed set is permitted but ovules are not discounted by self fertilisation when cross pollen is available (i.e. when mixed pollen loads are deposited onto stigmas or when cross pollen arrives shortly after the deposition of self pollen).

2.2.3 *Plasticity in Self-Fertility*

External environmental conditions, especially increases in temperature, have long been known (and frequently exploited in breeding programs) to increase self-fertility in self-incompatible plants (Stephenson and Bertin 1983; Ascher 1984; Levin 1996 and references therein). For example, when clones of four individuals (genotypes) of *Lolium perenne* were grown at 14, 18, 22 and 26°C, pollen tube penetration of stigma and seed production following self-pollinations increased with temperature (Elgersma et al. 1989). Moreover, the increase in self-fertility with temperature differed among the four genotypes.

Self-fertility is also known to change with the age of flowers and the number of developing fruits on the plant (e.g., East 1934; Owen 1942; Pandey 1960; Ascher and Peloquin 1966; Lawson and Williams 1976; Pundir and Al-Attar 1982; Vogler et al. 1998; Stephenson et al. 2000, 2003; Goodwillie et al. 2004; Travers et al. 2004). For example, in *C. rapunculoides*, a series of cross and self pollinations on young (first day of the female phase) flowers and old (fourth day of the female phase) flowers on more than 120 plants from two populations revealed that self-fertility increases on old flowers (Stephenson et al. 2000). In addition, in a study using clonal replicates of eight genotypes of *C. rapunculoides*, self-fertility was increased in plants in which there were no developing outcrossed fruits compared to plants in which outcrossed fruits were already developing on the plant prior to the self pollinations (Vogler et al. 1998). In both cases, the plants in which the young flowers were the most self fertile showed the greatest increases in self-fertility with floral age and low prior fruit production. Similarly, in *Solanum carolinense*, most genotypes produce some self seed upon self pollination of older flowers and some self seed production occurs on most plants in which no fruits have been produced on the first four inflorescences (Stephenson et al. 2003; Travers et al. 2004). In both cases, the plants with the S_9 allele, which was leaky, experienced the greatest increase in self-fertility with floral age and low prior fruit production. In both *C. rapunculoides* and *S. carolinense*, the number of self pollen tubes in the transmitting tissue of the style and the growth rate of self pollen tubes increases in older flowers (Richardson et al. 1990; Travers et al. 2004). Curiously, cross pollen tubes in *C. rapunculoides* grow more slowly in old styles than cross tubes in younger styles, suggesting that the condition of the transmitting tissue has degenerated somewhat with floral age.

We suspect that the increase in self-fertility on plants with no developing fruit is due to an increase in floral longevity that occurs when no resources are directed to developing fruit which, in turn, provides the slower growing self pollen tubes a longer time frame to enter the ovary before the flower wilts. Together, these studies indicate that environmental conditions (both internal and external) not only affect self-fertility but that the magnitude of the environmental effect depends on the genotype of the individual plants (i.e. there are genotype by environment interactions that determine self-fertility), indicating that self-fertility is, at least in some species, a plastic trait. It should be noted that both the presence of older unpollinated flowers and flowers on plants with few or no developing fruit are indicative of conditions of low availability of cross pollen (i.e. low pollinator activity or few *S*-alleles in the population).

2.2.4 Summary of Genetics of Self-Fertility

Many populations of self-incompatible species are polymorphic for self-fertility. Studies of the self-fertile plants in these populations reveal that self-fertility can occur in many ways, through mutations that affect either the pollen or the pistil part of the *S*-locus (S_f alleles); through linked modifiers of the *S*-locus that confer complete or partial (i.e. leaky *S*-alleles) self-fertility; and through any of a wide variety of unlinked modifiers of SI. These unlinked modifiers differ in the size of their impact on self-fertility and these modifiers can be dominant, recessive or epistatic in their mode of action (Levin 1996; Good-Avila and Stephenson 2002). Moreover, many species, including species in the Brassicaceae and Solanaceae, are known to have two or more different types of mutations that confer self-fertility (e.g. S_f alleles and polygenic unlinked modifiers conferring partial self-fertility, as discussed earlier). For example, there appears to be at least three distinct genetic causes for self-fertility circulating in two otherwise self-incompatible populations of *P. axillaris* (a pollen part mutation on the S_{17} allele; a linked modifier of the S_{13}-*S*-RNase gene; and an unlinked modifier that has a large effect on self-fertility) (Tsukamoto et al. 1999; Tsukamoto et al. 2003a, b). In *Leptosiphon jepsonii*, different unlinked polygenic modifiers of self-fertility are circulating in different populations (Weber and Goodwillie 2007). These findings show that there are not only many different genetic mechanisms to enhance self-fertility, but that mutations conferring self-fertility frequently arise in natural populations.

Although each of the genetic mechanisms described above will create populations that are polymorphic for self-fertility, they differ in their implications for inbreeding depression, pollen and seed discounting. This is due to differences among these mechanisms in their effects on the relative germination and growth rates of self and cross pollen tubes and their association with *S*-linked (sheltered) load. As noted by Uyenoyama (2003), deleterious recessive alleles are expected to accumulate around the *S*-locus because of the reduced levels of recombination in this region (see Chap. 6, this volume) and because the *S*-locus is always heterozygous (i.e. the deleterious recessive alleles are not exposed to selection). Consequently, the *S*-homozygous progeny produced by selfing would not only suffer from the deleterious recessives that segregate within that maternal lineage but would also suffer from the sheltered load. To date only two studies have experimentally calculated sheltered load and, in both cases, they found that *S*-homozygotes have greater inbreeding depression than their *S*-heterozygous siblings Moreover, the amount of sheltered load varied among S alleles (Stone 2004; Mena-Alí 2006). Consequently, S_f alleles and *S*-linked modifiers of SI would always be tightly associated with sheltered load. Moreover, the amount of *S*-linked load will vary depending upon the amount of load that was sheltered by the *S*-allele at the time that the mutation arose. Because the germination and growth rate of pollen tubes from S_f alleles is similar to that of cross pollen (Levin 1996), deposition of self pollen shortly after anthesis or the deposition of a mixed (self plus cross pollen) onto the stigma may also result in the production of selfed progeny, which will be less fit, at the expense of outcrossed progeny (seed discounting). Moreover, because floral wilting is often accelerated by fertilisation (O'Neill 1997), early self-fertilisation would also reduce the opportunities for donating pollen to conspecifics (pollen discounting). In contrast, unlinked modifiers of SI tend to slow the germination and/or growth of self pollen tubes relative to cross pollen tubes, which handicaps self pollen performance. Moreover, modifiers often exhibit genotype by environment interactions that enhance self-fertility in older flowers and flowers on plants with few developing outcrossed fruits. Consequently, self fertilisation tends to be delayed relative to cross fertilisation, thereby reducing both the seed and pollen discount. Finally, the *S*-linked load of inbred progeny in species with unlinked modifiers of SI would vary with the specific *S*-alleles in the population.

2.3 Fate of Self-Fertility Genes

If mutations that enhance self-fertility commonly occur in populations of SI species, then what are their fates? Presumably, most of these mutations are eliminated by genetic drift or by purifying selection brought on by the high levels of inbreeding depression commonly associated with out-crossing species (Husband and Schemske 1996). On the other hand, mutations that enhance self-fertility are expected to increase in frequency when inbreeding depression is less than 0.5 (due to the transmission advantage of selfing) and/or when the availability of cross pollen limits seed set. A broad survey of the factors correlated with pollen limitation in angiosperms revealed that self-incompatible species are significantly more pollen-limited than those that are self-compatible and that this association holds across species with different life histories (Larson and Barrett 2000). This suggests that a condition (pollen limitation) for the invasion of self-fertility mutations may be relatively common. If a mutation conferring self-fertility arises in a population exhibiting low numbers of *S*-alleles, pollen limitation and/or low-to-intermediate levels of inbreeding depression, then the mutation may become fixed, resulting in the loss of SI.

The transition from SI to SC is one of the most frequently travelled roads in the evolution of plant mating systems (Stebbins 1974; Goodwillie 1999; Takebayashi and Morrell 2001; Igic et al. 2006). Traditionally, populations of SI species with genetic polymorphisms for self-fertility have been viewed as populations in transition to SC or temporarily harbouring some self-fertility genes. However, evolutionary biologists have begun to ask whether genetic polymorphisms for self-fertility are the product of selection for the maintenance of a mixed (both cross and fertilisation permitted) mating system (Lloyd and Schoen 1992; Levin 1996; Stephenson et al. 2000) and recent theoretical models have explored the range of parameters that could lead to stable polymorphisms for mixed mating systems in SI species (Vallejo-Marín and Uyenoyama 2004; Porcher and Lande 2005; see also Chaps. 3, 5, and 6, this volume).

2.3.1 Conditions for Stable Polymorphisms

Porcher and Lande (2005) modelled the conditions for the loss of GSI by examining the evolutionary dynamics between inbreeding depression (caused by both a large number of nearly recessive lethals and variable background levels of inbreeding depression, including sheltered load) and the invasion of mutations enhancing self-fertility that are either linked or unlinked to the *S*-locus. Their model also incorporates pollen limited seed set. Considering multiple scenarios, they find broad conditions for the invasion of self-fertility mutations if self-fertilisation always occurs after opportunities for cross-fertilisation (i.e. delayed selfing and little or no seed discounting). Self-fertility can always invade and lead to the complete loss of SI in late-stage flowers just as delayed selfing can easily invade a primarily outcrossing self-compatible population (Lloyd 1992). Even though modifiers permitting delayed self-fertilisation would only increase the self-fertility of late-stage flowers, theory predicts that they could become fixed in populations and this would result in a stable polymorphism of pseudo self-fertility (PSF) even if self-fertilisation were infrequent. Thus, theory suggests that a breakdown of SI with floral age or flowering season should be relatively common, especially in populations with few *S*-alleles or pollen limitation. Breakdown of SI with floral age or over the flowering season has been found in several plant species (see Sect. 2.2.3). Moreover, it is likely that delayed self-fertility is present but undetected in many species, because hand pollinations on late stage flowers have not been conducted. Obviously, the utility of a breakdown of SI with floral or reproductive age as a mechanism to provide reproductive assurance is enhanced when it is coupled with additional floral mechanisms that enable autogamous self-fertilisation in old but not young flowers (Kalisz and Vogler 2003). For example, in *C. rapunculoides* when the pollen is not removed from the flowers during the male phase, the fully reflexed stigmas of older female phase flowers come into close proximity to the pollen and autogamous self fertilisation occurs on many of the plants (Stephenson et al. 2000). Recently, we (Mena-Alí et al. unpublished data) have obtained data showing that there is heritable genetic variation for autogamy within populations and that there are strong genetic correlations between self-fertility in old female phase flowers, the length of stigmatic lobes and autogamy. Moreover, there is evidence that a breakdown of SI in late-stage female flowers of *C. rapunculoides* actually increases the rate of self-fertilisation under field conditions utilising natural pollination when plants are subjected to restricted pollinator access (Good-Avila and Stephenson 2003, Mena-Alí et al. unpublished data).

When cross and self pollen compete for fertilisation (i.e. there is the potential for seed discounting), Porcher and Lande (2005) found that there are at least five important parameters that determine whether self-fertility mutations will invade and lead to stable polymorphisms for PSF. These are (1) whether the mutation is linked or unlinked to the S-locus, (2) the amount of S-linked load, (3) the amount of inbreeding depression, (4) the number of S-alleles in the population and (5) the availability of outcross pollen. In general, the conditions for the invasion of a self-fertility mutation are narrower if the mutation is linked to the S-locus. However, stable polymorphisms for S-linked mutations can occur if the mutations allow only a low rate of self-fertilisation ($s < 0.2$) and if there are few S-alleles and high inbreeding depression (including sheltered load) in the population. Thus if PSF is caused by S-linked modifiers, it may go unnoticed, since a low selfing rate is associated with such polymorphisms and the selfing rate may not be detected as significantly different from zero. At present, there are so few families for which the S-locus is known that it is difficult to know whether S-linked mutations may be associated with stable self-fertility polymorphisms (Stone 2002; Mena-Alí and Stephenson 2007).

If mutations conferring self-fertility are unlinked to the S-locus, the conditions for stable polymorphisms when cross and self pollen compete for ovules are somewhat broader (Porcher and Lande 2005). Polymorphisms can evolve in the presence of sheltered load and lead to stable polymorphisms with high selfing rates. Stable polymorphism in the presence of non S-linked mutations occurs if self fertile genotypes have higher seed set and benefit from reproductive compensation (resource savings due to self fertilisation) even in the presence of substantial inbreeding depression. This result is similar to that obtained by Vallejo-Marín and Uyenoyama (2004), who found that unlinked self-fertility mutations can be evolutionarily stable because positive genetic associations develop between S-locus heterozygotes and enhancers of SI, but these associations are balanced by a loss of reproductive opportunities in populations harbouring few S-alleles. The two most important factors determining the stability of self-fertility polymorphisms due to unlinked modifiers of SI are the levels of inbreeding depression (including S-linked load) and the extent of pollen limitation caused by either low S-allele number or poor availability of outcross pollen. The relationship between these factors is essential for determining both the success of modifiers conferring self-fertility and their potential stability, because they represent the balance between the advantages and disadvantages of reproducing via self-fertilisation. With a large number of S-alleles and no pollen limitation, mutations conferring self-fertility can invade whenever inbreeding depression is less than 0.5 (Charlesworth and Charlesworth 1979; Vallejo-Marín and Uyenoyama 2004; Porcher and Lande 2005). However, even if inbreeding depression is greater than 0.5, mutations conferring self-fertility can invade depending on the number of S-alleles in the population and the degree of pollen limitation (Porcher and Lande 2005). This suggests that self-fertility mutations will persist in very small or isolated populations even in the presence of high levels of inbreeding depression. If there is substantial sheltered load, the invasion of mutations conferring enhanced self-fertility is considerably more difficult and occurs primarily when self-fertile individuals practice relatively high selfing rates (Porcher and Lande 2005). This is because when self-fertility is conferred by non-S-linked mutations, individuals only purge genetic load when rates of self-fertilisation are high enough to purge load at the S-locus. In general, the presence of S-linked load impedes the invasion of self-fertility particularly in small populations where inbreeding depression is expected to be smaller. Glemin et al. (2001) proposed that high levels of S-linked load may explain why there are not consistent differences in the amount of self-fertility between the central and the peripheral populations of some species (as found by, e.g., Karron 1987). Some studies of endangered species and small populations have found evidence for a breakdown of SI in response to small population size (Les et al. 1991; Reinartz and Les 1994; Gilblin and Hamilton 1999), while others have shown that small populations retain SI and suffer from poor sexual recruitment of offspring. In these latter cases, S-allele diversity is typically low enough that individuals rely on uniparental reproduction such as clonal growth [e.g., *Elliottia racemosa* (Godt and Hamrick 1999), *Rutidosis leptorrhynchoides* (Young et al. 2002), and *Coreopsis rosea* (Wood unpublished data)].

2.3.2 Summary and Conclusions Concerning Stable Polymorphisms

In short, the circumstances that appear to offer the broadest conditions for stability in the amount of self-fertility are (1) self-fertility alleles that promote delayed self-fertilisation, (2) S-linked modifiers of SI that confer only a small increase in the rate of self-fertilisation and (3) non S-linked modifiers of SI that lead to large variation in the degree of self-fertility among individuals. All of these mutations are more likely to occur in populations with low S-allele numbers and/or high rates of pollen limitation and they are more likely to lead to the stable polymorphisms in self-fertility when there are high levels of inbreeding depression and/or S-linked load. These characteristics appear to describe many SI populations but particularly those occurring in low density or in highly fragmented habitats and those prone to repeated bouts of colonisation and extinction such as those with a weedy life history. Thus, a prediction is that stable self-fertility polymorphisms should be more common in species exhibiting metapopulation dynamics. Pannell and Barrett (1998) compared the seed productivity of obligate outcrossers and selfers in a metapopulation and found that the advantage of reproductive assurance was greatest when the number of occupied patches was small. As patch occupancy increased, obligate outcrossers had seed production similar to or higher than selfers. Although this model did not address the conditions for stable polymorphism in self-fertility genes, it showed that metapopulation dynamics can favour reproductive assurance particularly in annuals and in species with no seed bank, but that obligate outcrossers can exist in metapopulations. An interesting question is whether the presence of genes that enhance self-fertility in SI species allows weedy SI species to maximise both fitness and genetic diversity via outcrossing while also allowing reproductive assurance during episodes of colonisation and expansion (Mena-Alí et al. 2008).

Lastly, if pseudo-self-fertility represents a stable mating system, then we would expect such species to be phylogenetically clustered and for pseudo-self-fertility to be reconstructed as the ancestral state at internal nodes. A recent analysis of the rates of gain and loss of SI, pseudo-self-fertility and SC was carried out in the family Asteraceae, a family in which 10% of the species have pseudo-self-fertility. This showed that rates of transition from pseudo-self-fertility to SC were much higher than from pseudo-self-fertility to SI, suggesting that pseudo-self-fertility is most often part of a breakdown of SI (Ferrer and Good-Avila 2007). However, pseudo-self-fertility was found to be clustered in several clades and was the ancestral state in three tribes (the Asterae, Calenduleae and Senecioneae) known to harbour many partially SI taxa. Thus, although pseudo-self-fertility may be more frequently associated with the loss of SI, it may persist for extended periods of time. This begs the question, at what point do we consider a mating system to be evolutionarily stable? Examination of the variation in SI, SC and pseudo-self-fertility in families reveals tremendous variation in the phylogenetic clustering and persistence of SI and SC lineages (Takebayashi and Morrell 2001; Igic et al. 2006; Ferrer and Good-Avila 2007; see also Chap. 4, this volume). Within the Asteraceae, several clades were identified as being predominantly SI, but most clades harbour mixtures of SI, pseudo-self-fertility and SC taxa. Although there is strong evidence that SI systems have broken down repeatedly in angiosperm evolution (Stebbins 1974; Goodwillie 1999; Igic et al. 2006; see also Chap. 5, this volume), it is also increasingly apparent that self-fertilisation can be a reproductive dead end (Takebayashi and Morrell 2001) and that SI has arisen de novo repeatedly in angiosperm history (Weller et al. 1995; Steinbachs and Holsinger 2002; Good-Avila and Ferrer submitted).

Thus, examined from a higher level, the evolution of SI appears more as a struggle between genes favouring self-sterility and those favouring self-fertility. More experimental work is needed to determine if genetic polymorphisms that enhance self-fertility provides species with the flexibility to navigate the ever-changing dynamics of population and environmental variables while allowing them a discrete method to maximise offspring fitness.

Acknowledgments This work was supported by a USDA CSREES grant No. 2005-35320-15351 to A.G.S; NSERC Discovery Grant to SVG-A; and an NSF Minority Postdoctoral Research Fellowship DBI-0706721 to JIM.

References

Aizen MA, Searcy KB, Mulcahy DL (1990) Among- and within-flower comparisons of pollen tube growth following self and cross pollinations in *Dianthus chinensis* L. (Caryophyllaceae). Am J Bot 77:671–676
Ascher P (ed) (1984) Self-incompatibility. Petunia: Monographs on theoretical and applied genetics. Springer, Berlin Heidelberg New York
Ascher PD, Peloquin SJ (1966) Effect of floral ageing on the growth of compatible and incompatible pollen in *Lilium longiflorum*. Am J Bot 74:471–476
Baker HG (1955) Self-incompatibility and establishment after 'long-distance' dispersal. Evolution 9:347–349
Barrett SCH (2003) Mating strategies in flowering plants: The outcrossing-selfing paradigm and beyond. Phil Trans R Soc Lond B 358:991–1004
Bateman AJ (1956) Cryptic self-incompatibility in the wall flower: *Cheiranthus cheiri* L. Heredity 10:257–261
Billiard S, Castric V, Vekemans X (2007) A general model to explore complex dominance patterns in plant sporophytic self-incompatibility systems. Genetics 175:1351–1369
Bixby PJ, Levin DA (1996) Response to selection on autogamy in Phlox. Evolution 50:892–899
Bowman RN (1987) Cryptic self-incompatibility and the breeding system of Clarkia unguiculata (Onagraceae). Am J Bot 74:471–476
Busch JW (2005) The evolution of self-compatibility in geographically peripheral populations of *Leavenworthia alabamica* (Brassicaceae). Am J Bot 92:1503–1512
Byers DL, Waller DM (1999) Do plant populations purge their genetic load? Effects of population size and mating history on inbreeding depression. Annu Rev Ecol Syst 30:479–513
Casper BB, Sayigh LS, Lee SS (1988) Demonstration of cryptic incompatibility in distylous *Amsinckia douglasiana*. Evolution 42:248–253
Charlesworth D, Charlesworth B (1979) The evolution and breakdown of S-allele systems. Heredity 43:41–55
Charlesworth D, Charlesworth B (1987) Inbreeding depression and its evolutionary consequences. Annu Rev Ecol Syst 18:237–268
Coleman CE, Kao Th (1992) The flanking regions of two *Petunia inflata* S-alleles are heterogeneous and contain repetitive sequences. Plant Mol Biol 18:725–737
Cruz-Garcia F, Hancock CN, McClure B (2003) *S*-RNase complexes and pollen rejection. J Exp Bot 54:123–130
Cruzan MB, Barrett SCH (1993) Contribution of cryptic incompatibility to the mating system of *Eichhornia paniculata* (Pontederiaceae). Evolution 47:925–934
Darwin C (1876) The effects of cross and self-fertilisation in the vegetable kingdom. John Murray, London
Darwin C (1877) The different forms of flowers on plants of the same species. D. Appleton, New York
de Nettancourt D (1977) Incompatibility in angiosperms. Springer, Berlin Heidelberg New York
de Nettancourt D (1997) Incompatibility in Angiosperms. Sex Plant Reprod 10:185–199

East EM (1934) Norms of pollen-tube growth in incompatible matings of self-sterile plants. Proc Natl Acad Sci USA 20:225–230
Eckert CG, Allen M (1997) Cryptic self-incompatibility in tristylous *Decodon verticillatus* (Lythraceae). Am J Bot 84:1391
Elgersma A, Stephenson AG, Nijs APM (1989) Effects of genotype and temperature on pollen tube growth in perennial ryegrass (*Lolium perenne* L.). Sex Plant Reprod 2:225–230
Ferrer MM, Good-Avila SV (2007) Macroevolutionary analyses of the gain and loss of self-incompatibility in the Asteraceae. New Phytol 173:401–414
Fisher RA (1941) Average excess and average affect of a gene substitution. Ann Eugen 11:53–63
Gilblin DE, Hamilton CW (1999) The relationship of reproductive biology to the rarity of endemic *Aster curtas* (Asteraceae). Can J Bot 77:140–149
Glemin S, Bataillon T, Ronfort J, Mignot A, Olivieri I (2001) Inbreeding depression in small populations of self-incompatible plants. Genetics 159:1217–1229
Godt MJW, Hamrick JL (1999) Population genetic analysis of *Elliottia racemosa* (Ericaceae), a rare Georgia shrub. Mol Ecol 8:75–82
Goldraij A, Kondo K, Lee CB, Hancock CN, Sivaguru M, Vasquez-Santana S, Kim S, Phillips TE, Cruz-Garcia F, McClure B (2006) Compartmentalization of S-RNase and HT-B degradation in self-incompatible *Nicotiana*. Nature 439:805–810
Good-Avila SV, Stephenson A (2002) The inheritance of modifiers conferring self-fertility in *Campanula rapunculoides* L. (Campanulaceae). Evolution 56:263–272
Good-Avila SV, Stephenson AG (2003) Parental effects in a partially self-incompatible herb *Campanula rapunculoides* L. (Campanulaceae): Influence of variation in the strength of self-incompatibility on seed set and progeny performance. Am Nat 161:615–630
Good-Avila SV, Ferrer MM (submitted) The origin and evolutionary dynamics of self-incompatibility systems: Re-inventing self avoidance again and again
Goodwillie C (1999) Multiple origins of self-compatibility in *Linanthus* section *Leptoshiphon* (Polemoniaceae): Phylogenetic evidence from internal-transcribed-spacer sequence data. Evolution 53:1387–1395
Goodwillie C, Knight MC (2006) Inbreeding depression and mixed mating in *Leptosiphon jepsonii*: A comparison of three populations. Ann Bot 98:351–360
Goodwillie C, Partis KL, West JW (2004) Transient self-incompatibility confers delayed selfing in *Leptosiphon jepsonii* (Polemoniaceae). Int J Plant Sci 165:387–394
Hatakeyama K, Watanabe M, Takasaki T, Ojima K, Hinata K (1998) Dominance relationships between S-alleles in self-incompatible *Brassica campestris* L. Heredity 80:241–247
Hegedus A, Szabo Z, Nyeki J, Halasz J, Pedryc A (2006) Molecular analysis of S-haplotypes in peach, a self-compatible *Prunus* species. J Am Soc Hort Sci 131:738–743
Hinata K, Okasaki K (eds) (1986). Role of the stigma in the expression of self-incompatibility in crucifers in view of genetic analysis. Biotechncology and ecology of pollen. Springer, Berlin Heidelberg New York
Holsinger KE, Feldman MW, Christiansen FB (1984) The evolution of self-fertilization in plants: A population genetic model. Am Nat 124:446–453
Hua Z, Kao Th (2006) Identification and characterization of components of a putative *Petunia* S-Locus F-Box-containing E3 ligase complex involved in S-RNase-based self-incompatibility. Plant Cell 18:2531–2553
Husband BC, Schemske DW (1996) Evolution of the magnitude and timing of inbreeding depression in plants. Evolution 50:54–70
Igic B, Bohs L, Kohn JR (2006) Ancient polymorphism reveals unidirectional breeding system shifts. Proc Natl Acad Sci USA 103:1359–1363
Jones KN (1994) Nonrandom mating in *Clarkia gracilis* (Onagraceae): A case of cryptic self-incompatibility. Am J Bot 81:195–198
Kao Th, Tsukamoto T (2004) The molecular and genetic bases of S-RNase-based self-incompatibility. Plant Cell 16:S72–S83
Karron JD (1987) A comparison of levels of genetic polymorphism and self-compatibility in geographically restricted and widespread plant congeners. Evol Ecol 1:47–58

Kondo K, Yamamoto M, Matton DP, Sato T, Hirai M, Norioka S, Hattori T, Kowyama Y (2002) Cultivated tomato has defects in both S-RNase and HT genes required for stylar function of self-incompatibility. Plant J 29:627–636

Kowyama Y, Takahashi H, Muraoka K, Tani T, Hara K, Shiotani I (1994) Number, frequency and dominance relationships of S-alleles in diploid *Ipomoea trifida*. Heredity 73:275–283

Kruszewski L, Galloway LF (2006) Explaining outcrossing rate in *Campanulastrum americanum* (Campanulaceae): Geitonogamy and cryptic self-incompatibility. Int J Plant Sci 167:455–461

Larson BMH, Barrett SCH (2000) A comparative analysis of pollen limitation in flowering plants. Biol J Linn Soc 69:503–520

Lawson J, Williams W (1976) Environmental and genetic effects on pseudo-compatibility in relation to the production of hybrid seed. J Hort Sci 51:359–365

Les DH, Reinartz JA, Esselman EJ (1991) Genetic consequences of rarity in *Aster furcatus* (Asteraceae), a threatened, self-incompatible plant. Evolution 45:1641–1650

Levin DA (1996) The evolutionary significance of pseudo self-fertility. Am Nat 148:321–332

Lipow SR, Wyatt R (2000) Single gene control of postzygotic self-incompatibility in poke milkweed, *Asclepias exaltata* L. Genetics 154:893–907

Lloyd DG (1992) Self- and cross-fertilization in plants. II. The selection of self-fertilization. Int J Plant Sci 153:358–369

Lloyd DG, Schoen DJ (1992) Self- and cross-fertilization in plants. I. Functional dimensions. Int J Plant Sci 153:341–357

Mena-Alí JI (2006). Dynamics of the self-incompatibility alleles in populations of *Solanum carolinense*. Doctoral Dissertation, Graduate Program in Biology. University Park, The Pennsylvania State University

Mena-Alí JI, Stephenson AG (2007) Segregation analyses of partial self-incompatibility in self and cross progeny of *Solanum carolinense* reveal a leaky S-allele. Genetics 177:501–510

Mena-Alí JI, Kesser LH, Stephenson AG (2008) Inbreeding depression in *Solanum carolinense* (Solanaceae), a species with a plastic self-incompatibility response. BMC Evol Biol 8:10 doi:10.1186/1471-2148-8-10

Nagylaki T (1976) A model for the evolution of self-fertilization and vegetative reproduction. J Theor Biol 58:55–58

Nasrallah ME (1974) Genetic control of quantitative variation in self-incompatibility proteins detected by immunodiffusion. Genetics 76:45–50

Nasrallah ME (1989). The genetics of self-incompatibility reactions in *Brassica* and the effects of supressor genes. In: Lord EM, Bernier G (eds) Plant reproduction: from floral induction to pollination, vol 1. American Society of Plant Physiology Symposium Series, Rockville, MD, pp 146–155

Nasrallah ME, Kandasamy MK, Nasrallah JB (1992) A genetically defined trans-acting locus regulates S-locus function in *Brassica*. Plant J 2:497–506

Newbigin E, Uyenoyama MK (2005) The evolutionary dynamics of self-incompatibility systems. Trends Genet 21:500–505

Nielsen LR, Siegismund HR, Philipp M (2003) Partial self-incompatibility in the polyploid endemic species *Scalesia affinis* (Asteraceae) from the Galápagos: Remnants of a self-incompatibility system? Bot J Linn Soc 142:93–101

O'Neill SD (1997) Pollination regulation of flower development. Annu Rev Plant Physiol Plant Mol Biol 48:547–574

Ockendon DJ, Currah L (1978) Time of cross- and self-pollination affects the amount of self-seed set by partially self-incompatible plants of *Brassica oleracea*. Theor Appl Genet 52:233–237

Owen FV (1942) Inheritance of cross- and self-sterility and felf-fertility in *Beta vulgaris*. J Agric Res 64:678–698

Pandey KK (1960) Incompatibility in Abutilon 'Hybridum'. Am J Bot 47:877–883

Pannell JR, Barrett SCH (1998) Baker's law revisited: Reproductive assurance in a metapopulation. Evolution 52:657–668

Porcher E, Lande R (2005) Loss of gametophytic self-incompatibility with evolution of inbreeding depression. Evolution 59:46–60

Pundir NS, Al-Attar AA (1982) Physiological stress over self-incompatibility in *Raphanus sativus* L. Zeitschrift fur Pflanzenzuchtung 89:344–346

Reinartz JA, Les DH (1994) Bottleneck-induced dissolution of self-incompatibility and breeding system consequences in *Aster furcatus* (Asteraceae). Am J Bot 81:446–455

Richardson TE, Hrincevich A, Kao T-h, Stephenson AG (1990) Preliminary studies into age-dependent breakdown of self-incompatibility in *Campanula rapunculoides*: Seed set, pollen tube growth, and molecular data. Plant Cell Incompat Newslett 22:41–47

Rigney L, Thomson J, Cruzan M, Brunet J (1993) Differential success of pollen donors in a self-compatible lily. Evolution 47:915–924

Sato Y, Okamoto S, Nishio T (2004) Diversification and alteration of recognition specificity of the pollen ligand SP11/SCR in self-incompatibility of *Brassica* and *Raphanus*. Plant Cell 16:3230–3241

Schemske DW (1978) Evolution of reproductive characteristics in *Impatiens* (Balsaminaceae): The significance of cleistogamy and chasmogamy. Ecology 59:596–613

Schierup MH, Vekemans X, Christiansen FB (1997) Evolutionary dynamics of self-incompatibility alleles in plants. Genetics 147:835–846

Schierup MH, Bechsgaard JS, Nielsen LH, Christiansen FB (2006) Selection at work in self-incompatible *Arabidopsis lyrata*: Mating patterns in a natural population. Genetics 172:477–484

Schoen DJ, Lloyd DG (1984) The selection of cleistogamy and heteromorphic diaspores. Biol J Linn Soc 23:303–322

Schoen DJ, Morgan MT, Bataillon T (1996) How does self-pollination evolve? Inferences from floral ecology and molecular genetic variation. Phil Trans R Soc Lond B 351:1281–1290

Sherman-Broyles S, Boggs N, Farkas A, Liu P, Vrebalov J, Nasrallah ME, Nasrallah JB (2007) *S*-locus genes and the evolution of self-fertility in *Arabidopsis thaliana*. Plant Cell 19:94–106

Stebbins GL (1957) Self-fertilization and population viability in the higher plants. Am Nat 91:337–354

Stebbins GL (1974) Flowering plants: Evolution above the species level. Belknap, Cambridge, MA

Steinbachs JE, Holsinger KE (2002) S-RNase-mediated gametophytic self-incompatibility is ancestral in eudicots. Mol Biol Evol 19:825–829

Stephenson AG, Bertin RI (1983). Male competition, female choice, and sexual selection in plants. In: Real L (ed) Pollination biology. Academic Press, Orlando, pp 109–149

Stephenson AG, Winsor JA, Richardson TE, Singh A, Kao Th (1992). Effects of style age on the performance of self and cross pollen in *Campanula rapunculoides*. In: Ottaviano, E, Mulcahy, DL, Gorla, MS, Mulcahy GB (eds) Angiosperm pollen and ovules. Springer, Berlin Heidelberg New York, pp 117–121

Stephenson AG, Good SV, Vogler DW (2000) Interrelationships among inbreeding depression, plasticity in the self-incompatibility system, and the breeding system of *Campanula rapunculoides* L. (Campanulaceae). Ann Bot 85:211–219

Stephenson AG, Travers SE, Mena-Alí JI, Winsor JA (2003) Pollen performance before and during the autotrophic-heterotrophic transition of pollen tube growth. Phil Trans R Soc Lond B 358:1009–1018

Stone JL (2002) Molecular mechanisms underlying the breakdown of gametophytic self-incompatibility. Quart Rev Biol 77:17–32

Stone JL (2004) Sheltered load associated with *S*-alleles in *Solanum carolinense*. Heredity 92:335–342

Stone JL, Sasuclark MA, Blomberg CP (2006) Variation in the self-incompatibility response within and among populations of the tropical shrub *Witheringia solanacea* (Solanaceae). Am J Bot 93:592–598

Takebayashi N, Morrell PL (2001) Is self-fertilization an evolutionary dead end? Revisiting an old hypothesis with genetic theories and a macroevolutionary approach. Am J Bot 88:1143–1150

Tang C, Toomajian C, Sherman-Broyles S, Plagnol V, Guo Y-L, Hu TT, Clark RM, Nasrallah JB, Weigel D, Nordborg M (2007) The evolution of selfing in *Arabidopsis thaliana*. Science 317:1070–1072

Tao R, Watari A, Hanada T, Habu T, Yaegaki H, Yamaguchi M, Yamane H (2007) Self-compatible peach (*Prunus persica*) has mutant versions of the *S*-haplotypes found in self-incompatible *Prunus* species. Plant Mol Biol 63:109–123

Travers SE, Mena-Alí JI, Stephenson AG (2004) Plasticity in the self-incompatibility of *Solanum carolinense*. Plant Species Biol 19:127–135

Tsukamoto T, Ando T, Kokubun H, Watanabe H, Masada M, Zhu X, Marchesi E, Kao T-h (1999) Breakdown of self-incompatibility in a natural population of *Petunia axillaris* (Solanaceae) in Uruguay containing both self-incompatible and self-compatible plants. Sex Plant Reprod 12:6–13

Tsukamoto T, Ando T, Takahashi K, Omori T, Watanabe H, Kokubun H, Marchesi E, Kao T-h (2003a) Breakdown of self-incompatibility in a natural population of *Petunia axillaris* caused by loss of pollen function. Plant Physiol 131:1903–1912

Tsukamoto T, Ando T, Kokubun H, Watanabe H, Sato T, Masada M, Marchesi E, Kao T-h (2003b) Breakdown of self-incompatibility in a natural population of *Petunia axillaris* caused by a modifier locus that suppresses the expression of an S-RNase gene. Sex Plant Reprod 15:255–263

Ushijima K, Yamane H, Watari A, Kakehi E, Ikeda K, Hauck NR, Iezzoni AF, Tao R (2004) The *S*-haplotype-specific F-box protein gene, SFB, is defective in self-compatible haplotypes of *Prunus avium* and *P. mume*. Plant J 39:573–586

Uyenoyama MK (2003) Genealogy-dependent variation in viability among self-incompatibility genotypes. Theor Popul Biol 63:281–293

Uyenoyama MK, Holsinger KE, Waller DM (1993) Ecological and genetic factors directing the evolution of self-fertilization. Oxford Surv Evol Biol 9:327–381

Vallejo-Marín M, Uyenoyama MK (2004) On the evolutionary costs of self-incompatibility: Incomplete reproductive compensation due to pollen limitation. Evolution 58:1924–1935.

Vekemans X, Schierup MH, Christiansen FB (1998) Mate availability and fecundity selection in multi-allelic self-incompatibility systems in plants. Evolution 52:19–29

Vilanova S, Badenes ML, Burgos L, Martinez-Calvo J, Llacer G, Romero C (2006) Self-compatibility of two apricot selections is associated with two pollen-part mutations of different nature. Plant Physiol 142:629–641

Vogler DW, Stephenson A (2001) The potential for a mixed mating system *Campanula*. Int J Plant Sci 162:801–805

Vogler DW, Das C, Stephenson AG (1998) Phenotypic plasticity in the expression of self-incompatibility in *Campanula rapunculoides*. Heredity 81:546–555

Waller DM (1979) The relative costs of self- and cross-fertilized seeds in *Impatiens capensis* (Balsaminaceae). Am J Bot 66:313–320

Weber JJ, Goodwillie C (2007) Timing of self-compatibility, flower longevity, and potential for male outcross success in *Leptosiphon jepsonii* (Polemoniaceae). Am J Bot 94:1338–1343

Weller S, Ornduff R (1977) Cryptic self-incompatibility in *Amsinckia grandiflora*. Evolution 31:47–51

Weller SG, Donoghue MJ, Charlesworth D (1995). The evolution of self-incompatibility in flowering plants: A phylogenetic approach. Experimental and molecular approaches to plant biosystematics. Missouri Botanical Garden, St. Louis

Willi Y, Buskirk JV, Fischer M (2005) A threefold genetic allele effect: Population size affects cross-compatibility, inbreeding depression and drift load in the self-incompatible *Ranunculus reptans*. Genetics 169:2255–2265

Young AG, Hill JH, Murray BG, Peakall R (2002) Mating system, genetic diversity and clonal structure in the alpine herb *Rutidosis leiolepis* F. Muell. (Asteraceae). Biol Conserv 106:71–78

Chapter 3
On the Evolutionary Modification of Self-Incompatibility: Implications of Partial Clonality for Allelic Diversity and Genealogical Structure

M. Vallejo-Marín and M.K. Uyenoyama

Abstract Experimental investigations of homomorphic self-incompatibility (SI) have revealed an unanticipated level of complexity in its expression, permitting fine regulation over the course of a lifetime or a range of environmental conditions. Many flowering plants express some level of clonal reproduction, and phylogenetic analyses suggest that clonality evolves in a correlated fashion with SI in *Solanum* (Solanaceae). Here, we use a diffusion approximation to explore the effects on the evolutionary dynamics of SI of vegetative propagation with SI restricted to reproduction through seed. While clonality reduces the strength of frequency-dependent selection maintaining *S*-allele diversity, much of the great depth typical of *S*-allele genealogies is preserved. Our results suggest that clonality can play an important role in the evolution of SI systems, and may afford insight into unexplained features of allele genealogies in the Solanaceae.

Abbreviations

GSI	Gametophytic self-incompatibility
MRCA	Most recent common ancestor
PSI	Pseudo self-incompatibility
S-RNase	*S*-locus ribonuclease
SC	Self-compatibility

M. Vallejo-Marín
Department of Ecology and Evolutionary Biology, 25 Willcocks Street, University of Toronto, Toronto, ON M5S 3B2, Canada, e-mail: m.vallejo@utoronto.ca

M.K. Uyenoyama
Department of Biology, Box 90338, Duke University, Durham, NC 27708–0338, USA, e-mail: marcy@duke.edu

SCR	*S*-locus cysteine rich (also termed SP11) – the pollen determinant in Brassica
SI	Self-incompatibility
SRK	S-Receptor Kinase (the pistil determinant in Brassica)
SSI	Sporophytic self-incompatibility
ΨSCR	SRK-like sequences in Arabidopsis

3.1 Introduction

A watershed era in the study of mating systems began upon the cloning of the *S*-locus over two decades ago (Nasrallah et al., 1985; Anderson et al., 1986). Within evolutionary biology, the revelation of the genetic basis of self-incompatibility (SI) in two model systems precipitated new explorations of the origins, evolutionary modification, and genomic consequences of its expression. In addition to illuminating physiological and regulatory mechanisms of SI, the nucleotide sequence of the *S*-locus region itself afforded insight into the evolutionary process over the extraordinary time depth of *S*-allele genealogies.

Virtually every system subjected to close examination has now been found to depart from canonical SI, defined as a one-factor system that both defines compatibility classes and excludes incompatible pollen from fertilization in every reproductive episode. These empirical discoveries invite a broadened concept of SI. Here, we characterize SI as a versatile component of a diversified strategy of reproduction, the evolutionary modification of which spans the full spectrum from total disablement to plastic regulation of its expression.

Section 3.2 begins with a consideration of recent phylogenetic analyses of the transition to self-compatibility (SC). These studies provide evidence of multiple independent departures from canonical SI, and do not exclude subsequent restoration of SI. Further, some transitions from canonical SI may not in fact reflect a loss of function, but rather modification of its expression: a plastic response to physiological state or external environment, for example (see also Chap. 2). Within the Solanaceae, an exemplar for studies of the genetic basis and evolution of SI, a number of species express SI in combination with partial clonal reproduction. Because partial clonality and SI show correlated evolution (Vallejo-Marín and O'Brien, 2007), we develop a diffusion approximation of partial vegetative propagation with SI restricted to reproduction through seed (Sect. 3.3).

We find that while partial clonality reduces the number of *S*-alleles maintained at stochastic equilibrium, it preserves much of the great depth characteristic of *S*-allele genealogies. Our analysis illustrates that the nature of expression of SI influences the inferential framework extending from gene genealogies based on the *S*-locus region to reconstruction of the evolutionary history of mating systems.

3.2 Mating System Dynamics

A characteristic of SI that prevails across the plant kingdom is its evolutionary lability (Barrett, 1988). Stebbins (1974, p 51) described the transition from SI to self-fertilization as one of the evolutionary paths most often taken by flowering plants. While it seems clear that many SC species exist in families in which SI appears to be ancestral (see Chap. 4), distinguishing the modification of SI from its complete disablement requires direct experimental study. In this section, we review some recent analyses of transitions from canonical SI.

3.2.1 Relative Transition Rates

In the Solanaceae, Igic et al. (2006) found that transitions from complete SI have occurred at rates at least several-fold higher than the reverse. They interpreted the failure to find greater support for a model that permits both gain and loss of SI over to a model that precludes gains as evidence of unidirectional loss of gametophytic SI (GSI) in the Solanaceae (see also Chap. 5). However, even the analysis in which they specified the SI status of ancestral nodes indicated virtually identical likelihoods of the two models, and the analysis without assigned ancestral states gave significantly more support to the model that allowed positive rates of gain of SI.

Some cases of apparent loss of SI may in fact constitute transitions from canonical SI to some form of modified expression, as has been described, for example, by Levin (1996). In their survey of the literature on breeding systems in the Asteraceae, Ferrer and Good-Ávila (2007) recognized not only full SI and full SC, but also partial or pseudo-SI (PSI), broadly defined as variation in the strength of SI among individuals or populations in response to floral age, temperature, or various environmental factors (see also Chap. 2). Their phylogenetically independent maximum-likelihood analysis indicated substantial rates of gain of SI from both SC and PSI.

3.2.2 Multiple Origins of SC in **Arabidopsis**

SRK/SCR-based sporophytic self-incompatibility (SSI, see Chaps. 6–8) has not been detected outside the Brassicaceae. From an analysis of nucleotide variation at cytoplasmic and nuclear loci, Koch et al. (2001) estimated the emergence of this family about 40 million years ago (mya). This figure appears to coincide with an estimate for the origin of this form of SSI (40–50 mya; Uyenoyama, 1995). A key case study is the derivation of SC *Arabidopsis thaliana* from a presumed SI ancestor in the past 5–6 million years, a divergence time supported by analyses of both nucleotide substitutions (Koch et al., 2001) and chromosomal rearrangements (Koch and Kiefer, 2005).

Shimizu et al. (2004) interpreted the low variation at $\Psi SCR1$, the pseudogene homologous to the determinant of pollen specificity (*SCR*), as evidence of a very recent (<0.32 mya) transition to SC through the selective sweep of an *S*-allele with impaired pollen function. Substitution of a loss-of-function *S*-allele has been proposed as a possible route from SI to SC (Charlesworth and Charlesworth, 1979). However, contrary to expectation under the scenario, Tang et al. (2007) found multiple, highly divergent *S*-locus haplotypes in a survey of 96 accessions of *A. thaliana*. Shimizu et al. (2008) confirmed these results in accessions from Europe (haplogroup A) and African island populations (haplogroup B). Haplogroup A, found in 96% of world wide accessions, is associated with $\Psi SCR1$, but *SCR* in haplogroup B shows no obvious disabling mutations. In haplotype B, it is *SRK*, the regulator of the stylar response to incompatible pollen, that shows a frameshift mutation (Tang et al., 2007; Shimizu et al., 2008).

Nasrallah et al. (2004) showed that *A. thaliana* accessions differ in their response to the acquisition by transformation of functional *SRK* and *SCR* genes from its SI congener *A. lyrata*. The only accession among the seven tested that showed full expression of SI upon transformation (C24) contains a highly rearranged *S*-locus region, the apparent product of recombination between *S*-haplotypes (Sherman-Broyles et al., 2007). Other accessions show pseudo-SC, with expression of SI breaking down in later stages of flower development. Liu et al. (2007) showed that a locus distinct from both *SCR* and *SRK* modified the age-dependent pattern of SI expression in a PSC transformant.

These studies demonstrate that the transition of *A. thaliana* from SI to SC derives from multiple origins and genetic mechanisms (see also Chap. 6). Of particular relevance to our discussion is the possibility that at least some of these transitions may have involved partial or modified expression of SI rather than the immediate and total loss of function.

3.2.3 Modified Forms of SI

Mounting empirical evidence suggests that in many taxa the expression of SI within individuals or populations may be restricted to certain developmental stages or environmental conditions (Levin 1996; Good-Ávila et al., in Chap. 2).

3.2.3.1 Partial SI

Partial expression of SI occurs in many taxa representing taxonomic groups across the plant kingdom (Levin, 1996). In one of the few phylogenetic analyses of partial SI, Ferrer and Good-Ávila (2007) found that in the Asteraceae, 10% of the 571 species studied expressed some form of PSI. Good-Ávila and Stephenson (2002) found a heritable increase in self-fertility with floral age in *Campanula rapunculoides* (Campanulaceae). In *Solanum carolinense* (Solanaceae), Travers et al. (2004) used direct experimental manipulation to demonstrate an increase in self-fertility with floral age and in plants that were prevented from setting fruit earlier in the reproductive season. Moreover, by measuring fruit and seed set following self-pollination across clonally replicated genets, they detected a genetic component to the variation in self-fertility. Also in *S. carolinense*, Mena-Alí and Stephenson (2007) found *S*-allele-specific variation in the level of self-fertility, suggesting that this effect reflects a change in the rate of translation or turnover or in the enzymatic activity level of the particular *S-RNase* allele, perhaps due to a linked modifier gene.

3.2.3.2 SI Under Partial Clonal Reproduction

Natural populations of a number of species, including many in the typically colonizing genus *Solanum* (Solanaceae), comprise individuals derived from asexual propagation together with individuals derived from seeds generated through complete SI expression. The capacity for clonal reproduction may reduce the selective advantage of the loss or modification of SI by permitting the extended persistence of isolated SI colonizers. In this case, one might expect correlated evolution of SI status and clonality status to follow from Baker's (1955; 1967) rule.

In the only phylogenetic study of the correlated evolution of SI and life-history to date, Vallejo-Marín and O'Brien (2007) used the Pagel method (Pagel, 1994; Pagel and Meade, 2006) to address the joint evolution of SI and clonality. They detected significant rates of transition from SI to SC under clonality and from clonality to non-clonality under SC (see Fig. 2 of Vallejo-Marín and O'Brien, 2007). The loss of clonality in colonizing SI *Solanum* species may be strongly deleterious, perhaps reflecting that having an alternative means of reproduction can become essential when seed set is limited by an insufficiency of compatible pollen.

3.3 *S*-Locus Evolution Under Partial Clonality

In this section, we use an extension of the one-dimensional diffusion equation approximation of *S*-locus evolution introduced by Wright (1939) to explore the consequences of partial clonal reproduction on the evolutionary dynamics of GSI (see Appendix 1).

3.3.1 Diffusion Approximation

Figure 3.1 depicts a population in which zygotes derived from seeds generated under full expression of SI constitute a fraction γ of reproductives, with the complement derived by clonal propagation. For example, a proportion c of the reproductive resources of an individual, irrespective of *S*-locus genotype, may be devoted to clonal reproduction, with clonally derived offspring (ramets) surviving at rate τ relative to sexually derived offspring (genets).

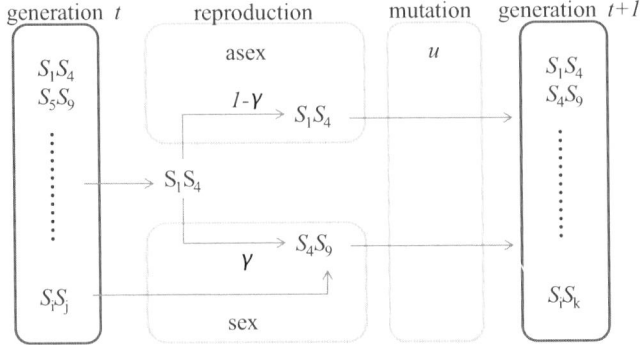

Fig. 3.1 Flow chart of a single generation in the simulations

To develop a one-dimensional diffusion approximation of the high-dimensional dynamics of arbitrary numbers of S-alleles, we assume exchangeability among S-alleles (Wright, 1939; Fisher, 1958, Chap. 4). Over a single generation, the expected change in frequency of a particular S-allele (q) corresponds to

$$\Delta q = \frac{\gamma q(F-q)}{(1-F)(1-2F)}, \qquad (3.1)$$

for F denoting homozygosity, approximated by the inverse of the number of S-alleles maintained in the population ($1/n$); and γ the sexual fraction, defined as the proportion of reproductive individuals that are themselves sexually derived:

$$\gamma = \frac{1-c}{c\tau + 1 - c}. \qquad (3.2)$$

Novel (functionally distinct) S-alleles enter the population at rate u per gene per generation, and genetic drift is incorporated through random sampling of N zygotes to form the next generation, reflecting that compatibility reflects interactions between zygotes and gametes rather than between gametes (Fisher, 1958; Wright, 1960).

These expressions give rise to a diffusion approximation with infinitesimal mean and variance coefficients

$$\mu(x) = -\frac{\gamma x(x-F)}{(1-F)(1-2F)} - ux, \qquad (3.3a)$$

$$\sigma^2(x) = \frac{x(1-2x)}{2N}. \qquad (3.3b)$$

Our diffusion coefficients reduce under full sexual reproduction ($\gamma = 1$) to those of the Wright (1960) model. Partial clonal reproduction decreases the force of SI selection on the infinitesimal mean change (3.3a) by a factor corresponding to the proportion of the population derived by sexual reproduction (3.2).

Drift generates change on the order of the inverse of the effective population size (3.3b). With the exception of $\gamma x(x-F)$ on the order of $1/2N$, the intense frequency-dependent selection induced by SI overwhelms drift (Neuhauser, 1999). Like the Wright (1939) model, our diffusion approximation holds only very close to extinction of the focal allele ($x \approx 1/2N$) or very close to the deterministic equilibrium frequency ($x \approx F = 1/n$). The process jumps between the domains of extinction and common frequencies at rate

$$\lambda = \theta \eta, \tag{3.4}$$

in which

$$\theta = 2Nu \tag{3.5}$$

represents the number of novel S-alleles produced in each generation by mutation and η the probability that an S-allele present in a single copy (frequency $1/2N$) will jump to common frequencies before extinction. Because drift reflects sampling of N zygotes rather than $2N$ genes, θ here is half the scaled mutation rate that customarily arises under random union of gametes. Under our model,

$$\eta = \frac{2n}{(n-1)(n-2)} \left[\gamma - \frac{u(n-1)(n-2)}{n} \right]. \tag{3.6}$$

Under complete clonal reproduction ($\gamma = 0$), mutation and drift of clones alone govern S-allele dynamics. Sexual reproduction ($\gamma > 0$) permits SI selection, but if mutation dominates selection ($u \gg \gamma$), the invasion probability (3.6) converges to zero for $\theta > 1$ (Chap. 15, Sect. 4, Example D of Karlin and Taylor, 1981). Here, we restrict attention to cases where the frequency of sexually derived individuals at reproduction ($M = \gamma N$) is sufficiently large relative to the scaled mutation rate (θ) to ensure a positive jump probability η (3.6) and that sexual reproduction through SI can occur ($n \geq 3$); a necessary condition is

$$\gamma > \frac{u(n-1)(n-2)}{n}.$$

At approximate steady state, the number of common S-alleles (n) segregating in the population reflects a balance between the rate at which S-alleles jump to common frequencies λ (3.4) and the rate at which common alleles jump to extinction (Uyenoyama, 2003). In our model, n is implicitly determined by

$$1 = \theta G e^{\frac{2Mn}{(n-1)(n-2)} - \theta} \sqrt{\frac{2\pi(n-1)}{Mn + \theta(n-1)}} \left\{ \left[1 + \frac{\theta(n-1)}{Mn} \right] \frac{n-2}{n} \right\}^{\theta + Mn/(n-1)}, \tag{3.7}$$

in which

$$G = \frac{2n\gamma}{\eta(n-1)(n-2)} = \frac{2Mn}{2Mn - \theta(n-1)(n-2)}$$

and

$$M = \gamma N. \tag{3.8}$$

These expressions indicate that the number of sexually derived individuals at reproduction (M) and the scaled mutation rate (θ) jointly determine allelic diversity at the S-locus. In particular, the effect of partial clonal reproduction on the evolutionary dynamics at the S-locus cannot be explained simply by a reduction in effective population size by the factor γ. The analogous expression derived by Yokoyama and Hetherington (1982) for the Wright model ($\gamma = 1$) also depends on both u and N, and not only their product.

The Wright–Fisher formula for the stationary distribution provides an expression for the density of the frequency of the focal S-allele at approximate steady-state. In this case, the probability that the focal S-allele segregates in the population at a frequency lying in a small interval near x is proportional to

$$\Phi(x) = 2\theta e^{\frac{2Mn^2 x}{(n-1)(n-2)}} (1-2x)^{Mn/(n-1)+\theta-1} x^{-1}. \tag{3.9}$$

3.3.2 S-Allele Number and Frequency

Self-incompatibility induces an intense form of frequency-dependent selection (Vekemans and Slatkin, 1994; Neuhauser, 1999) that typically maintains scores of S-alleles. Wright (1939) reduced the description of this high-dimensional evolutionary process to a one-dimensional diffusion equation through a series of inspired approximations. To explore the robustness of our model to similar approximations, we compared theoretical expectations of the number of common S-alleles maintained and their frequency spectrum to the results of numerical simulations (see Appendix 2).

Figure 3.2 presents box plots of the total number of segregating S-alleles over a range of values of γ in a population of $N = 250$ zygotes under a rate of mutation to new S-alleles of $u = 10^{-4}$. Also shown is the prediction from our model, obtained by numerically solving (3.7) for n and determining the largest integer not exceeding the expectation (floor). The median total number of S-alleles observed generally exceeds by 1 or 2 the expected number of common S-alleles, those occurring in approximately equal frequency.

To study the spectrum of allele frequencies, we determined the frequency in the population of each segregating S-allele in 100 independent simulations of a population of 250 zygotes under mutation rate $u = 10^{-4}$. Figure 3.3 shows the total number of alleles segregating in frequencies indicated on the abscissa under two assignments of the proportion of sexually derived zygotes (γ). Also shown is the theoretical stationary distribution (3.9) of the frequency of an S-allele. For a given set of population parameters (N, u, γ), we first determined the expected number of S-alleles as the largest integer not exceeding the root of (3.7) and then assigned this value to n in (3.9) to obtain the solid curves in Fig. 3.3.

In accordance with a prediction (3.9) of our model, a number of S-alleles hover at the edge of extinction, with the frequencies of most alleles clustered around a higher mode. Arrows indicating the expected frequency of a common allele (inverse of the

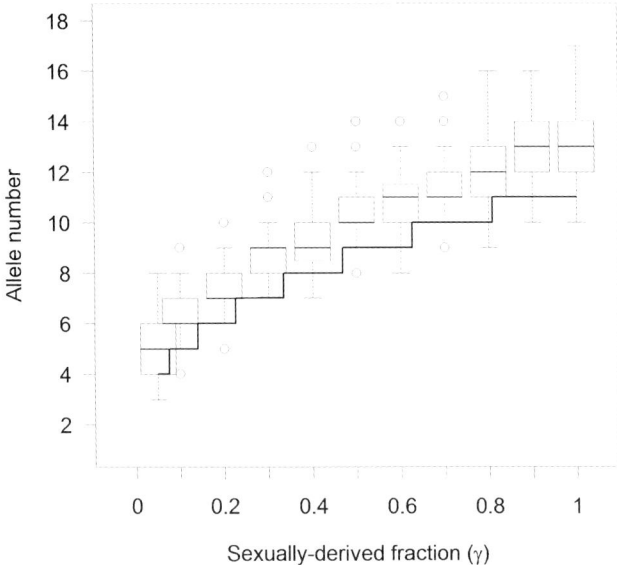

Fig. 3.2 Total number of S-alleles segregating in a population of 250 zygotes with a mutation rate of $u = 10^{-4}$. Each box plot summarizes the results of 100 independent simulations, censused in generation 15,000 following the burn-in period (median, first and third quartiles, whiskers at approximate 95% confidence limits, and *circles* for observations beyond the whiskers). Each *horizontal line* represents the expected number of S-alleles, corresponding to the largest integer not exceeding the solution for n in (3.7)

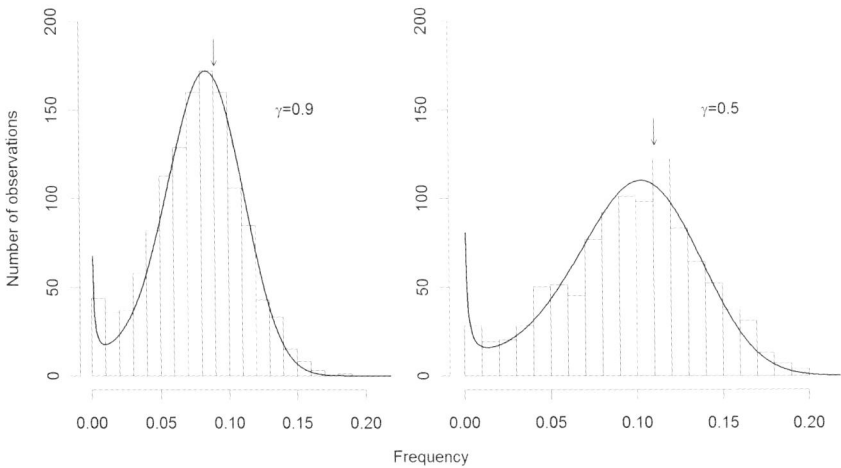

Fig. 3.3 Histograms of S-allele frequencies observed in 100 independent simulations compared to expected stationary distributions (3.9) with proportions $\gamma = 0.9$ (*left*) and $\gamma = 0.5$ (*right*) of reproductives sexually derived. Arrows indicate the expected frequency of a common allele ($1/n$). The *vertical* positions of the *solid curves* are determined by eye. Other parameters as in Fig. 3.2

expected number) lie close to this mode. As γ declines, fewer S-alleles segregate; the modal frequency increases and allele frequency vary over a greater range.

3.3.3 Age of the Root

Having established that our diffusion equation approximation provides a good description of a number of classical characteristics of S-allele evolution, we now use the model to address the evolutionary process from a genealogical perspective, aspects of which are more difficult to explore using forward-in-time simulations alone.

3.3.3.1 Expansion of Time Scale

Upon the advent of nucleotide sequences sampled from the S-locus and adjacent genomic regions, explorations of genealogical relationships among S-allele lineages revealed an extensive pattern of deep divergence (Ioerger et al., 1990; Dwyer et al., 1991), on the order of tens of millions of years.

Takahata (1990) characterized genealogies of functionally distinct classes of alleles maintained by exchangeable forms of balancing selection as similar to neutral genealogies, but on a time scale expanded by the scaling factor

$$f = \frac{n^2}{4N\lambda},$$

for n the number of common alleles maintained at stochastic steady state, N the effective population size, and λ (3.4) the rate at which rare allelic lineages become common. This characterization of the process of divergence of S-allele lineages suggests that the expected age of the most recent common ancestor (MRCA) of n S-alleles corresponds to

$$4Nf(1-1/n) \tag{3.10}$$

(see, for example, Tajima, 1983). Under SI with clonal reproduction (3.4), the scaling factor corresponds to

$$f = \frac{n^2(n-1)(n-2)}{4\theta[2Mn - \theta(n-1)(n-2)]}. \tag{3.11}$$

3.3.3.2 Reduction in the Sexually Derived Fraction

To explore whether the effects of clonality on the evolutionary dynamics are equivalent to a reduction in effective population size alone, we compared the number of common S-alleles maintained under partial clonal reproduction (closed symbols in Fig. 3.4) to that maintained under pure non-clonality in a population comprising the

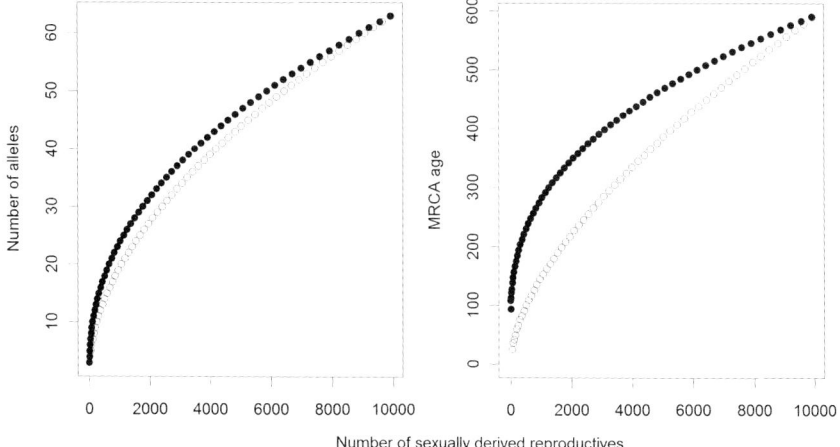

Fig. 3.4 Number of segregating S-alleles (n, *left panel*) and expected age of the root of the S-allele tree ($4Nf(1 - 1/n)$, *right panel*) under a rate of mutation to new S-alleles of $u = 10^{-5}$. *Closed symbols* correspond to a population of effective size $N = 10^4$ reproducing under partial clonality ($\gamma > 0$). *Open symbols* correspond a population practicing exclusive sexual reproduction with effective population size equal to the sexual fraction ($N\gamma$) in the partially clonal population

same number of sexually derived parents ($N\gamma$; open symbols in Fig. 3.4). The left panel of Fig. 3.4 confirms the positive relationship between the steady-state number of common S-alleles (n) and the proportion of sexually derived reproductives (γ) as indicated in Fig. 3.2. Further, it shows that partial clonality maintains somewhat more S-alleles than does pure sexuality in a smaller population, possibly reflecting that mutations to new S-alleles arise in both the sexual and clonal fractions.

To explore the effect of clonality on the depth of S-allele genealogies, we made a similar comparison of the expected age of the MRCA (3.10) using the scaling factor (3.11) induced by SI expression under partial clonality. The right panel of Fig. 3.4 indicates that the age of the root of the genealogy of the segregating S-alleles under partial clonality can exceed that under pure sexuality.

3.3.3.3 Discrepancy Between Allele Number and Genealogical Depth

Figure 3.4 suggests that even as the number of S-alleles (n) declines with the rate of sexual reproduction (γ), genealogical depth can remain large. To address the implications of an incorrect presumption of the absence of clonality, we conducted a thought experiment. Given a rate of mutation to new S-alleles (u) and a number of S-alleles (n) segregating in a population undergoing partial clonal reproduction, how great would be the discrepancy between the actual depth of the S-allele genealogy and the depth expected under complete sexual reproduction? We simulated the process by comparing, for a given assignment of effective population size (N), mutation rate (u), and number of S-alleles (n), the expected age of the MRCA under partial clonal reproduction (3.10) to the age expected in a purely sexual population with an identical number of common S-alleles with an effective size N^*.

To find the expected age of the MRCA (3.10) in the partially clonal population, we determined, for values of γ over its entire range $(0, 1)$, the expected numbers of common S-alleles (n). Because n is restricted to integer values, a range of γ values is consistent with each value of n; we chose the minimum value of γ to represent the range. For each pair of values for n and γ, we obtained the expected age of the MRCA from (3.10). To find the age of the MRCA in a purely sexual population, given the same values of n and u, we first solved the Yokoyama and Hetherington (1982) equation, which corresponds to (3.7) with $\gamma = 1$, for the range of effective population sizes consistent with n, and then obtained the corresponding scaling factor (f^*) from (3.11) that would be appropriate for a completely sexual population of size equal to the number of sexually-derived reproductives in the partially clonal population.

Figure 3.5 shows the ratio the ages of the MRCA $(N^* f^* / N f)$ as a function of the sexual proportion γ. Each S-allele number n determines a range of N^* values, giving rise to the range of $N^* f^* / N f$ ratios. Except for values of γ very close to 1 (absence of clonality), where the two models are identical, $N^* f^* / N f$ lies below unity. This implies that wrongly presuming the absence of clonal reproduction would cause the depth of the S-allele genealogy to appear too deep relative to the number of segregating S-alleles (n). Figure 3.5 indicates that under high levels of clonality

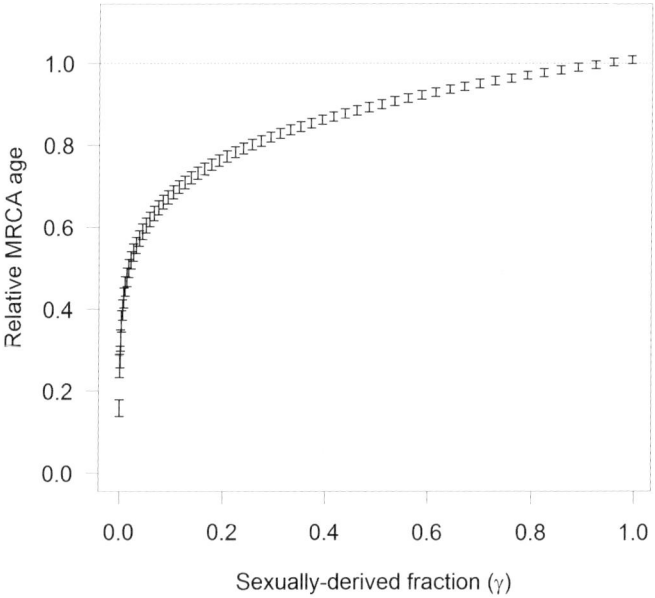

Fig. 3.5 Expected age of the MRCA in a purely sexual population as a proportion of the expected age of the MRCA in a partially clonal population of effective size $N = 10^4$, under a mutation rate $u = 10^{-5}$

(low γ), the actual genealogical depth can exceed the depth expected under non-clonality by several-fold.

3.4 Discussion

3.4.1 Clonality in the Solanaceae

Our diffusion approximation suggests that partial clonal reproduction can have profound effects on S-allele diversity and their genealogical span. We expect that clonality would constitute an important influence on the evolutionary dynamics at the S-locus only if its history had a genealogical depth comparable to that of S-alleles. Clonal reproduction characterized all (33/33) SI *Solanum* species studied by Vallejo-Marín and O'Brien (2007) and 68.5% (37/54) of SC species. While no formal treatment exists of the prevalence or the ancestral clonality status within the Solanaceae, the widespread occurrence of herbaceous perennials in this family (*e.g.*, *Solanum*), together with the observation that the majority of herbaceous perennial can reproduce vegetatively (Richards, 1986), suggests that clonality may be a common feature of this group. Groenendale et al. (1996) estimate that 50–75% of species in this family are clonal, and the closest related family in their analysis (Convolvulaceae) also shows a high incidence of clonality. These observations are consistent with the view that clonality has played a substantial role in the history of groups in the Solanaceae, including the genus *Solanum* in particular.

While extensive trans-specific sharing of S-allele lineages appears to be characteristic of all systems of SI, differences in genealogical structure are apparent among groups (Uyenoyama, 1997; Richman and Kohn, 1999). In particular, the analysis of Uyenoyama and Takebayashi (2004) suggested that a difference in effective population size alone between *Solanum carolinense* (with an estimated 13 segregating S-alleles) and three species of *Physalis* (with an estimated 72 segregating S-alleles) is unlikely to account for the differences in tempo and mode of S-allele diversification in these solanaceous taxa. Even though *S. carolinense* shows the lowest number of segregating S-alleles among species for which estimates are available (Lawrence, 2000), its S-allele lineages span almost the entire genealogical range contained in the Solanaceae (Uyenoyama and Takebayashi, 2004). Our analysis raises the possibility that clonal reproduction might contribute to this pattern of great genealogical depth in spite of low S-allele number.

3.4.2 Evolutionary Stability of Partial SI

Although partial expression of SI has traditionally been characterized as a transient state between complete SI and complete SC (Goodwillie et al., 2005), PSI can also represent a persistent condition.

Uyenoyama et al. (2001) explored the generation of new S-allele specificities through the successive loss and restoration of recognition between the pollen and stylar components of SI. Despite the shared evolutionary fate of factors under absolute linkage to the S-locus, evolutionary conflict between the male and female components can promote both mutations that impair pollen SI function, permitting an escape from rejection, and compensating mutations in the stylar component, restoring rejection (Newbigin and Uyenoyama, 2005). Under certain conditions, haplotypes with impaired pollen SI function can persist in stable polymorphism with full-function haplotypes (Uyenoyama et al., 2001).

Beyond the persistence of a partial breakdown until the restoration of full function, partial expression of SI can in fact constitute an evolutionarily stable state, resistant to the invasion of modifier alleles that increase as well as decrease the level of expression (Vallejo-Marín and Uyenoyama, 2004). In that model, we considered incomplete reproductive compensation of rejected mating opportunities through the expression of SI, perhaps as a consequence of low pollinator service or restriction of the domain of pollen transfer to the local neighborhood (Wilcock and Neiland, 2002). Selection on a modifier locus that controls SI expression level comprises two components: a direct effect (offspring number weighted by relatedness at the modifier locus) and a disequilibrium effect (greater rates of rejection and lower rates of reproductive compensation in S-locus heterozygotes than homozygotes). In a restricted parameter range, exact cancellation of these two components gives rise to an evolutionary stable intermediate level of SI expression.

3.4.3 Paradoxical Effects on Mating Systems

By reducing the intensity of frequency-dependent S-locus selection (3.2), clonal reproduction permits the maintenance of fewer segregating S-alleles (Fig. 3.2). Further, by promoting the spatial clumping of individuals with identical genotypes, clonality may exacerbate pollen limitation by reducing the fraction of compatible pollen received (Charpentier, 2002). Both effects serve to increase the evolutionary cost of SI under clonality.

Paradoxically, clonality may also preserve SI under conditions of pollen limitation by promoting reproductive assurance through vegetative reproduction (Baker, 1955, 1967; Vallejo-Marín and O'Brien, 2007). For example, Pannell and Barrett (1998) demonstrated that perennial SI taxa or those that maintain a persistent seed bank are less likely to become locally extinct. Similarly, asexual reproduction may permit isolated colonists to propagate themselves until conditions favorable for sexual reproduction arise (Baker, 1955; Baker and Cox, 1984).

These considerations suggest that under conditions of pollen limitation, clonality can promote the maintenance or the breakdown of SI: a contrast Vallejo-Marín (2007) has dubbed the SI-clonality paradox (see also Honnay and Jacquemyn, 2008). Similarly, SI itself has paradoxical effects on genetic load: by reducing the expression and purging of recessive deleterious mutations, especially in regions tightly linked to the *S*-locus, SI can shelter and increase genetic load (Uyenoyama, 2003; Stone, 2004). Because clonality and SI affect multiple components of reproduction, the resolution of the various conflicting evolutionary pressures may depend on many ecological factors (including plant density, spatial distribution of genotypes, and pollinator type and availability) as well as genetic factors (including epistasis among deleterious mutations and genomic structure in the vicinity of the *S*-locus).

3.5 Conclusions

Adoption of an unambiguous definition of the SI phenotype has been essential throughout the long history of the study of SI, and especially in the identification and cloning of the cardinal components of the *S*-locus. At the present juncture, following two remarkable decades of molecular-level analysis in several systems of homomorphic SI, a picture has emerged of the *S*-locus as a dynamic genomic region replete with evolutionary conflict, death, and rebirth.

Homology can now be defined at the genetic as well as phenotypic level (Kusaba et al., 2001), permitting the reconstruction of the history of SI even in taxa that do not express the SI phenotype (Tang et al., 2007; Shimizu et al., 2008). These developments suggest the possibility of now belaying the study of SI on the genetics to explore a wider diversity of SI phenotypes.

Recent experimental studies have revealed an array of departures from canonical SI, consistent with a characterization of SI systems as highly adaptable and responsive to the exigencies of survival and reproduction. As yet, rather few studies have attempted to explore the evolutionary processes through which such diversification arise. Here, we have addressed a simple departure from canonical SI, incorporation of clonal propagation together with full SI expression in reproduction through seeds. We found that the addition of a single parameter to the classical Wright (1939) model of GSI gives rise to new evolutionary dynamics. In particular, while clonality reduces the number of segregating *S*-alleles, much of the extraordinary depth of *S*-allele genealogies persists.

Recognition of SI as a dynamic component of the reproductive phenotype invites experimental investigation of its diverse manifestations. A deepening understanding of the genetic mechanisms of modification in turn demands the theoretical explorations of their implications for the evolutionary process and empirical reconstructions of genealogical history.

Acknowledgements MVM was supported during part of this study by a post-doctoral fellowship from the Canada Research Chair program to SCH Barrett. US Public Health Service grant GM 37841 (MKU) provided partial support for this research.

Appendix 1: Diffusion Equation Approximation

At reproduction, the frequency in the population of zygotes bearing S-locus genotype $S_i S_j$ is

$$TP'_{ij} = c\tau P_{ij} + (1-c) \sum_{k \neq i,j} \left[\frac{q_i P_{jk}}{1-q_j-q_k} + \frac{q_j P_{ik}}{1-q_i-q_k} \right]/2, \qquad (3.12)$$

for c is the proportion of reproductive effort invested in clonal reproduction, τ the relative viability of clonally derived offspring, and T a normalizer to be determined. We address the change in frequency of a specific S-allele, arbitrarily designated S_i, imposing the assumption of equality among the frequencies of any of the other $n-1$ S-alleles presently segregating in the population.

For q the frequency of the focal allele S_i, the frequency of any other allele ($\hat{q}$) satisfies

$$(n-1)\hat{q} + q = 1.$$

Similarly, let P denote the frequency of the genotype comprising S_i together with any other allele, and $\hat{P}$ the frequency of any given genotype comprising alleles other than S_i:

$$\binom{n-1}{2} \hat{P} + (n-1)P = 1. \qquad (3.13)$$

Because the allelic and genotypic frequencies are related by

$$q = (n-1)P/2, \qquad (3.14)$$

a description of the population requires only a single free variable (q).

From (3.12) and an analogous recursion in the frequency of any given genotype that does not carry S_i, we obtain

$$TP' = c\tau P + (1-c)(n-2) \left[\frac{q\hat{P}}{1-2\hat{q}} + \frac{\hat{q}P}{1-q-\hat{q}} \right]/2$$

$$T\hat{P}' = c\tau \hat{P} + (1-c)\hat{q} \left[\frac{(n-3)\hat{P}}{1-2\hat{q}} + \frac{P}{1-q-\hat{q}} \right].$$

Together with (3.13), these expressions determine the normalizer T:

$$T = c\tau + 1 - c.$$

Using (3.14), the expected change in allele frequency (3.1) derives from the expression for the change in P. From (3.1), the derivation of the diffusion approximation follows as described by Uyenoyama (2003). Analysis of the diffusion equation derived here follows that of Uyenoyama (2003), with the replacement of a and b in that article by

$$a^* = \frac{\gamma}{(1-F)(1-2F)}$$
$$b^* = \frac{\gamma}{2(1-F)} + u.$$

Appendix 2: Simulations

Each simulation was started with a population of N individuals, heterozygous at the S-locus. This initial population had k S-alleles in equal frequencies. During each reproductive cycle, a new set of N individuals was created from the parental generation (Fig. 3.1). Each new offspring was formed by randomly selecting a parental individual as the maternal plant, and then allowing it to reproduce either clonally or sexually with probabilities c and $1-c$, respectively. We assumed no differences in any components of fitness among individuals or among sexual and clonal offspring (i.e. $\tau = 1$), which in turn implies $c = 1 - \gamma$. For sexual reproduction, a new zygote was created by mating the maternal plant with a randomly selected pollen grain. All individuals, regardless of whether they reproduced by seed or vegetatively, contributed to the pollen pool. We assumed the complete expression of SI during the sexual component of the reproductive cycle. If the selected pollen was incompatible with the maternal plant, a new pollen grain was drawn. This process was continued until a compatible pollen grain was found, or until 900 pollination attempts were tried, after which we selected a new maternal parent. For the clonal component of reproduction, the maternal parent was used to generate a genetically identical offspring. New alleles at the S-locus were introduced by mutation at rate u. Mutations were introduced after the zygote was formed, i.e., we assumed that new mutations are not expressed in the gametophyte. Mutations occurred either in the maternal or paternal S-allele copy but not in both, and were equally likely to occur in sexually or clonally produced offspring.

To allow the system to stabilize before measurements were taken, we introduced a burn-in period that lasted until the initial k S-alleles coalesced to a single ancestor. After the burn-in period, we iterated the system for 15,000 more reproductive cycles before censusing the population. We repeated this process 100 times for each parameter combination to generate the simulation results.

References

Anderson MA, Cornish EC, Mau SL, Williams EG, Hoggart R, Atkinson A, Bonig I, Grego B, Simpson R, Roche PJ, Haley JD, Penschow JD, Niall HD, Tregear GW, Coghlan JP, Crawford RJ, Clarke AE (1986) Cloning of cDNA for a stylar glycoprotein associated with expression of self-incompatibility in *Nicotiana alata*. Nature 321:38–44

Baker HG (1955) Self-compatibility and establishment after "long-distance" dispersal. Evolution 9:347–349

Baker HG (1967) Support for Baker's Law – as a rule. Evolution 21:853–856

Baker HG, Cox PA (1984) Further thoughts on dioecism and islands. Ann Mo Bot Gard 71:244–253
Barrett S (1988). The evolution, maintenance, and loss of self-incompatibility systems. In: Lovett Doust J, Lovett Doust L (eds) Plant reproductive ecology: Patterns and strategies. Oxford University Press, Oxford, pp 98–124
Charlesworth D, Charlesworth B (1979) The evolution and breakdown of *S*-allele systems. Heredity 43:41–55
Charpentier A (2002) Consequences of clonal growth for plant mating. Evol Ecol 15:521–530
Dwyer KG, Balent MA, Nasrallah JB, Nasrallah ME (1991) DNA sequences of self-incompatibility genes from *Brassica campestris* and *B. oleracea*: Polymorphism predating speciation. Plant Mol Biol 16:481–486
Ferrer MM, Good-Ávila SV (2007) Macrophylogenetic analyses of the gain and loss of self-incompatibility in the Asteraceae. New Phytol 173:401–414
Fisher RA (1958) The genetical theory of natural selection, 2nd edn. Oxford University Press, Dover, New York
Good-Ávila SV, Stephenson AG (2002) The inheritance of modifiers conferring self-fertility in the partially self-incompatible perennial, *Campanula rapunculoides* L. (Campanulaceae). Evolution 56:263–272
Goodwillie C, Kalisz S, Eckert CG (2005) The evolutionary enigma of mixed mating systems in plants: Occurrence, theoretical explanations, and empirical evidence. Annu Rev Ecol Evol Syst 36:47–79
Groenendale JMv, Klimes L, Klimesova J, Hendriks RJJ (1996) Comparative ecology of clonal plants. Philos Trans R Soc Lond B 351:1331–1339
Honnay O, Jacquemyn H (2008) Mating system evolution under strong clonality: towards self-compatibility or self-incompatibility? Evol Ecol DOI 10.1007/s10682-007-9207-3
Igic B, Bohs L, Kohn JR (2006) Ancient polymorphism reveals unidirectional breeding system shifts. Proc Natl Acad Sci USA 103:1359–1363
Ioerger TR, Clark AG, Kao Th (1990) Polymorphism at the self-incompatibility locus in Solanaceae predates speciation. Proc Natl Acad Sci USA 87:9732–9735
Karlin S, Taylor HM (1981) A second course in stochastic processes. Academic, New York
Koch MA, Kiefer M (2005) Genome evolution among cruciferous plants: A lecture from the comparison of the genetic maps of three diploid species – *Capsella rubella*, *Arabidopsis lyrata* subsp. *petraea*, and *A. thaliana*. Am J Bot 92:761–767
Koch MA, Haubold B, Mitchell-Olds T (2001) Molecular systematics of the Brassicaceae: Evidence from coding plastidic *matK* and nuclear *Chs* sequences. Am J Bot 88:534–544
Kusaba M, Dwyer K, Hendershot J, Vrebalov J, Nasrallah JB, Nasrallah ME (2001) Self-incompatibility in the genus *Arabidopsis*: Characterization of the *S*-locus in the outcrossing *A. lyrata* and its autogamous relative *A. thaliana*. Plant Cell 13:627–643
Lawrence MJ (2000) Population genetics of the homomorphic self-incompatibility polymorphisms in flowering plants. Ann Bot (Lond) 85(Suppl. A):221–226
Levin DA (1996) The evolutionary significance of pseudo-self-fertility. Am Nat 148:321–332
Liu P, Sherman-Broyles S, Nasrallah ME, Nasrallah JB (2007) A cryptic modifier causing transient self-incompatibility in *Arabidopsis thaliana*. Curr Biol 17:734–740
Mena-Alí JI, Stephenson AG (2007) Segregation analyses of partial self-incompatibility in self and cross progeny of *Solanum carolinense* reveal a leaky s-allele. Genetics 177:501–510
Nasrallah JB, Kao Th, Goldberg ML, Nasrallah ME (1985) A cDNA clone encoding an *S*-locus-specific glycoprotein from *Brassica oleracea*. Nature 318:263–267
Nasrallah ME, Liu P, Sherman-Broyles S, Nasrallah JB (2004) Natural variation in expression of self-incompatibility in *Arabidopsis thaliana*: Implications for the evolution of selfing. Proc Natl Acad Sci USA 101:16070–16074
Neuhauser C (1999) The ancestral graph and gene genealogy under frequency-dependent selection. Theor Popul Biol 56:203–214
Newbigin E, Uyenoyama MK (2005) The evolutionary dynamics of self-incompatibility systems. Trends Genet 21:500–505

Pagel M, Meade A (2006) Bayesian analysis of correlated evolution of discrete characters by reversible-jump Markov chain Monte Carlo. Am Nat 167:808–825

Pagel MD (1994) Detecting correlated evolution on phylogenies: A general method for the comparative analysis of discrete characters. Philos Trans R Soc Lond B 255:37–45

Pannell JR, Barrett SCH (1998) Baker's law revisited: Reproductive assurance in a metapopulation. Evolution 52:657–668

Richards AJ (1986) Plant breeding systems. George Allen & Unwin, London

Richman AD, Kohn JR (1999) Self-incompatibility alleles from *Physalis*: Implications for historical inference from balanced genetic polymorphisms. Proc Natl Acad Sci USA 96:168–172

Sherman-Broyles S, Boggs N, Farkas A, Liu P, Vrebalov J, Nasrallah ME, Nasrallah JB (2007) *S*-locus genes and the evolution of self-fertility in *Arabidopsis thaliana*. Plant Cell 19:94–106

Shimizu KK, Cork JM, Caicedo AL, Mays CA, Moore RC, Olsen KM, Ruzsa S, Coop G, Bustamante CD, Awadalla P, Purugganan MD (2004) Darwinian selection on a selfing locus. Science 306:2081–2084

Shimizu KK, Shimizu-Inatsugi R, Tsuchimatsu T, Purugganan MD (2008) Independent origins of self-compatibility in *Arabidopsis thaliana*. Mol Ecol 17:704–714

Stebbins GL (1974) Flowering plants: Evolution above the species level. Harvard University Press, Cambridge, MA

Stone JL (2004) Sheltered load associated with S-alleles in *Solanum carolinense*. Am J Bot 92:335–342

Tajima F (1983) Evolutionary relationship of DNA sequences in finite populations. Genetics 105:437–460

Takahata N (1990) A simple genealogical structure of strongly balanced allelic lines and transspecies evolution of polymorphism. Proc Natl Acad Sci USA 87:2419–2423

Tang C, Toomajian C, Sherman-Broyles S, Plagnol V, Guo YL, Ho TT, Clark RM, Nasrallah JB, Weigel D, Nordborg M (2007) The evolution of selfing in *Arabidopsis thaliana*. Science 317:1070–1072

Travers SE, Mena-Alí JI, Stephenson AG (2004) Plasticity in the self-incompatibility of *Solanum carolinense*. Plant Species Biol 19:127–135

Uyenoyama MK (1995) A generalized least-squares estimate for the origin of sporophytic self-incompatibility. Genetics 139:975–992

Uyenoyama MK (1997) Genealogical structure among alleles regulating self-incompatibility in natural populations of flowering plants. Genetics 147:1389–1400

Uyenoyama MK (2003) Genealogy-dependent variation in viability among self-incompatibility genotypes. Theor Popul Biol 63:281–293

Uyenoyama MK, Takebayashi N (2004) Genus-specific diversification of mating types. In: Singh R, Uyenoyama MK (eds) The evolution of population biology. Cambridge University Press, New York, pp 254–271

Uyenoyama MK, Zhang Y, Newbigin E (2001) On the origin of self-incompatibility haplotypes: Transition through self-compatible intermediates. Genetics 157:1805–1817

Vallejo-Marín M (2007) The paradox of clonality and the evolution of self-incompatibility. Plant Signal Behav 2:265–266

Vallejo-Marín M, O'Brien HE (2007) Correlated evolution of self-incompatibility and clonal reproduction in *Solanum* (Solanaceae). New Phytol 173:415–421

Vallejo-Marín M, Uyenoyama MK (2004) On the evolutionary costs of self-incompatibility: Incomplete reproductive compensation due to pollen limitation. Evolution 58:1924–1935

Vekemans X, Slatkin M (1994) Gene and allelic genealogies at a gametophytic self-incompatibility locus. Genetics 137:1157–1165

Wilcock C, Neiland R (2002) Pollination failure in plants: Why it happens and when it matters. Trends Plant Sci 7:270–277

Wright S (1939) The distribution of self-sterility alleles in populations. Genetics 24:538–552

Wright S (1960) On the number of self-incompatibility alleles maintained in equilibrium by a given mutation rate in a population of a given size: A reexamination. Biometrics 16:61–85

Yokoyama S, Hetherington LE (1982) The expected number of self-incompatibility alleles in finite plant populations. Heredity 48:299–303

Chapter 4
Evolution and Phylogeny of Self-Incompatibility Systems in Angiosperms

A.M. Allen and S.J. Hiscock

Abstract The evolution of self-incompatibility (SI) has been debated since its genetic bases were first fully described over half a century ago. The diversity of SI systems and their scattered distribution among different flowering plant lineages suggest that SI has evolved independently on many different occasions. Recent advances in our understanding of the evolutionary relationships among angiosperms together with a growing understanding of the genetic, cellular, and molecular control of SI systems have now allowed objective insights to be made into the evolution of SI. Mapping the presence of the various SI systems within robust angiosperm phylogenies allows us to view the phylogenetic distribution of SI as a whole, and compare the distribution and phylogenetic relationships of the different forms of SI. Here we discuss the diversity of SI systems in flowering plants and current hypotheses on their evolution and phylogenetic relationships. Using the APG II tree we map the distribution of SI and self-sterility and show: (1) Self-sterility and SI are well represented among basal angiosperm lineages, suggesting that self-sterility/SI rather than self-compatibility (SC) was the ancestral angiosperm mating system; (2) late-acting ovarian SI is the basal SI state; (3) gametophytic SI systems evolved independently in the earliest diverging angiosperm lineages; (4) heteromorphic sporophytic SI evolved on numerous different occasions in eudicots and probably only once in monocots; and (5) different forms of homomorphic sporophytic SI evolved relatively recently in eudicots.

Abbreviations

ANITA	Amborellales, Nymphaeales, and the ITA clade
APG	Angiosperm Phylogeny Group
G-locus	A gametophytic locus associated with SSI

A.M. Allen and S.J. Hiscock
School of Biological Sciences, University of Bristol, Woodland Road, Bristol, BS8 1UG, UK,
e-mail: simon.hiscock@bristol.ac.uk

GSI	Gametophytic SI
OSI	Ovarian SI
PSC	Pseudo-self-compatibility
S-locus	Self-incompatibility locus
S-RNase	*S*-ribonuclease
SC	Self-compatibility
SCR/SP11	*S*-locus cysteine-rich protein/*S*-locus protein 11 (the pollen *S*-determinant in Brassica)
SI	Self-incompatibility
SLF/SFB	*S*-locus *F*-box (*SFB/SLF* – *S*-*Locus F-box*/*S-haplotype-specific F-box* – the pollen *S*-determinant in many GSI systems; *SLF* in *Antirrhinum* and *Petunia*; *SFB* in *Prunus*)
SRK	*S*-locus Receptor Kinase (the pistil *S*-determinant in *Brassica*)
SSI	Sporophytic SI
Z-locus	A second self-incompatibility locus in grasses

4.1 Introduction

Self-incompatibility (SI), defined as 'the inability of a fertile seed plant to produce zygotes after self-pollination' (de Nettancourt 1977), is the most important and widespread mechanism promoting outcrossing in angiosperms. It has been suggested that the early evolution of SI was a key factor contributing to the rapid diversification of angiosperms during the Cretaceous (Whitehouse 1950). Certainly, possession of a genetic mechanism to prevent inbreeding would have allowed the first flowering plants to capitalise on the use of newly available pollinators by removing the risks of self-fertilisation. The utilisation of biotic pollination systems during this period is likely to have been an important driving force of speciation and the spread of newly evolved angiosperm taxa (Crepet 1985) and possession of a mechanism for self-sterility (SI) would have increased the efficiency of biotic cross-pollination. Indeed, recent studies of the mating systems of basal angiosperms are providing increasing evidence that the first flowering plants may have been self-sterile (Sage and Sampson 2003). SI systems and self-sterility are distributed widely throughout all the principal lineages of angiosperms, being present in approximately 19 orders, 71 families and 250 genera comprising approximately 60% of all angiosperm species (East 1940; Brewbaker 1957; de Nettancourt 1977; Charlesworth, 1985; Richards 1997; Hiscock and Kues 1999).

The great variety of SI systems and their scattered distribution in angiosperms have made it difficult to elucidate the ancestral SI state or the evolutionary relationships between the different SI systems. In recent years, the extensive use of molecular phylogenetics has allowed the construction of robust phylogenetic trees and identification, for the first time, of the most basal angiosperm lineages (Soltis et al. 1999; Qiu et al. 1999, 2005). Among these basal lineages, several mechanisms promoting outbreeding have been identified, most notably possession of unisexual flowers and self-sterility (possibly SI). We use the term self-sterility instead of SI in this context, because the site of incompatibility is frequently unclear, and often ovarian or post-zygotic, and SI *sensu strictu* is usually considered pre-zygotic (Heslop-Harrison 1975). Nevertheless, the broad definition of SI used by de Nettancourt (above) can encompass many forms of post-zygotic self-sterility (de Nettancourt 1977). Whatever the mechanism(s) involved, reduced levels of seed set, or its absence after selfing now appear to be widespread among basal angiosperm taxa (see later).

4.1.1 Diversity of SI Systems in Angiosperms

Most genetically well-characterised SI systems are controlled by a single highly polymorphic locus, the *S*-locus (de Nettancourt 1977, 2001; Hiscock and McInnis 2003). Nevertheless, a number of multi-locus systems have also been described (Lundqvist 1990a, b, 1991), most notably the two-locus (*S*- and *Z*-locus) SI system found in the grasses (Poaceae) (Li et al. 1997; Sect. 3.2.2). The *S*-locus contains a minimum of two tightly linked (non-recombining) polymorphic genes, one controlling pollen identity and the other pistil identity. In addition to these primary male and female determinants, many other genes are known or predicted to reside at the *S*-locus (Watanabe et al. 2000; McCubbin and Kao 1999). Characterised *S*-loci are large, and often contain other genes not involved in SI. For instance, the *S*-locus of *Petunia* is estimated to span 4.4 Mb and comprise a huge multi-gene complex where recombination is suppressed (Wang et al. 2003). Genetic data also show that genes not linked to the *S*-locus are important for fully functional SI (Boyes et al. 1991; Umbach et al. 1990). These so-called modifier loci have been identified in many different species and encode proteins of diverse function, many of which are involved in signalling pathways downstream of the *S*-protein-mediated self-recognition machinery (Good-Avila and Stephenson 2002; Hiscock et al. 2003).

Genetically characterised SI systems fall into one of the two broad categories: gametophytic self-incompatibility (GSI) and sporophytic self-incompatibility (SSI), which are defined by the genetic control of the incompatibility phenotype in pollen. In GSI, the incompatibility phenotype of the pollen is determined by the pollen's own haploid genome. In SSI, the incompatibility phenotype of the pollen grain is determined by the diploid genome of the parental plant, thereby allowing *S*-alleles to display dominance interactions in both pollen and pistil. GSI and SSI can be distinguished by reciprocal crossing experiments (diallels) that produce compatibility/incompatibility patterns characteristic of either GSI or SSI. GSI systems always show half compatibility between individuals that share one *S*-allele in common. However, in SSI systems, crosses are always fully compatible or fully incompatible and may show reciprocal differences in compatibility if there are dominance interactions between *S*-alleles (de Nettancourt et al. 2001; Hiscock and McInnis 2003).

Important correlations have been made between the possession of GSI or SSI and other reproductive features (Brewbaker 1959; Zavada 1984; Heslop-Harrison and Shivanna 1977). For instance, GSI has been correlated with possession of a wet stigma and bicellular pollen, whereas possession of a dry stigma and tricellular pollen is frequently associated with SSI; the grasses (GSI with tricellular pollen and dry stigma) and *Papaver* (GSI with dry stigma) being the most notable exceptions. SI systems can also be distinguished at a crude level on the basis of the site of their incompatibility reaction. All the well-characterised SSI systems show pollen inhibition on the stigma surface, whereas in GSI systems inhibition of incompatible pollen tubes frequently occurs within the style (but with exceptions in the grasses and *Papaver*, which have stigmatic inhibition). Such correlations are useful for generating hypotheses about the nature of a particular SI system in the absence of genetic data (Gibbs 1986).

In addition to GSI and SSI, there are ovarian or late-acting SI systems, so called because inhibition of incompatible pollen tubes takes place in the ovary, either just before fertilisation or at some stage after fertilisation (de Nettancourt 1977, 2001; Seavey and Bawa 1986; Barrett 1998; Gibbs and Bianchi 1999). Unless the cytology of the incompatibility response has been investigated it is impossible to differentiate late-acting pre-zygotic SI from post-zygotic self-sterility, or early acting inbreeding depression (Kenrick et al. 1986). For this reason, late-acting SI is often used synonymously with self-sterility. Here we will only use the term 'SI' for such systems when there is clear evidence for a pre-zygotic or pre-mitotic barrier to successful fertilisation. Compared to GSI and SSI the genetics and physiology of late-acting SI are poorly understood. Some late-acting SI systems have been proposed to be under gametophytic control, whereas others have been deduced to be under sporophytic control, and some, most notably cocoa (*Theobroma cacao*) and passion fruit (*Passiflora edulis*), appear to be under joint gametophytic and sporophytic control (Hasenstein and Zavada 2001; Suassuna et al. 2003). The presence of OSI has often been correlated with a woody perennial habit, and gynoecia with hollow styles, or the absence of a true style, where there is little contact between the male and female reproductive tissues (Kenrick et al. 1986). These traits are often features of the basal angiosperms and many tropical and sub-tropical trees. Indeed, the woody perennial nature of many plants with putative OSI makes genetic analysis very time consuming or impossible (Seavey and Bawa, 1984).

SI systems can also be differentiated on the basis of floral morphology. In heteromorphic SI systems incompatibility phenotypes are correlated with distinct floral morphological differences, most notably long (pin) or short (thrum) styles characteristic of distyly and first described in detail by Darwin (Darwin 1862; see Chap. 1, this volume). By contrast, homomorphic SI systems show no distinct floral morphologies associated with incompatibility phenotype and comprise all GSI systems and all multi-allelic SSI systems.

4.1.2 Evolutionary Origin(s) of SI Systems

It has often been proposed that SI systems evolved from pathogen defence mechanisms (de Nettancourt 1977; Hodgkin et al. 1988; Elleman and Dickinson 1999;

Dickinson 1995). As well as the immediate morphological similarities between a host response to an invading fungal hypha and the pistil's rejection of an 'invading' incompatible pollen tube, there are a number of important biochemical and molecular similarities between the two processes. In *Brassica*, for instance, the female determinant of SSI is a serine–threonine receptor kinase that belongs to a large gene family of *S*-like receptor kinase genes, members of which are involved in host–pathogen defence responses (Cock et al. 1997; Martin et al. 1993). Similarly, the *S*-RNases that mediate GSI in the Solanaceae and Rosaceae are related to RNases involved in defence against pathogens (Kao and McCubbin 1996).

Another theory proposes that SI evolved through the elaboration of ancient inter-specific incompatibility systems that first appeared in the gymnosperm-like ancestors of angiosperms (Pandey 1960). Evidence from studies in *Brassica* and *Nicotiana* suggests that genes controlling SI are involved in inter-specific incompatibility (Hiscock and Dickinson 1993; Murfett et al. 1996). Inter-specific incompatibility appears widespread among gymnosperms (Hagman 1975; McWilliam 1959) and there is compelling evidence for a rudimentary pre-zygotic SI system operating in the nucellus of *Picea glauca* ovules (Runions and Owens 1998). Other species of gymnosperm, such as *Pseudotsuga menziesii, Pinus sylvestris* and *Larix laricina*, show various degrees of embryo abortion after self-fertilisation, suggesting that certain forms of self-sterility and also perhaps late-acting SI most probably have their origins in the gymnosperm ancestor of angiosperms (Pandey 1960).

The aim of this review is to survey the distribution of the various SI (and self-sterility) systems in angiosperms, combining current SI data with the most recent angiosperm phylogenies, to develop a picture of the evolution and phylogeny of self-incompatibility.

4.2 Was Self-Incompatibility Present in the First Angiosperms?

There is conflicting evidence over whether SI (or self-sterility) or self-compatibility (SC) was the ancestral condition of the earliest angiosperms. Stebbins (1957) believed that SI was ancestral and evolved at the point of angiosperm diversification, while SC evolved at a later stage. However, Bateman (1952) suggested that SI evolved progressively and independently many times, but not at the origin of the group, a view shared by East (1940), who argued that the prevalence of SI systems in herbaceous plants indicated a more recent origin. Current molecular data supports a polyphyletic origin of SI, owing to the diversity of the different SI systems discovered in a relatively small number of families (Charlesworth et al. 2005; Silva and Goring 2001; Hiscock and McInnis 2003). This conclusion is strengthened further by the pattern of distribution of SI systems among angiosperm families: Some closely related families possess completely different and unrelated SI systems, while other more distantly related families share the same type of SI system (Matton et al. 1994).

Many authors have proposed that outbreeding, particularly in the form of SI/self-sterility, may have facilitated the rapid early diversification of angiosperms

(Whitehouse 1950; Zavada 1984; Zavada and Taylor 1986). If this was the case, one would expect to find SI/self-sterility in basal extant angiosperm lineages (Friis et al. 2000). Interestingly one of the few phylogenetic studies of the evolution of SI in angiosperms (Weller et al. 1995) predicted the ancestral angiosperm mating system to be SC, with the various SI systems arising polyphyletically. Since this study, however, there has been a surge of interest in the mating systems of basal angiosperms. This has been driven by the availability since 1999 of well-supported phylogenies that have resolved, for the first time, the identity and relationships between the most basal angiosperm groups (Soltis et al. 1999; Qiu et al. 1999, 2005; Zanis et al. 2003). Many detailed studies of cross and self-pollinations in basal angiosperms are now available (Sage and Sampson 2003; Hristova et al. 2005; Bernhardt et al. 2003; Koehl et al. 2004; Pontieri and Sage 1999; Sage et al. 2001), making it possible to formulate a far clearer picture on the basal condition of angiosperm mating systems and importantly to make more accurate predictions about the evolution and phylogenetic relationships of the various different forms of SI.

4.2.1 Self-Incompatibility in Basal Angiosperms

According to current phylogenies (APG II, 2003), the three most basal angiosperm clades, collectively referred to as the ANITA grade, consist of the Amborellales, Nymphaeales, and the ITA clade (Illiciaceae, Trimeniaceae and Austrobaileyaceae). The next most basal lineages include the Chloranthaceace, the Magnoliids (Canellales, Piperales, Laurales and Magnoliales) and the monocots. Fossil evidence has shown that these groups were all well represented in the Early Cretaceous, when most floras were dominated by basal angiosperms and Magnoliids (Friis et al. 2000). *Amborella trichopoda* (Amborellales), the most basal extant angiosperm, is dioecious, and so outcrossing is obligate. However, female flowers usually contain staminodes, and occasionally produce fully developed stamens that lack pollen (Thien et al. 2003; Endress 2001), indicating that its relatives (extinct now) were certainly co-sexual. Indeed dioecy in *Amborella* is probably a relatively recent consequence of its island isolation, because dioecy is known to evolve readily in island floras (Thomson and Barrett 1981). Unfortunately, we will never know the ancestral mating system of co-sexual Amborellales. Flowers of Nymphaeales, however, are always co-sexual and strongly protogynous (carpels maturing before stamens), but there are no conclusive reports of SI/self-sterility in this order (Bernhardt and Thien 1987). Nevertheless, species of *Nuphar* have been reported to produce deformed fruits after self-pollination (Schneider and Moore 1977), suggesting the presence of self-sterility or a late-acting SI system.

Reports of SI in *Illicium floridanum* (Illiciaceae) have proved controversial. One study (Thien et al. 1983) reported inhibition of pollen tube growth on the stigma surface and the degeneration of ovules following self-pollination, suggesting SI. However more recent studies of cross and self pollinations found no evidence for pre-zygotic SI, concluding that self-sterility is acting at a late stage of embryo development, either through late acting post-zygotic self-sterility or through inbreeding depression (Koehl et al. 2004). Self-sterility in the form of parthenocarpy (fruit production without seed) has been observed in other species of Illiciaceae, and also in species of Austrobaileyaceae and several Magnoliid families. Despite not containing viable seed, it is possible for the self-pollinated carpels of these taxa to persist for long periods of time and swell as if seed had been set (Bernhardt and Thien 1987), thereby giving a false impression of SC. As with self-sterility in *I. floridanum*, it has proved difficult to determine whether parthenocarpy is due to late acting SI/self-sterility or early acting inbreeding depression. SI/self-sterility has been inferred in *Austrobaileya scandens* (Austrobaileyaceae) from population level studies of genetic diversity (Williams and Kennard 2006), but as with *I. floridanum*, earlier reports of pre-zygotic SI (Prakash and Alexander 1984) were not confirmed by recent studies, and the basis of self-sterility in *A. scandens*, late-acting self-sterility or early acting inbreeding depression, has yet to be determined (Williams and Kennard 2006).

The final family of the ANITA group, Trimeniaceae, does contain at least one species exhibiting classic pre-zygotic SI, *Trimenia moorei*, a species possessing bicellular pollen and a dry stigma (Bernhardt et al. 2003). Whilst there is limited knowledge of the breeding systems of other species in this monotypic family, stigmatic SI has been demonstrated conclusively in *T. moorei* using controlled cross and self-pollination followed by observations of pollen tube growth and the fertilisation process (Bernhardt et al. 2003).

Stigmatic SI on a dry-type stigma has also been described in the Chloranthaceae, the next diverging basal lineage close to ANITA (Hristova et al. 2005). The position of this family within the phylogeny is still uncertain, but molecular data has placed it between the ANITA grade and the Magnoliids, as a monophyletic group (APG II 2003), a placement supported by morphological (Balthazar and Endress 1999) and fossil data (Crane and Lidgard 1989). Species belonging to the genera *Chloranthus* and *Sarcandra* show almost no pollen germination ($<1\%$) following controlled self-pollinations, but overall pollen viability in these species is low because in controlled cross (compatible) pollinations pollen germination was only $\sim7\%$ (Hristova et al. 2005).

The finding of stigmatic SI in basal angiosperms is intriguing because stigmatic SI (together with the dry-type of stigma) is traditionally viewed as associated with more advanced forms of SSI (Dickinson 1995). The presence of a dry stigma and stigmatic SI in the Trimeniaceae (Bernhardt et al. 2003), Chloranthaceae (Hristova et al. 2005) and species of Magnoliid (Pontieri and Sage 1999) indicates that this assumption now needs to be revised. Indeed, the identification of a dry-type stigma in *Illicium floridanum* (Koehl et al. 2004) indicates that the dry stigma rather than the wet stigma (Dickinson 1995) may be the true basal condition of the angiosperm stigma.

Self-sterility and SI are widespread among the Magnoliids, being reported in species from all four orders, Canellales, Piperales, Laurales and Magnoliales (Sage and Sampson 2003; Pontieri and Sage 1999; Gibson and Wheelright 1996; Bernhardt and Thien 1987). In the Saururaceae, an early diverging family in the

Piperales, pre-zygotic (stigmatic) SI has been confirmed in three out of the five species (Pontieri and Sage 1999). Members of the Canellales, Laurales and Magnoliales (e.g. species of Winteraceae, Lauraceae, Magnoliaceae and Annoneaceae) also exhibit full or partial self-sterility, often accompanied by parthenocarpy, where self-pollinated carpels swell, but do not contain seeds with a functional embryo (Bernhardt and Thien 1987). A more recent study found convincing evidence for post-zygotic ovarian self-sterility in *Pseudowintera axillaris* (Winteraceae) – following self-pollination, embryos were consistently aborted at the same stage of development, indicating the presence of an active self-sterility (SI) system as opposed to early acting inbreeding depression (Sage and Sampson 2003).

4.2.2 Self-Incompatibility in the Monocots

Within the monocotyledons, a diverse range of SI systems is predicted based on variation in the site of the incompatibility reaction observed among different species, which range from stigmatic inhibition, through stylar inhibition, to late acting incompatibility within the ovary (Sage et al. 2000). The majority of genetically characterised SI systems in the monocots show gametophytic control (e.g. Poaceae, Liliaceae, Commelinaceae; de Nettancourt 2001), but members of one family, Pontederiaceae, exhibit heteromorphic SSI (Bianchi et al. 2000). Heterostyly associated with a separate late-acting self-sterility system has also been identified in the Amaryllidaceae, in the daffodil *Narcissus triandrus* (Sage et al. 1999, 2000). Interestingly, there is no evidence for the existence of homomorphic SSI in the monocotyledons (Sage et al. 2001).

The best-studied SI system in the monocots is the two-locus GSI system of the grasses (Poaceae) (discussed in detail in Chap. 13), but other well-studied taxa include *Narcissus* (Sage et al. 1999), *Trillium* (Sage et al. 2001), *Tradescantia* (Owens and McGrath 1984), *Lilium* (Lundqvist 1991; Tezuka et al. 2007), *Gasteria* (Naarborg and Willemse 1992) and *Eichhornia azurea* (Bianchi et al. 2000), and reflect the spectrum of different SI/self-sterility systems present in this major group. Members of the largest angiosperm family, the Orchidaceae, employ a vast range of mechanisms to promote outbreeding, including many examples of SI (presumed to be GSI) and self-sterility (Borba et al. 2001). However, despite predictions of the presence of such mechanisms in over 750 species (East 1940), little is known about the physiology and genetic control of SI in the Orchidaceae on account of their very long lifecycles, which effectively preclude genetic analysis.

In the Liliaceae at least two different GSI systems have been identified, a seemingly widespread single-locus system and a four-locus system in *Lilium martagon* (Lundqvist 1991). An additional late acting self-sterility system has also been reported in *Clintonia borealis*, but the genetic control of this system is unknown (Dorken and Husband 1999). *Lilium longiflorum* has proved to be a useful species for the study of single-locus GSI, because of its large flowers and clear-cut SI reaction (Ascher and Peloquin 1968). Studies of SI have therefore focused on the physiology of the SI reaction in the pistil. Typically, pollen tubes enter the pistil through three crevices on the stigma before growing through a hollow stylar canal in which incompatible pollen tubes are inhibited (Hiratsuka et al. 1983). Recent work suggests that SI is associated with low levels of cAMP (cyclic AMP) and acetylcholine (ACh) in the pistil, which in turn regulate the levels of ChAT (ACh-forming enzyme) and AChE (ACh-decomposing enzyme) (Tezuka et al. 2007). Nothing, however, is known about the *S*-genes that mediate recognition.

A large survey of 110 species in 22 genera of the Commelinaceae revealed SI to be widespread in this family, with 55 species showing SI, of which the majority 53 exhibited stigmatic arrest of self-pollen and two showed later stylar inhibition (Owens 1981). At least four species (*Gibasis oaxacana*, *G. karwinskyana*, *Tradescantia paludosa* and *T. ohiensis*) were demonstrated to possess single-locus GSI, characterised by stigmatic pollen inhibition in association with a wet stigma and binucleate pollen (Annerstedt and Lundqvist 1967; Owens and McGrath 1984). SI has also been studied in detail in the genus *Trillium* (Melanthiaceae), which comprises ~42 species. *T. grandiflorum* and *T. erectum* both possess stigmatic SI, in combination with a dry-type stigma and bicellular pollen (Sage et al. 2001).

4.3 Phylogenetic Distribution of SI Systems

It is over a decade since Weller et al. (1995) carried out the first detailed phylogenetic analysis of the distribution of SI systems in angiosperms. Since then there have been significant advances in our understanding of the phylogenetic relationships between the flowering plants (Soltis et al. 1999; Qiu et al. 1999, 2005), giving us a far better understanding of the phylogenetic relationships amongst the most basal angiosperm lineages. Studies of the mating systems of basal angiosperms have mushroomed over the past 12 years, identifying novel systems of self-sterility and SI in many of these basal groups (Sect. 4.3).

There is now evidence for the presence of self-sterility and SI in four of the five most basal angiosperm lineages: self-sterility is present in the Austrobaileyaceae and Illiciaceae, and SI is present in the Trimeniaceae and Chloranthaceae (Williams and Kennard 2006; Koehl et al. 2004; Bernhardt et al. 2003). There is also evidence for the presence of self-sterility in at least one member (*Nuphar*) of the Nymphaeaceae (Bernhardt and Thien 1987). Among basal lineages Trimeniaceae, Chloranthaceae and Saururaceae, there is clear evidence for the presence of true pre-zygotic SI which, perhaps surprisingly, takes the form of stigmatic inhibition of self-pollen. Nevertheless, reports of late-acting forms of self-sterility (or SI) are more widespread among the basal groups, suggesting that this is the ancestral condition of self-sterility/SI. This apparent diversity of self-sterility and SI systems among basal angiosperms suggests a rapid parallel evolution of SI/self-sterility in the earliest diverging angiosperm lineages (Sage and Sampson 2003). Among the monocots, a similar picture of diversity among SI/self-sterility systems is emerging, with different forms of SI/self-sterility occurring both between and within families, suggesting complex and ongoing evolution of SI within this ancient monophyletic group.

Given the extent of recent advances, it is therefore timely to reassess the distribution of the various SI/self-sterility systems within a phylogenetic context. We therefore used the same approach as Weller et al. (1995) to map the presence of all known SI and self-sterility systems onto the most recent angiosperm phylogeny (APG II 2003) to determine whether advances over the past 12 years have affected the phylogenetic relationships between SI/self-sterility and SC (Fig. 4.1). The most significant finding is at the base of the tree, where there is now strong support for the presence of self-sterility (possibly in the form of late-acting SI) in many of the most basal angiosperm groups. This places real uncertainty on SC being the ancestral mating system state of flowering plants and raises the possibility that self-sterility

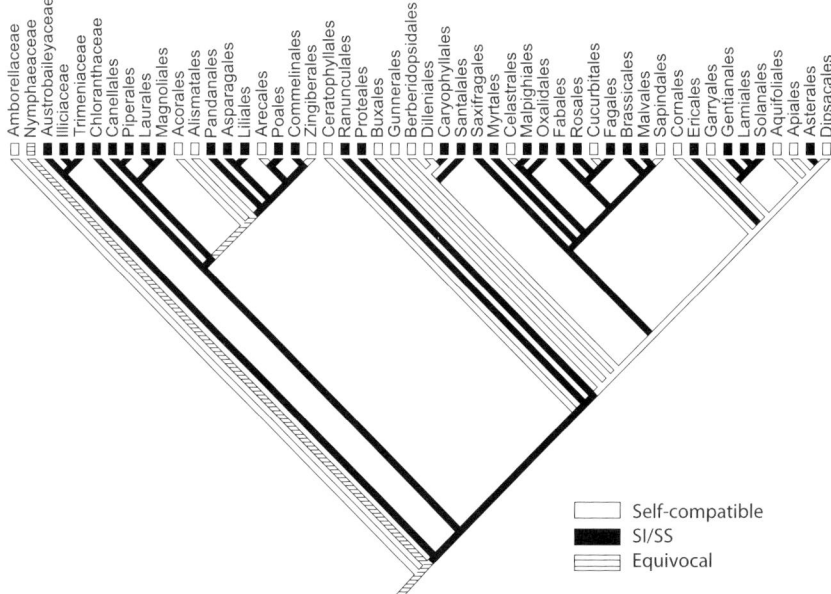

Fig. 4.1 Phylogenetic distribution of SI and self-sterility (SS) systems within the angiosperms Phylogenetic relationships between taxa are based on the current angiosperm phylogeny (APG II, 2003). The presence of SI/SS was mapped using MacClade 4.0. There is evidence for the presence of SI/SS in the following families: Nymphaeaceae (equivocal, see Sect. 3.2.1), Austrobaileyaceae, Illiciaceae, Trimeniaceae, Chloranthaceae, Winteraceae, Saururaceae, Lauraceae, Magnoliaceae, Annoneaceae, Magnoliaceae Degeneriaceae, Orchidaceae, Amaryllidaceae, Hemerocallidaceae, Iridaceae, Asphodelaceae, Liliaceae, Melanthiaceae, Poaceae, Bromeliaceae, Commelinaceae Pontederiaceae, Ranunculaceae, Papaveraceae, Nelumbonaceae, Chenopodiaceae, Caryophyllaceae, Polygonaceae, Plumbaginaceae, Olacaceae, Santalaceae, Saxifragaceae, Grossulariaceae, Myrtaceae, Onagraceae, Lythraceae, Passifloraceae Linaceae, Clusiaceae, Erythroxylaceae, Turneraceae, Oxalidaceae, Connaraceae, Fabaceae, Fagaceae, Nothofagaceae, Betulaceae, Rosaceae, Resedaceae, Brassicaceae, Capparidaceae, Malvaceae, Bixaceae, Sterculiaceae, Cistaceae, Ericaceae, Polemoniaceae, Primulaceae, Boranginaceae, Asclepidaceae, Rubiaceae, Loganiaceae, Gentainaceae, Rubiaceae, Bignoniaceae, Plantaginaceae, Gesneriaceae, Oleaceae, Acanthaceae, Solanaceae, Convolvulaceae, Campanulaceae, Menyanthaceae, Asteraceae

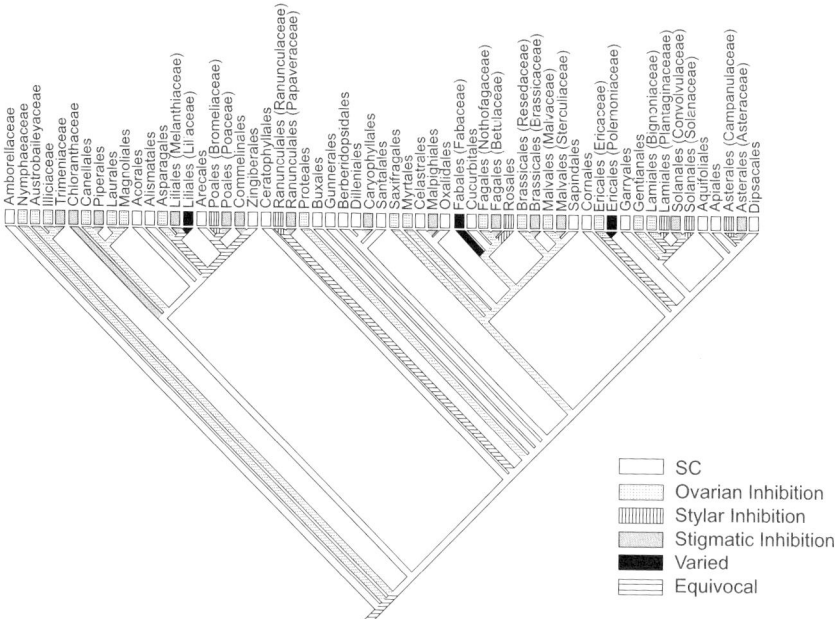

Fig. 4.2 Phylogenetic distribution of SI systems classified by the different sites of pollen tube inhibition in the pistil. Sites of pollen tube inhibition were predicted using evidence from all relevant literature. Heteromorphic SI systems were excluded from this analysis due to their characteristic variability in site of inhibition. Where there is evidence for different sites of inhibition in the same order, the order is differentiated at the family level, and the family indicated in brackets. Variation of pollen tube inhibition was seen in two families; Ovarian and stylar inhibition have been reported in different species of the Liliaceae (Dorken and Husband 1999; Ascher and Peloquin 1968), and different members of the Polemoniaceae exhibited ovarian and stigmatic inhibition (Levin 1993; Goodwillie 1997; LaDoux and Friar 2006). These families have been assigned a different character, as the ancestral condition is not known. The site of inhibition is indicated in the key

or SI was the ancestral angiosperm condition. The only basal lineage for which there is no suggestion of self-sterility/SI is *Amborella*, which is dioecious, but there is compelling evidence for self-sterility/SI in other members of the ANITA grade, Chloranthaceae, Magnoliids and Monocots. Furthermore, mapping of stigmatic SI in basal groups predicts that either this early-acting system of SI evolved independently in three different early diverging lineages, or it evolved once in their common ancestor and was subsequently lost by certain groups (Fig. 4.2). It is difficult to draw too many conclusions about the gain and loss of SI/self-sterility and SC among basal lineages, because of the incomplete sampling of taxa, but current data produces a phylogeny consistent with very early evolution and diversification of self-sterility/SI systems followed by multiple losses and gains of self-sterility/SI during angiosperm diversification.

More work is needed to define the genetic basis of the self-sterility and SI systems of basal angiosperms and future studies must take the form of quantitative comparisons (diallels) of self and cross pollen tube growth, and seed development following controlled cross and self-pollinations. This will be critical for distinguishing late acting SI systems and genetically determined self-sterility systems from the action of early acting inbreeding depression (Koehl et al. 2004).

For the remainder of this review we focus on the distribution of the main systems of SI for which genetic and/or cell biological data is available. By examining the phylogenetic distribution of each system type in turn, we will present an overall picture of the phylogenetic distribution of the different SI systems and their possible evolutionary relationships.

4.3.1 Late-Acting Ovarian Self Incompatibility (OSI)

Late-acting ovarian SI (OSI) occurs throughout the angiosperms, being present in nearly all major clades, from basal taxa to the more derived eudicot lineages (Fig. 4.3). The combination of self-pollen tube entry into the ovary followed by

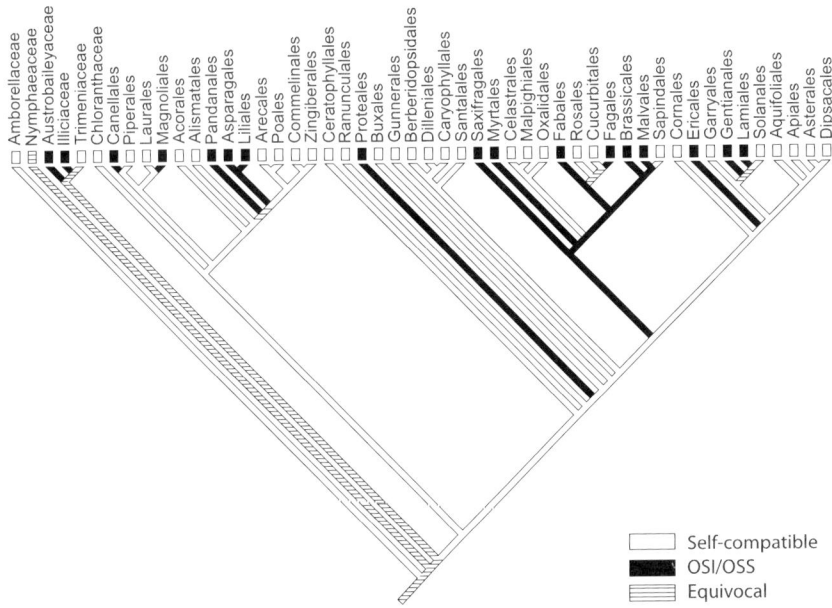

Fig. 4.3 Phylogenetic distribution of late-acting ovarian SI/SS systems within the angiosperms. Evidence of OSI/OSS has been reported in the following taxa: Nymphaeaceae, equivocal, see Sect. 3.2.1 (Nymphaleales); Austrobaileyaceae, Illiciaceae (Austobaileyales); Winteraceae (Canallales); Magnoliaceae (Magnoliales); Amaryllidaceae, Hemerocallidaceae, Asphodelaceae (Asparagales); Liliaceae (Liliales); Nelumbonaceae (Proteales); Grossulariaceae (Saxifragales); Myrtaceae, Onagraceae (Myrtales); Fabaceae (Fabales); Fagaceae, Nothofagaceae (Fagales); Resedaceae, Capparidaceae (Brassicales); Malvaceae, Malvaceae, Bixaceae, Sterculiaceae, Cistaceae (Malvales); Ericaceae, Polemoniaceae (Ericales); Asclepidaceae, Rubiaceae (Gentianales); Bignoniaceae (Lamiales).

little or no seed set after self-pollinations has been reported in woody perennial species from many different families, including Sterculiaceae (*Theobroma cacao, Sterculia chicha* and *Cola nitida*); Ericaceae (*Rhododendron sp.*); Fabaceae (*Acacia retinodes*); Winteraceae (*Pseudowintera colorata*); Myrtaceae (*Eucalyptus sp.*); Bignoniaceae (*Tabebuia sp.*); Malvaceae (*Chorisia sp.*) (Kenrick et al. 1986; Gibbs and Bianchi 1999; Sage and Sampson 2003). OSI has also been described in a few bulbous herbaceous taxa, including *Narcissus sp.* and *Hemerocallis sp.* (Amaryllidaceae), and *Lilium sp.* (Liliaceae) (de Nettancourt 1977).

The scattered distribution of late-acting OSI throughout the angiosperms suggests that these systems evolved independently on numerous occasions at different times during angiosperm diversification. It is less likely that their distribution reflects multiple losses of a common ancestral system (Sage et al. 1994). The presence of OSI and late-acting self-sterility in basal angiosperms suggests that these self-sterility systems were present in the first angiosperms and probably represent the basal state of SI. Indeed their phylogenetic distribution indicates that they predate stylar and stigmatic SI systems. This is consistent with current thinking, which views OSI as ancient, even predating the evolution of the closed carpel (Kenrick et al. 1986; Endress and Igersheim 2000; Endress 2001; Sage and Sampson 2003). These ancestral systems may then have evolved into or been superseded by more advanced stigmatic or stylar mechanisms of SI. This transformation could have occurred through a change in the timing and site of expression of the incompatibility gene(s) as the gynoecium increased in complexity (Sage et al. 1994).

Among the various documented cases of OSI, there is much variation in the timing and site of the pollen rejection reaction. OSI systems have been described as pre- or post-zygotic (Sage et al. 1994; Sage and Sampson 2003). Examples of pre-zygotic OSI occur in *Theobroma cacao*, where gametes fail to fuse in the ovule (Knight and Rogers 1955), and *Acacia retinodes*, where the pollen tube is arrested in the nucellus (Kenrick et al. 1986). Post-zygotic systems have been reported in *Rhododendron* species, where the ovule is fertilised, but the resulting zygote and endosperm is aborted (Williams et al. 1986), and *Gasteria* species, where self-fertilisation fails to stimulate ovule development (Sears 1937).

The genetic basis of most OSI systems is unknown but is frequently hypothesised to be gametophytic (Sage et al. 1994). Several attempts have been made to characterise the genetic control of late acting SI systems on the basis of the controlled pollination experiments. In *Acacia retinodes*, controlled crosses suggested gametophytic control (Knox and Kenrick 1983) and more extensive crossing experiments in *Theobroma cacao*, indicated joint gametophytic/sporophytic control, mediated by three unlinked loci (Cope 1958; Knight and Rogers 1955). Controlled crossing experiments in *Gasteria spp.* suggested control of OSI by two or more gametophytic loci (Naarborg and Willemse 1992). These limited studies suggest that OSI systems are most likely controlled by multiple loci. Nevertheless, more extensive controlled reciprocal crossing studies (diallels) are clearly needed in these systems to establish their genetic control unequivocally and rule out the potential effects of early acting inbreeding depression (Charlesworth 1985). This approach was taken in a study of the post-zygotic self-sterility in *Asclepias exalta* (Asclepiadaceae) (Lipow and

Wyatt 2000). Three diallel crosses were performed between full-sibs generated from self-sterile plants from one wild population. The plants segregated into four compatibility classes, indicating single locus control of self-sterility, but the authors were unable to assign either gametophytic or sporophytic control to the system (Lipow and Wyatt 2000).

Recently, OSI has been studied extensively in *Narcissus triandrus* (Hodgins and Barrett 2006, 2007). *N. triandrus* is unique in possessing trimorphic heterostyly and an unlinked late-acting self-sterility system (Sage et al. 1999). Examination of cross and self-pollen tube growth and ovule development following cross and self-fertilisation, revealed that the OSI mechanism is acting pre-zygotically. After self-fertilisation, ovules ceased to develop prior to pollen tube entry, but no difference was seen in pollen tube growth. This has been hypothesised to be due to the absence of stimuli required for normal ovule development, implying long-distance messaging between pollen tubes and developing ovules (Sage et al. 1999). Within the genus *Narcissus*, the diversity of mating systems indicates that mating system evolution is ongoing (Barrett and Harder 2005; Perez-Barrales et al. 2006), suggesting that the OSI system in *N. triandrus* may be of relatively recent origin. This supports phylogenetic predictions that different forms of late-acting OSI/self-sterility have evolved (and been lost) recurrently since they first appeared in the earliest angiosperms.

4.3.2 Gametophytic Self-Incompatibility (GSI)

GSI systems are the most abundant SI mechanism in the flowering plants. The phylogenetic distribution of GSI systems within the angiosperms illustrates the prevalence of this system in all major clades of the monocots and eudicots (Fig. 4.4). The striking mechanistic differences between the two GSI systems characterised at a molecular level, the *S*-RNase-based system of the Solanaceae, Plantaginaceae and Rosaceae, and the Ca^{2+}-dependent system of the Papaveraceae (reviewed by McClure and Franklin-Tong (2006); see also Chaps. 9–11, this volume), indicate that single-locus GSI has evolved at least twice. Phylogenetic analysis of *S*-RNase gene sequences (Igic and Kohn 2001; Steinbachs and Holsinger 2002; see Chap. 5, this volume) indicates that *S*-RNase based GSI is an ancient form of SI present in the common ancestor of the eurosids and euasterids (indicated in Fig. 4.4) and there fore ancestral to ~75% of eudicots. The absence of *S*-RNase-based GSI in so many eudicot orders indicates that it has been lost many times during eudicot diversification, while other forms of SI, most notably SSI, evolved subsequently. It will be interesting to establish whether homologues of *S*-RNase genes are present in GSI species of the core eudicots and monocots to determine whether the origin of this SI system can be traced back even further.

The Ca^{2+}-based GSI system identified in *Papaver rhoeas* (poppy) has yet to be identified in any other plant species, and so is currently unique to the Papaveraceae (Ranunculales, see Fig. 4.4). As this molecular mechanism is so different from the

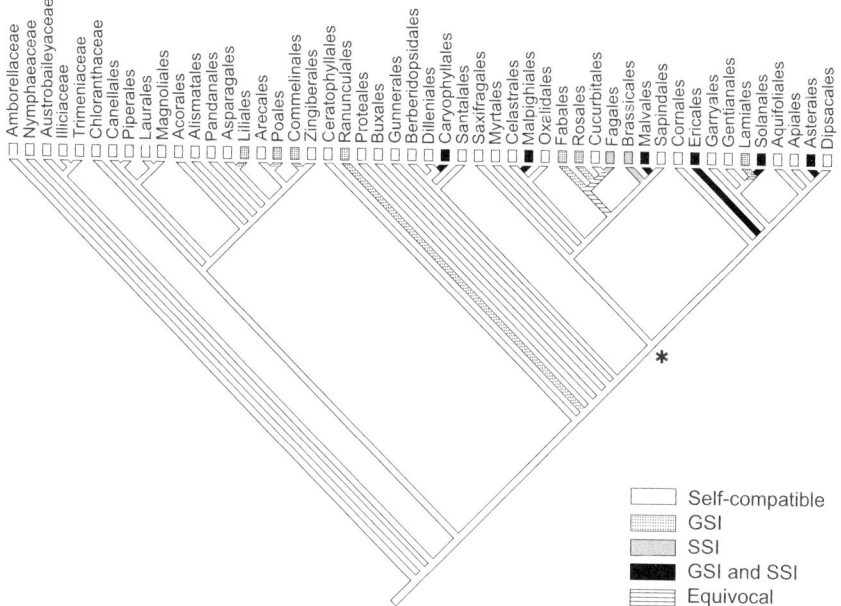

Fig. 4.4 Phylogenetic distribution of homomorphic GSI and SSI systems within the angiosperms The predicted point of evolution of RNase-based GSI is indicated by a star. Gametophytic control of SI has been reported in the following families: Liliaceae (Liliales); Poaceae (Poales); Commeliniaceae (Commelinales); Ranunculaceae, Papaveraceae (Ranunculales); Chenopodiaceae (Caryophyllales); Passifloriaceae (Malphigiales); Fabaceae (Fabales); Rosaceae (Rosales); Sterculiaceae, Cistaceae (Malvales); Ericaceae, Polemoniaceae (Ericales); Plantaginaceae (Lamiales); Solanaceae (Solanales); Campanulaceae (Asterales). Evidence of sporophytic control of SI has been found in the following families: Caryophyllaceae (Caryophyllales), Passifloraceae (Malphigiales), Betulaceae (Fagales), Brassicaceae (Brassicales), Sterculiaceae (Malvales), Polemoniaceae (Ericales), Convolvulaceae (Solanales) and Asteraceae (Asterales)

S-RNase-based system, it is highly unlikely that the two GSI systems are related. Diversification of the Ranunculales occurred early in the evolution of the eudicots, making this order the most basal eudicot lineage (Worberg et al. 2007). This diversification predates the origin of the common ancestor of the Solanaceae, Rosaceae and Plantaginaceae. This suggests that the Papaveraceae-type GSI system most probably predates S-RNase-based GSI and may represent the ancestral state of SI in eudicots, although alternatively it may be unique to the Papaveraceae, in which case it probably evolved just once. Molecular-based studies in basal eudicots and monocots with GSI (and SI basal angiosperms like *Trimenia*) would be extremely useful in helping to resolve the evolutionary relationships between these two distinct GSI systems.

A multi-locus GSI system has been described in three species of *Ranunculus* (Østerbye 1977; Lundqvist 1990a), a genus in the Ranunculales and so related to poppy. The physiology of this system is unknown, but incompatibility is stigmatic, making it a good system in which to explore the possibility of a Ca^{2+}-based GSI in another member of the Ranunculales. Stigmatic GSI has also been described in the monocot genus *Tradescantia* (Owens and McGrath 1984; see Sect. 3.3.2) and other basal species, like the monocot *Trillium erectum*, and basal angiosperms *Trimenia* and *Saururus* exhibit stigmatic SI that is believed to be under gametophytic control. It would be extremely interesting to determine whether these SI systems have the same GSI mechanism as poppy, or whether they represent yet further diversity among GSI systems.

In addition to the grasses and *Ranunculus*, GSI controlled by multiple loci has also been described in the Chenopodiaceae [*Beta vulgaris* (Lundqvist et al. 1973; Larsen 1978)], Liliaceae [*Lilium martagon* (Lunqvist 1991)] and Fabaceae [*Lotus tenuis* (Lundqvist 1993)], where, as with *Ranunculus* spp., GSI is controlled by four linked loci, which must be matched in the pollen and pistil for an incompatibility reaction to occur. Mating is therefore possible between close relatives that share three out of four alleles. For this reason multi-locus systems are thought to represent a more primitive and less efficient type of GSI from which single-locus GSI systems may be derived (Lundqvist, 1990a).

4.3.3 Sporophytic Self-Incompatibility (SSI)

4.3.3.1 Homomorphic SSI

Homomorphic SSI is much less widespread than GSI, and appears to be restricted to the eudicots (Fig. 4.4). SSI typically occurs in families at the ends of branches in the more advanced eudicot lineages, indicating that it is of more recent origin than GSI. Homomorphic SSI has been confirmed in six eudicot families: Asteraceae, Betulaceae, Brassicaceae, Caryophyllaceae, Convolvulaceae and Polemoniaceae (reviewed in Hiscock and McInnis 2003; Hiscock and Tabah 2003). It has also been reported in the Passifloraceae, Malvaceae and Sterculiaceae (Hiscock and McInnis 2003; Suassuna et al. 2003; Hiscock and Tabah 2003). However, it is not known how many mechanistically different types of SI systems are acting in these taxa because molecular data is only available for the Brassicaceae *S*-locus determinants.

Like the *Papaver* GSI system, *Brassica* SSI is controlled by a ligand/receptor interaction, but the ligand (SCR/SP11) is carried by the pollen grain and the receptor (SRK) resides in the epidermal (papilla) cells of the stigma (Takayama et al. 2001 and Chaps. 7 and 8, this volume). Attempts to identify the *Brassica*-type SSI system in other angiosperm families have so far proved fruitless, suggesting that, as with GSI, SSI operates through a variety of different molecular mechanisms. SRK-like sequences have been identified in representative SSI species from both the Convolvulaceae (*Ipomoea trifida*, see (Kowyama et al. 2000; Chap. 12, this volume) and Asteraceae (*Senecio squalidus*) (Tabah et al. 2004), where they are considered unlikely to play a role in SI because they are not linked to the *S*-locus. SSI in the Convolvulaceae and Asteraceae therefore appear to utilise a different molecular mechanism to the Brassicaceae, but it remains to be determined whether a common mechanism of SSI is present in *Ipomoea* and *Senecio*.

4.3.3.2 Heteromorphic SSI

Heteromorphic SSI has a scattered distribution among different angiosperm lineages, being recorded in 18 orders, 25 families and 155 genera (Barrett 1992). This suggests that heteromorphic SSI has evolved on numerous occasions during angiosperm diversification. Most of these taxa are distylous, with tristyly found in just four families: Lythraceae, Oxalidaceae, Pontederiaceae and Connaraceae (de Nettancourt et al. 2001; see Chap. 1).

It is unknown whether heteromorphic SI evolved from a homomorphic SSI system or independently from a self-compatible ancestor. It is likely that they have independent origins, as they consistently occur in different families. One exception to this rule is the distylous species *Melochia tomentosa*, which belongs to the Sterculiaceae; a family also containing *T. cacao,* a homomorphic species with late-acting ovarian SI reported to be under joint sporophytic/gametophytic control (Martin 1967; Cope, 1958). Interestingly, in heterostylous *Narcissus triandrus*, the SI reaction operates through a late-acting ovarian self-sterility system, the genetic control of which is unknown, but is unlikely to be a simple sporophytic system (Sage et al. 1999).

Compared to homomorphic SSI less is known about the physiology of heteromorphic SSI. The site of the incompatibility reaction can be the stigma, the style or even the ovary, depending on the species and morph type (de Nettancourt et al. 2001). This variation has been proposed to be the result of recognition molecules being present within the pollen grain and pollen tube (Stevens and Murray 1982). Several floral characters have been correlated with the site of pollen inhibition, including pollen morphology, stigma type (typically wet in thrum flowers and dry in pin flowers) and style type (solid or hollow). Stigmatic rejection has been reported to occur on wet stigmas, a feature that has not been detected in homomorphic SSI (de Nettancourt et al. 2001). Genes that regulate heteromorphic SSI have yet to be identified in any species although concerted efforts to identify morph-specific genes are ongoing in *Primula* (McCubbin et al. 2006), *Turnera* (Shore et al. 2006) and *Fagopyrum* (Milijuš-Đukić et al. 2004, 2007; see Chaps. 1 and 14, this volume, for further details).

4.4 The Relationship Between GSI and SSI

The phylogenetic distribution of GSI and SSI systems indicates that gametophytic systems predate sporophytic systems. This poses interesting questions about the relationship between the two types of SI. Molecular evidence suggests that SSI systems are unlikely to have evolved from GSI systems due to the strikingly different molecular mechanisms characterised to date. SSI therefore most likely evolved in SC lineages derived from GSI ancestors of SSI systems and became superimposed upon existing GSI systems. Evidence of joint gametophytic/sporophytic SI systems, especially in species usually considered exclusively SSI (reviewed by Lewis 1994), suggests that SSI can evolve and be maintained in the presence of cryptic GSI.

There is evidence of cryptic GSI acting in the Brassicaceae, Asteraceae and most recently the Passifloraceae (Lewis et al. 1988; Hiscock 2000; Lewis 1994, Saussuna et al. 2003). Crossing anomalies have consistently been reported in SSI species of these families, where a cross between individuals sharing a particular *S*-haplotype is compatible when it is predicted to be incompatible (Lewis 1994; Hiscock 2000). Lewis et al. (1988) proposed and provided conclusive genetic evidence that these anomalous compatibilities arise due to the action of a second gametophytic locus (*G*) acting alongside the sporophytic system, and further suggested that the *G*-gene was likely to be a feature of all sporophytic systems. This theory is strengthened by the absence of anomalous cross results in GSI systems, compared to the 2–10% anomalous cross results usually encountered with genetic analyses of SSI (Suassuna et al. 2003). In the Brassicaceae, recombination tests indicated that the *G*-locus and the *S*-locus are linked, with recombination rates ranging from ~10 to ~28% (Lewis 1994). For an incompatible reaction to occur, both *G*- and *S*-alleles must match (although dominance of *S*-alleles may affect this). In *Passiflora edulis*, reciprocal crossing experiments indicate dual action of the *S* and *G*-loci in SI, and also implicate the involvement of further loci in SI (do Rêgo et al. 1999; Suassuna et al. 2003).

The *G*-gene resides at a diallelic gametophytic locus that usually exists in a homozygous state (Lewis 1994). In this state the *G*-gene is silent, and does not affect the incompatibility response. However, when present in a heterozygous state, different *G*-alleles may affect the outcome of a cross by enabling a compatible reaction to occur when incompatibility is expected. In this way, possession of a cryptic GSI system would increase the mating potential of SSI individuals with shared *S*-alleles (Hiscock 2000; Brennan et al. 2003). Another indicator of the presence of a cryptic GSI system is variation in the site of the incompatibility reaction. In SSI species such as *Senecio squalidus* (Asteraceae), pollen development is most commonly arrested at the stigma surface, but can penetrate the stigma and grow through stigmatic tissue before being arrested. This later-acting response occurs only between individuals sharing an *S*-allele, and never in self-pollinations (Hiscock 2000; Hiscock et al. 2002) and may reflect the action of the cryptic *G*-gene (Lewis 1994).

Other examples of joint gametophytic/sporophytic control of SI include *Theobroma cacao* (Sterculiaceae), *Cerastium arvense* (Caryophyllaceae) and *Stellaria holostea* (Caryophyllaceae) (Cope 1958; Knight and Rogers 1955; Lundqvist, 1990b, 1994). The late-acting OSS system of *T. cacao* (see Sect. 3.3.1) is hypothesised to be controlled by three unlinked loci: *A*, *B* and *S*, where *A* and *B* act as accessory loci to *S* (Knight and Rogers 1955; Cope 1962). In this system incompatibility factors determine both the pollen and ovule phenotype, meaning that the interaction occurs directly between gametophytes, not as in other systems between the male gametophyte (pollen) and the sporophyte (pistil). The *T. cacao S*-locus is multi-allelic, and alleles show dominance relationships, as seen in SSI systems (de Nettancourt 2001). There is evidence of gametophytic control of the pollen phenotype in the sporophytic system of the Caryophyllaceae (Lundqvist 1990b, 1994). This system is thought to reflect the action of a single gene in transition between gametophytic and sporophytic control (de Nettancourt 2001).

4.5 Discussion

The diversity of SI systems in flowering plants is reflected in differences in the genetic control, the numbers of controlling loci and differences in the site and timing of the incompatibility response. These differences predict variation in the molecular mechanisms of SI, a prediction born out by molecular characterisation of S-genes in species from the Solanaceae, Papaveraceae and Brassicaceae. More recent molecular analyses of SI in species of Poaceae, Convolvulaceae and Asteraceae predict further novel molecular mechanisms of SI (Hiscock and McInnis 2003). The evolution of such diversity in the modes of recognition and rejection of self-pollen is likely to be a consequence of intense selection for more effective mechanisms to prevent self-fertilisation and promote outcrossing in co-sexual flowers (Hiscock and McInnis 2003).

As well as evolving de novo many times during angiosperm diversification, SI systems have clearly been lost many times, as shown by the widespread occurrence of SC in all major angiosperm lineages. Subsequent evolution of new SI systems must then have occurred in many lineages where former SI systems once existed. This is certainly true for the origin of SSI in lineages whose common ancestor possessed S-RNase-based GSI (Igic and Kohn 2001; see also Chap. 5, this volume). The frequent loss of SI during angiosperm diversification is emphasised in large families that possess a mixture of SI and SC lineages. The Asteraceae, the second largest angiosperm family, contains approximately 1,620 genera and 22,750 species (APG II 2003), of which 63% are estimated to be SI (presumably all SSI), with the remaining species possessing a mixture of pseudo-self-compatibility (PSC, 10%) and SC (27%) (Ferrer and Good-Avila 2007). This widespread occurrence of SI suggests that it is ancestral in this family. Using a phylogenetic approach, Ferrer and Good-Avila (2007) attempted to model the evolution of SI in the Asteraceae and, while they could not determine whether SI or SC was the ancestral state, they showed that neither SI nor SC is a terminal state, but part of a flexible and dynamic evolutionary process. This dynamic process can be seen in action in the genus *Senecio*, where hybridisation, polyploidisation and other speciation events have had a profound impact on SI, often resulting in the loss of SI (Abbott et al. in press; Hegarty et al. 2006) and possibly also the gain of SI. Indeed, de novo evolution of SI from SC has never been described, but in a recent study of seed set and cross-compatibility in resynthesised lines of SC allohexaploid *Senecio cambrensis*, fully SSI individuals were identified segregating out of second and third generation SC lines (Brennan and Hiscock, in preparation). The original SC *S. cambrensis* lines were generated by colchicine-treatment of an F1 *S. vulgaris* (SC) x *S. squalidus* (SI) hybrid and all were fully self-fertile, indicating the SI machinery present in the *S. squalidus* genome had been suppressed in the hybrid, only to be re-activated when parental traits segregated out in the F2 and F3 generations. Whilst this is clearly not de novo evolution of SI, it illustrates how hybridisation and polyploidy might impact on the loss and gain of SI.

When compared with the diversity of SI systems between families, the widespread conservation of SI mechanisms within families is striking, indicating that within most families SI is monophyletic (Charlesworth 1985; Weller et al. 1995). Nevertheless, possession of a single SI system is not true of every family. At least two different SI systems have been described in the Fabaceae, Liliaceae, Solanaceae, Primulaceae and Polemoniaceae. In the Fabaceae, the one-locus GSI system of *Trifolium* is thought to be derived from the four-locus system of *Lotus tenuis* (Lundqvist 1993). Ovarian SI is widespread in this family (Gibbs and Bianchi 1999), but stigmatic inhibition of pollen is also seen in *L. tenuis* (Lundqvist 1993). A similar theory has been proposed for an evolutionary link between the four-locus system of *Lilium martagon* and the single-locus GSI of *Lilium longiflorum* (Lundqvist 1991).

In addition to the well-studied single-locus *S*-RNase-based GSI system of the Solanaceae, a two-locus GSI system has been reported in *Physalis ixocarpa* (Pandey 1957) and a late-acting self-sterility system has been described *Lycium cestroides* (Aguilar and Bernardello 2001), although only the former has been well-characterised genetically. Like the two-locus GSI system of the Poaceae, the two loci of *Physalis ixocarpa* act independently but, unlike the Poaceae system, only one allele needs to match in the pollen and style for incompatibility to occur (Pandey 1957). It is thought that the *Physalis* two-locus system is the result of a duplication of the *S*-locus (Pandey 1957, 1980). Given the apparent ubiquity of the *S*-RNase-based system in this family, it would be extremely interesting to explore the role (if any) of *S*-RNases in these two species, particularly as the stylar incompatibility response in *P. ixocarpa* appears identical to that seen in other GSI species in the Solanaceae, (Pandey 1957).

Although heteromorphic SSI is present in the majority of species within the Primulaceae, and might therefore be safely assumed to be the only form of SI operating in this family, rigorous genetic diallel analyses identified a single-locus GSI system in *Anagallis monelli* (Talavera et al. 2001). This finding poses intriguing questions about the evolution of SI systems and offers an opportunity to explore evolutionary shifts of SI within a group of closely related species.

Of all angiosperm families investigated to date, the Polemoniaceae possesses the greatest diversity of SI/self-sterility systems. Stigmatic inhibition of self-pollen is a feature of three species: *Polemonium viscosum* (Galen et al. 1989), *Phlox drummondii* (Levin 1975) and *Linanthus parviflorus* (Goodwillie 1997), while *Ipomopsis tenuifolia* and *I. aggregata* possess a late-acting OSI mechanism (LaDoux and Friar 2006; Sage et al. 2006). Differences are also seen in the genetic control of SI in these species, with gametophytic control in *P. drummondii* (Levin 1975), sporophytic control in *L. parviflorus* (Goodwillie 1997) and multi-locus control in *I. tenuifolia* (LaDoux and Friar 2006). Several theories have been proposed for the existence of such diversity within the Polemoniaceae. The different SI/self-sterility systems could have evolved independently in different taxa, after diversification of the family (Barrett et al. 1996). Alternatively, an ancestral GSI system has been proposed to exist, underlying the more recent evolution of SSI in some taxa (Goodwillie 1997). A third theory extends this proposal further to specify the ancestral condition as a late-acting ovarian system, which may still be present in species showing stigmatic SI (Sage et al. 2006). The dynamic evolution of SI in this family provides a unique model system in which to study the relatively recent evolutionary relationships of the three basic forms of angiosperm SI as an allegory to the evolution of SI systems in flowering plants generally.

Phylogenetic studies of *S*-genes in the most well characterised SI systems have allowed conclusions to be drawn about the evolutionary history of these gene sequences. The reader is referred to Chaps. 3, 5 and 6 of this volume. *S*-RNases have been inferred to be extremely ancient, originating before the divergence of the rosids and asterids, making this form of GSI ancestral to ~75% of eudicots (Igic and Kohn 2001). The origin of SRK sequences in the Brassicaceae has been traced back to before the divergence of the *Brassica/Raphanus* and *Arabidopsis/Capsella* lineages, approximately 12.2–19.5 mya (Paetsch et al. 2006), but is probably at least 50 mya (Uyenoyama 1995). Further dating is constrained by the failure, so far, to identify the SRK/SCR SSI system in a family other than the Brassicaceae. Similarly, phylogenetic analysis of the *Papaver S*-pistil determinant is constrained because no studies have been carried out to date to examine whether the *Papaver S*-determinant sequences are present in other species, other than very limited studies within the Papaveraceae. Future studies of GSI in other species in the Ranunculales may shed light on the origin of the *Papaver* system. Interestingly, homologues of the poppy *S*-protein gene have been identified in *Arabidopsis* (Ride et al. 1999), where they appear to be involved in diverse cell signalling processes, indicating that like the *S*-RNase/SLF and SRK/SCR genes they are members of a multigene family, the origins of which are likely to predate a role in SI.

The diverse nature of angiosperm SI systems poses interesting questions about their evolutionary relationships. For instance, did they all arise de novo or did one system evolve into another? Molecular studies have so far provided no evidence for any links between the *S*-proteins that mediate self-recognition, even though commonalities exist between the downstream events leading up to pollen tube inhibition, for example kinase-mediated phosphorylation, Ca^{2+} signalling, actin depolymerisation, and involvement of F-boxes and ubiquitination (see reviews by Takayama and Isogai 2005; McClure and Franklin-Tong 2006; Chaps. 7–11, this volume). Although there is strong evidence for a polyphyletic origin of SI, the exact origin of each system is unknown. It is possible that at least one system may have been possessed by the first angiosperms. We still know little about the genetic basis of SI and self-sterility observed in extant basal angiosperms, and whether they share a common mechanism or mechanisms (Sage et al. 1994). If a genetically determined form of SI was present in the first angiosperms, it is most likely to have been a form of late-acting OSI/self-sterility, as this is the most common form of self-sterility detected among extant basal angiosperms (see Sect. 4.2.1). Indeed, because the first angiosperms are unlikely to have possessed a true stigma and style, or a closed ovary, any incompatibility reaction would presumably be centred in ovarian tissues or within the ovule.

Despite forming an effective barrier to seed set from self-fertilisation, late-acting ovarian SI/self-sterility systems also impose a cost on the plant, by often removing the chance of seed set from a cross-pollination – ovule discounting (Dorken and Husband 1999). By contrast, stylar and stigmatic SI systems are more efficient as outbreeding systems, because incompatible pollen tubes are halted at an earlier stage, thereby allowing compatible pollen tubes unimpeded opportunity to fertilise ovules. Indeed, the presence of stigmatic SI in Trimeniaceae suggests that selection favoured the evolution of pre-zygotic SI early during angiosperm diversification. Stylar incompatibility is typically slow, but occurs throughout a large amount of tissue, and compatible pollen tubes reach tissue free of arrested pollen tubes in the lower part of the style and in the ovary. Stigmatic SI completely eliminates the 'clogging' effect of incompatible pollen tubes in the gynoecium, and so is typically thought of as the most efficient and most advanced form of SI, because it prevents wastage of female resources on self-pollen tubes (Sage et al. 1994).

Our phylogenetic analysis of SI/self-sterility is consistent with that of Weller et al. (1995) in supporting a polyphyletic origin of SI in angiosperms. Our study, however, unlike that of Weller et al. (1995), predicts self-sterility or SI rather than SC as the most likely mating system of the earliest angiosperms. This difference between the two studies is a consequence of the greater resolution at the base of current angiosperm phylogenetic trees (APG II 2003), and greater availability of data on mating systems in basal groups. Other findings are largely concordant with those of Weller et al. (1995). To summarise, therefore, our findings show that (1) Self-sterility and SI are well represented among basal angiosperm lineages, including the ANITA grade, Chloranthaceae and Magnoliids; (2) late-acting ovarian SI/self-sterility is likely to be the basal state of SI in angiosperms; (3) GSI systems evolved independently in the earliest diverging lineages of monocots and eudicots; (4) heteromorphic SSI evolved on numerous different occasions during diversification of the eudicots and perhaps once in monocots and (5) different forms of homomorphic SSI evolved independently in eudicots relatively recently.

This survey, like that of Weller et al. (1995), has been confounded by our fragmentary knowledge of the occurrence of SI. Despite the availability of more detailed studies of SI/self-sterility in basal angiosperms, there are still many major angiosperm families for which little or no data on SI/self-sterility exists. More basic studies of incompatibility are needed using simple controlled cross and self-pollinations, and once self-sterility is confirmed, detailed cytological and genetic studies are needed to determine the site of incompatibility and its genetic control. For many groups this is no easy task because SI systems are often not easy to identify and define. Cryptic, partial and leaky SI systems have all been shown to exist in addition to the main types already discussed (Bowman 1987; Stephenson et al. 2000; Goodwillie et al. 2005; see also Chap. 2, this volume). Nevertheless, more and more complete studies of novel plant mating systems are appearing in the literature and a clearer picture is beginning to emerge of the evolution of SI. The ancestral state of the angiosperms can no longer be thought of as self-compatibility, because emerging data from studies of basal angiosperms strongly suggests that late-acting ovarian SI/self-sterility was present among the first hermaphrodite angiosperms. A major goal in the coming years will be to characterise these poorly understood systems of SI in more detail to determine their genetic control and the nature of the molecules involved in regulating the incompatibility response.

Acknowledgements We thank Patricia Sanchez-Baracaldo for help with using MacClade 4.0 and Noni Franklin-Tong, Matthew Hegarty, Adrian Brennan and Christopher Thorogood for useful comments on earlier versions of this manuscript. Work in SJH's lab is supported by the Natural Environment Research Council (NERC), the Leverhulme Trust and the Lady Emily Smyth Research Station (University of Bristol).

References

Abbott RJ, Brennan AC, James JK, Forbes DG, Hegarty MJ, Hiscock SJ (in press) Recent hybrid origin and invasion of the British isles by a self-incompatible species, Oxford ragwort (*Senecio squalidus* L., Asteraceae). Biol Invasions

Aguilar R, Bernardello G (2001) The breeding system of *Lycium cestroides*: A Solanaceae with ovarian self-incompatibility. Sex Plant Reprod 13:273–277

Angiosperm Phylogeny Group (2003) An update of the Angiosperm Phylogeny Group classification for the orders and families of flowering plants. Bot J Linn Soc 141:399–436

Annerstedt I, Lundqvist A (1967) Genetics of self-incompatibility in *Tradescantia paludosa* (Commelinaceae). Hereditas 58:13–30

Ascher PD, Peloquin SJ (1968) Pollen tube growth and incompatibility following intra- and interspecific pollinations in *Lilium longiflorum*. Am J Bot 55:1230–1234

Balthazar M, Endress PK (1999) Floral bract function, flowering process and breeding systems of *Sarcandra* and *Chloranthus* (Chloranthaceae). Plant Syst Evol 218:161–178

Barrett SCH (1992) Heterostylous genetic polymorphisms: Model systems for evolutionary analysis. In: Barrett SCH (ed) Evolution and function of heterostyly. Springer, Berlin Heidelberg New York, pp 1–29

Barrett SCH (1998) The evolution of mating strategies in flowering plants. Trends Plant Sci 3:335–341

Barrett SCH, Cruzan MB (1994) Incompatibility in heterostylous plants. In: Williams EG, Clarke AE, Knox RB (eds) Advances in cellular and molecular biology of plants. Genetic control of self incompatibility and reproductive development in flowering plants, vol 2. Kluwer, Dordrecht, pp 189–219

Barrett SCH, Harder LD (2005) The evolution of polymorphic sexual systems in daffodils (Narcissus). New Phytol 165:45–53

Barrett SCH, Harder LD, Worley AC (1996) The comparative biology of pollination and mating in flowering plants. Phil Trans R Soc Lond B 351:1271–1280

Bateman AJ (1952) Self-incompatibility in angiosperms: I. Theory. Heredity 6:285–310

Bernhardt P, Thien LB (1987) Self-isolation and insect pollination in the primitive angiosperms: New evaluations of older hypotheses. Plant Syst Evol 156:159–176

Bernhardt P, Sage TL, Weston P, Azuma H, Lam M, Thien LB, Bruhl J (2003) The pollination of *Trimenia moorei* (Trimeniaceae): Floral volatiles, insect/wind pollen vectors, and stigmatic self-incompatiblity in a basal angiosperm. Ann Bot 92:1–14

Bianchi M, Vesprini J, Barrett SCH (2000) Trimorphic incompatibility in Eichhornia azurea (Pontederiaceae). Sex Plant Reprod 12:203–208

Borba EL, Semir JO, Shepherd GJ (2001) Self-incompatibility, Inbreeding Depression and Crossing Potential in Five Brazilian *Pleurothallis* (Orchidaceae) Species. Ann. Bot 88:89–99

Bowman RN (1987) Cryptic Self-Incompatibility and the Breeding System of Clarkia unguiculata (Onagraceae). Am J Bot 74:471–476

Boyes DC, Chen C, Tantikanjana T, Esch JJ, Nasrallah JB (1991) Isolation of a second *S*-locus-related cDNA from *Brassica oleracea:* Genetic relationships between the *S*-locus and two related loci. Genetics 127:221–228

Brennan AC, Harris SA, Hiscock SJ (2003) Population genetics of sporophytic self incompatibility in *Senecio squalidus* L. (Asteraceae) II: A spatial autocorrelation approach to determining mating behaviour in the presence of low *S*-allele diversity. Heredity 91:502–509

Brewbaker, JL (1957) Pollen cytology and self-incompatibility systems in plants. J Hered 48:271–277

Charlesworth D (1985) Distribution of dioecy and self-incompatibility in angiosperms. In: Greenwood J, Slatkin M (eds) Evolution - essays in honour of John Maynard Smith. Cambridge University Press, Cambridge, pp 237–268

Charlesworth D, Vekemans X, Castric V, Glémin S (2005) Plant self-incompatibility systems: A molecular evolutionary perspective. New Phytol 168:61–69

Cock JM, Swarup R, Dumas C (1997) Natural antisense transcripts of the *S*-locus receptor kinase gene and related sequences in *Brassica oleracea*. Mol Gen Genet 255:514–524

Cope FW (1958) Incompatibility in *Theobroma cacao*. Nature 181:279

Cope FW (1962) The mechanism of pollen incompatibility in *Theobroma cacao*. Heredity 17:157–182

Crane PR, Lidgard S (1989) Angiosperm diversification and paleolatitudinal gradients in cretaceous floristic diversity. Science 246:675–678

Crepet WL (1985) Advanced (constant) insect pollination mechanisms: Patterns of evolution and implications vis-à-vis angiosperm diversity. Ann Mo Bot Gard 71:607–630

Darwin (1862) On the two forms, or dimorphic condition, in the species of *Primula* and on their remarkable sexual relation. J Linn Soc Bot 6:77–96

de Nettancourt D (1977) Incompatibility in angiosperms. Springer, Berlin Heidelberg New York

de Nettancourt D (2001) Incompatibility and incongruity in wild and cultivated plants. Springer, Berlin Heidelberg New York

Dickinson H (1995) Dry stigmas, water and self-incompatibility. Sex Plant Reprod 8:1–10

do Rêgo MM, Bruckner CH, da Silva EAM, Finger FL, de Siqueira DL, Fernandes AA (1999) Self-incompatibility in passion fruit: Evidence of two locus genetic control. Theor Appl Genet 98:564–568

Dorken ME, Husband BC (1999) Self-Sterility in the Understory Herb Clintonia borealis (Liliaceae). Int J Plant Sci 160:577–584

East EM (1940) The distribution of self-sterility in the flowering plants. Proc Am Philos Soc 82:449–518

Elleman CJ, Dickinson HG (1999) Commonalities between pollen/stigma and host/pathogen interactions: Calcium accumulation during stigmatic penetration by *Brassica oleracea* pollen tubes. Sex Plant Reprod 12:194–202

Endress PK (2001) The Flowers in Extant Basal Angiosperms and Inferences on Ancestral Flowers. Int J Plant Sci 162:1111–1140

Endress PK, Igersheim A (2000) Gynoecium Structure and Evolution in Basal Angiosperms. Int J Plant Sci 161:S211–S223

Ferrer MM, Good-Avila SV (2007) Macrophylogenetic analyses of the gain and loss of self-incompatibility in the Asteraceae. New Phytol 173:401–414

Friis EM, Pedersen KR, Crane PR (2000) Reproductive structure and organization of basal Angiosperms from the early Cretaceous (barremian or aptian) of western Portugal. Int J Plant Sci 161:S169–S182

Galen C, Gregory T, Galloway LF (1989) Costs of self pollination in a self-incompatible plant, *Polemonium viscosum*. Am J Bot 76:1675–1680

Gibbs PE (1986) Do Homomorphic and Heteromorphic Self-Incompatibility Systems Have the Same Sporophytic Mechanism? Plant Syst Evol 154:285–323

Gibbs PE, Bianchi MB (1999) Does Late-acting Self-incompatibility (LSI) Show Family Clustering? Two More Species of Bignoniaceae with LSI: *Dolichandra cynanchoides* and *Tabebuia nodosa*. Ann Bot 84:449–457

Gibson JP, Wheelright NT (1996) Mating systems of *Ocotea tenera* (Lauraceae), a gynodioecious tropical tree. Am J Bot 83:890–894

Good-Avila SV, Stephenson AG. (2002) The inheritance of modifiers conferring self-fertility in the partially self-incompatible perennial, *Campanula rapunculoides* L. (Campanulaceae). Evolution 56:263–272

Goodwillie C (1997) The genetic control of self-incompatibility in *Linanthus parviflorus* (Polemoniaceae). Heredity 79:424–432

Goodwillie C, Kalisz S, Eckert CG (2005) The evolutionary enigma of mixed mating systems in plants: Occurrence, theoretical explanations, and empirical evidence. Annu Rev Ecol Evol Syst 36:47–79

Hagman M (1975) Incompatibility in forest trees. Proc R Soc Lond B 188:313–326

Hasenstein KH, Zavada MS (2001) Auxin modification of the incompatibility response in *Theobroma cacao*. Physiol Plant 112:113–118

Hegarty MJ, Barker GL, Wilson ID, Abbott RJ, Edwards KJ, Hiscock SJ (2006) Transcriptome shock after interspecific hybridization in *Senecio* is ameliorated by genome duplication. Current Biol 16:1652–1659

Heslop-Harrison Y (1975) Incompatibility and the pollen stigma interaction. Annu Rev Plant Physiol 26:403–425

Heslop-Harrison Y, Shivanna KR (1977) The receptive surface of the Angiosperm stigma. Ann. Bot. 41:1233–1258

Hiratsuka S, Tezuka T, Yamamoto Y (1983) Use of longitudinally bisected pistils of *Lilium longiflorum* for studies on self-incompatibility. Plant Cell Physiol 24:765–768

Hiscock SJ (2000) Self-incompatibility in Senecio squalidus L. (Asteraceae). Ann Bot 85:181–190

Hiscock SJ, Dickinson HG (1993) Unilateral incompatibility within the Brassicaceae: Further evidence for the involvement of the self-incompatibility *S*-locus. Theor Appl Genet 86:744–753

Hiscock SJ, Kues U (1999) Cellular and molecular mechanisms of sexual incompatibility in plants and fungi. Int Rev Cytol 193:165–295

Hiscock SJ, McInnis SM (2003) The diversity of self-incompatibility systems in flowering plants. Plant Biol 5:23–32

Hiscock SJ, Tabah DA (2003) The different mechanisms of sporophytic self-incompatibility. Phil Trans R Soc Lond B 358:1037–1045

Hiscock SJ, Hoedemaekers K, Friedman WE, Dickinson HG (2002) The stigma surface and pollen-stigma interactions in *Senecio squalidus* L. (Asteraceae) following cross (compatible) and self (incompatible) pollinations. Int J Plant Sci 163:1–16

Hiscock SJ, McInnis SM, Tabah DA, Henderson CA, Brennan AC (2003) Sporophytic self-incompatibility in Senecio squalidus L. (Asteraceae) - the search for S. J Exp Bot 54:169–174

Hodgins KA, Barrett SCH (2006) Mating patterns and demography in the tristylous daffodil *Narcissus triandrus*. Heredity 96:262–270

Hodgins KA, Barrett SCH (2007) Population structure and genetic diversity in tristylous *Narcissus triandrus*: Insights from microsatellite and chloroplast DNA variation. Mol Ecol 16:2317–2332

Hodgkin T, Lyon GD, Dickinson HG (1988) Recognition in Flowering Plants: A Comparison of the Brassica Self-Incompatibility System and Plant Pathogen Interactions. New Phytol 110:557–569

Hristova K, Lam M, Feild T, Sage TL (2005) Transmitting tissue ECM distribution and composition, and pollen germinability in *Sarcandra glabra* and *Chloranthus japonicus* (Chloranthaceae). Ann Bot 96:779–791

Igic B, Kohn JR (2001) Evolutionary relationships among self incompatibility RNases. Proc Natl Acad Sci USA 98:13167–13171

Kao TH, McCubbin AG (1996). How flowering plants discriminate between self and non-self pollen to prevent inbreeding. Proc Natl Acad Sci USA 93:12059–12065

Kenrick J, Kaul V, Williams EG (1986) Self-incompatibility in *Acacia retinodes:* Site of pollen-tube arrest is the nucellus. Planta 169:245–250

Knight R, Rogers HH (1955) Incompatibility in *Theobroma cacao.* Heredity 9:69–77

Knox RB, Kenrick J (1983) Polyad function in relation to the breeding system of *Acacia.* In: Mulcahy DL, Ottaviano E (eds) Pollen: Biology and implications for plant breeding. Elsevier Biomedical, New York, pp 411–417

Koehl V, Thien LB, Heij EG, Sage TL (2004) The causes of self-sterility in natural populations of the relictual angiosperm, *Illicium floridanum* (Illiciaceae). Ann Bot 94:43–50

Kowyama Y, Tsuchiya T, Kakeda K (2000) Sporophytic self-incompatibility in *Ipomoea trifida,* a close relative of sweet potato. Ann Bot 85:191–196

LaDoux T, Friar EA (2006) Late acting self incompatibility in Ipomopsis tenuifolia (Gray) V. Grant (Polemoniaceae). Int J Plant Sci 167:463–471

Larsen K (1978) Oligoallelism in the multigenic incompatibility system of *Beta vulgaris* L. Incompat Newslett 10:23–28

Levin DA (1975) Gametophytic selection in phlox. In Mulcahy DL (ed), Gamete competition in plants and animals. North Holland, Amsterdam, Netherlands, pp 207–217

Lewis D (1994) Gametophytic-sporophytic incompatibility. In: Williams EG, Clarke AE, Knox RB (eds) Advances in cellular and molecular biology of plants. Genetic control of self incompatibility and reproductive development in flowering plants, vol 2. Kluwer, Dordrecht, pp 88–101

Lewis D, Verma SC, Zuberi MI (1988) Gametophytic-sporophytic incompatibility in the Cruciferae – *Raphanus sativus.* Heredity 61:355–366

Li X, Nicholas P, Nield J, Hayman D, Langridge P (1997) Self-incompatibility in the grasses: Evolutionary relationship of the S gene from *Phalaris coerulescens* to homologous sequences in other grasses. Plant Mol Biol 34:223–232

Lipow SR, Wyatt R (2000) Single Gene Control of Postzygotic Self-Incompatibility in Poke Milkweed, Asclepias exaltata L. Genetics 154:893–907

Lundqvist A (1990a) The complex S-gene system for control of self-incompatibility in the buttercup genus *Ranunculus.* Hereditas 113:29–46

Lundqvist A. (1990b). One-locus sporophytic *S*-gene system with traces of gametophytic pollen control in *Cerastium arvense* ssp. *strictum* (Carophyllaceae). Hereditas 113:203–215

Lundqvist A (1991) Four-locus S-gene control of self-incompatibility made probable in *Lilium martagon* (Liliaceae). Hereditas 114:57–63

Lundqvist (1993) The self-incompatibility system in *Lotus tenuis* (Fabaceae). Hereditas 119:59–66

Lundqvist A, Østerbye U, Larsen K, Linde-Laursen I (1973) Complex self-incompatibility systems in *Ranwnculus acris L. and Beta vulgaris L.* Hereditas 74:161–168

Martin FW (1967) Distyly, Self-Incompatibility, and Evolution in Melochia. Evolution 21:493–499

Martin G, Brommonschenkel S, Chunwongse J, Frary A, Ganal M, Spivey R, Wu T, Earle E, Tanksley S (1993) Map-based cloning of a protein kinase gene conferring disease resistance in tomato. Science 262:1432–1436

Matton DP, Nass N, Clarke AE, Newbigin E (1994) Self-Incompatibility: How Plants Avoid Ilegitimate Offspring. Proc Nat Acad Sci USA 91:1992–1997

McClure BA, Franklin-Tong V (2006) Gametophytic self-incompatibility: Understanding the cellular mechanisms involved in "self" pollen tube inhibition. Planta 224:233–245

McCubbin AG, Kao T (1999) The emerging complexity of self-incompatibility (S-) loci. Sex Plant Reprod 12:1–5

McCubbin AG, Lee C, Hetrick A (2006) Identification of genes showing differential expression between morphs in developing flowers of *Primula vulgaris.* Sex Plant Reprod 19:63–72

McWilliam JR (1959) Interspecific incompatibility in *Pinus.* Am J Bot 46:425–433

Milijuš-Đukić J, Ninković S, Radović S, Maksimović V, Brkljačić J, Nešković M (2004) Detection of proteins possibly involved in self-incompatibility response in distylous buckwheat. Biol Plant 48:293–296

Milijuš-Đukić J, Radović S, Maksimović V (2007) Treatment of isolated pistils with protease inhibitors overcomes the self-incompatibility response in buckwheat.Arch. Biol. Sci 59:45–49

Murfett J, Strabala TJ, Zurek DM, Mou B, Beecher B, McClure BA (1996) S-RNase and interspecific pollen rejection in the genus *Nicotiana*: Multiple pollen-rejection pathways contribute to unilateral incompatibility between self-incompatible and self-compatible species. Plant Cell 8:943–958

Naarborg AT, Willemse MTM (1992) The ovular incompatibility system in *Gasteria verrucosa*. Euphytica 58:231–240

Østerbye U (1977) Self-incompatibility in *Ranunculus acris* L.: Four loci in a German population. Hereditas 87:174–178

Owens SJ (1981) Self-incompatibility in the Commelinaceae. Ann Bot 47:567–581

Owens SJ, McGrath S (1984) Self-incompatibility and the pollen-stigma interaction in *Tradescantia ohiensis*. Protoplasma 121:209–213

Paetsch M, Mayland-Quellhorst S, Neuffer B (2006) Evolution of the self-incompatibility system in the Brassicaceae: Identification of *S*-locus receptor kinase (SRK) in self-incompatible *Capsella grandiflora*. Heredity 97:283–290

Pandey KK (1957) Genetics of incompatibility in *Physalis ixocarpa* Brot. A new system. Am J Bot 44:879–887

Pandey KK (1960) Incompatibility in *Abitulon hybridum*. Am J Bot 47:877–883

Pandey KK (1980) Evolution of incompatibility systems in plants: Origin of independent and complementary control of incompatibility in angiosperms. New Phytol 84:381–400

Perez-Barrales R, Vargas P, Arroyo J (2006) New evidence for the Darwinian hypothesis of heterostyly: Breeding systems and pollinators in *Narcissus* sect. Apodanthi. New Phytol 171:553–567

Pontieri V, Sage TL (1999) Evidence for stigmatic self-incompatibility, pollination induced ovule enlargement and transmitting tissue exudates in the paleoherb, *Saururus cernuus* L. (Saururaceae). Ann Bot 84:507–519

Prakash N, Alexander JH (1984) Self-incompatibility in *Austrobaileya scandens*. In: William EG, Knox RB (eds) Pollination "84. School of Botany, University of Melbourne, Melbourne, pp 214–216

Qiu Y, Dombrovska O, Lee J, et al. (2005) Phylogenetic analysis of basal angiosperms based nine plastid, mitochondrial, and nuclear genes. Int J Plant Sci 166:815–842

Qiu YL, Lee J, Bernasconi-Quadroni F, Soltis DE, Soltis PS, Zanis M, Zimmer EA, Chen Z, Savolainen V, Chase MW. (1999) The earliest angiosperms: evidence from mitochondrial, plastid and nuclear genomes. Nature 402:404–407

Richards AJ (1997) Plant breeding systems. Chapman and Hall, London

Ride JP, Davies EM, Franklin FCH, Marshall DF (1999) Analysis of Arabidopsis genome reveals a large new gene family in plants. Plant Mol Biol 39:927–932

Runions JD, Owens J (1998) Evidence of prezygotic self-incompatiblity in a conifer. In: Owens SJ, Rudall PJ (eds) Reproductive biology. Kew, Royal Botanic Gardens, pp 55–264

Sage TL, Sampson FB (2003) Evidence for ovarian self-incompatibility as a cause of self-sterility in the primitive woody angiosperm, *Pseudowintera axillaris* (Winteraceae). Ann Bot 91:1–10

Sage TL, Bertin R, Williams EG (1994) Ovarian and other late-acting self-incompatibility. In: Williams EG, Clarke AE, Knox RB (eds) Advances in cellular and molecular biology of plants. Genetic control of self incompatibility and reproductive development in flowering plants, vol 2. Kluwer, Dordrecht, pp 116–140

Sage TL, Strumas F, Cole WW, Barrett SCH (1999) Differential ovule development following self- and cross-pollination: The basis of self-sterility in *Narcissus triandrus* (Amaryllidaceae). Am J Bot 86:855–870

Sage TL, Pontieri V, Christopher R (2000) Self-incompatibility and mate recognition in monocotyledons. In: Monocots systematics and evolution. Proceedings of the second international conference on the comparative biology of Monocots. Aust. J. Bot. 2:268–275

Sage TL, Griffin SR, Pontieri V, Drobac P, Cole WW, Barrett SCH (2001) Stigmatic Self-incompatibility and mating patterns in *Trillium grandiflorum* and *Trillium erectum* (Melanthiaceae). Ann Bot 88:829–841

Sage TL, Price MV, Waser NM (2006) Self-sterility in *Ipomopsis aggregata* (Polemoniaceae) is due to prezygotic ovule degeneration. Am J Bot 93:254–262

Schneider EL, Moore LA (1977) Morphological studies of the *Nymphaeaceae*. VII. The floral biology of *Nuphar lutea* subsp. *macrophylla*. Brittonia 29:88–99

Sears ER (1937) Cytological phenomena connected with self-sterility in the flowering plants. Genetics 22:130–181

Seavey SR, Bawa KS (1986) Late-acting self-incompatibility in angiosperms. Bot Rev 52:195–219

Shore JS, Arbo MM, Fernández A (2006) Breeding system variation, genetics and evolution in the Turneraceae. New Phytol 171:539–551

Silva NF, Goring DR (2001) Mechanisms of self-incompatibility in flowering plants. Cell Mol Life Sci 58:1988–2007

Soltis PS, Soltis DE, Chase MW (1999) Angiosperm phylogeny inferred from multiple genes as a tool for comparative biology. Nature 402:358–359

Stebbins, GL (1957) Self fertilization and population variability in the higher plants. Am Nat 861:337–354

Steinbachs JE, Holsinger KE (2002) S-RNase-mediated gametophytic self-incompatibility is ancestral in Eudicots. Mol Biol Evol 19:825–829

Stephenson AG, Good SV, Vogler DW (2000) Inter-relationships among inbreeding depression, plasticity in the self-incompatibility system, and the breeding system of *Campanula rapunculoides* L. (Campanulaceae). Ann Bot 85:211–219

Stevens VAM, Murray BG (1982) Studies on heteromorphic self-incompatibility systems: Physiological aspects of the incompatibility system in *Primula obconica*. Theor Appl Genet 61:245–256

Suassuna T de MF, Bruckner CH, Carvalho CR de, Bor A (2003) Self-incompatibility in passionfruit: Evidence of gametophytic-sporophytic control. Theor Appl Genet 106:298–302

Tabah DA, McInnis SM, Hiscock SJ (2004) Members of the S-receptor kinase multigene family in *Senecio squalidus* L. (Asteraceae), a species with sporophytic self-incompatibility. Sex Plant Reprod 17:131–140

Takayama S, Isogai A (2005) Self-incompatibility in plants. Annu Rev Plant Biol 56:467–489

Takayama S, Shimosato H, Shiba H, Funato M, Che FS, Watanabe M, Iwano M, Isogai A (2001) Direct ligand–receptor complex interaction controls *Brassica* self-incompatibility. Nature 413:534–538

Talavera S, Gibbs PE, Fernândez-Piedra MP, Ortiz-Herrera (2001) Genetic control of self-incompatibility in *Anagallis monelli* (Primulaceae: Myrsinaceae). Heredity 87:589–597

Tezuka T, Akitab I, Yoshinoc N, Suzuki Y (2007) Regulation of self-incompatibility by acetylcholine and cAMP in *Lilium longiflorum*. J Plant Physiol 164:878–885

Thien LB, White DA, Yatsu LY (1983) The reproductive biology of a relict *Illicium floridanum* Ellis. Am J Bot 70:719–727

Thien LB, Sage TL, Jaffré T, Bernhardt P, Pontieri V, Weston P, Malloch D, Azuma H, Graham SW, McPherson MA, Rai HS, Sage RF, Dupre JL (2003) The population structure and floral biology of *Amborella trichopoda* Baillon (Amborellaceae). Ann Mo Bot Gard 90:466–490

Thomson JD, Barrett SCH (1981) Selection for outcrossing, sexual selection and the evolution of dioecy in plants. Am Nat 118:443–449

Umbach AL, Lalonde BA, Kandasamy MK, Nasrallah JB, Nasrallah ME (1990) Immunodetection of protein glycoforms encoded by two independent genes of the self-incompatibility multigene family of Brassica. Plant Physiol. 93:739–747

Uyenoyama MK (1995) A generalized least-squares estimate for the origin of sporophytic self-incompatibility. Genetics 139:975–992

Wang Y, Wang X, McCubbin AG, Kao T (2003) Genetic mapping and molecular characterization of the self-incompatibility *S*-locus in *Petunia inflata*. Plant Mol Biol 53:565–580

Watanabe M, Suzuki G, Takayama S, Isogai A, Hinata K (2000) Genomic organization of the SLG/SRK region of the *S*-locus in Brassica species. Ann Bot 85:155–160

Weller SG, Donoghue MJ, Charlesworth D (1995) The evolution of self-incompatibility in flowering plants: A phylogenetic approach. In: Hoch PC, Stephenson AG (eds) Experimental and molecular approaches to plant biosystematics. Missouri Botanical Garden, St Louis, pp 355–382

Whitehouse HLK (1950) Multiple allelomorph incompatibility of pollen and style in the evolution of angiosperms. Ann Bot N S 14:199–216

Williams EG, Kaul V, Rouse JL, Palser BF (1986) Overgrowth of Pollen Tubes in embryo sacs of *Rhododendron* following interspecific pollinations. Aust J Bot 34:413–423

Williams JH, Kennard KS (2006) Microsatellite loci for the basal angiosperm *Austrobaileya scandens* (Austrobaileyaceae). Mol Ecol Notes 6:201–203

Worberg A, Quandt D, Barniske AM, Löhne C, Hilu KW, Borsch T (2007) Phylogeny of basal eudicots: Insights from non-coding and rapidly evolving DNA. Org Divers Evol 7:55–77

Zanis MJ, Soltis PS, Qiu YL, Zimmer E, Soltis DE (2003) Phylogenetic analyses and perianth evolution in basal angiosperms. Ann Mo Bot Gard 90:129–150

Zavada MS (1984) The relation between pollen exine sculpturing and self-incompatibility mechanisms. Plant Syst Evol 147:63–78

Zavada MS, Taylor TN (1986) The Role of self-incompatibility and sexual selection in the gymnosperm-angiosperm transition: A hypothesis. Am Nat 128:538–550

Chapter 5
What Genealogies of S-alleles Tell Us

J.R. Kohn

Abstract Drawing on examples from S-RNase-based self-incompatibility (SI), *S*-locus genealogies are used to infer the demographic history of lineages, the history of mating-system transitions in entire plant families and aspects of the evolution of the *S*-locus itself. Two lineages of Solanaceae suffered severe restrictions of *S*-locus diversity evident after millions of years. Broadly shared ancestral *S*-locus polymorphism is evidence that loss of this form of incompatibility was irreversible in the Solanaceae. Frequent and irreversible loss implies incompatibility is either declining in frequency through time, or that it confers an increased diversification rate relative to self-compatibility (SC). Differences in diversification rate among self-incompatible and self-compatible lineages likely cause the failure of current phylogenetic methods to correctly reconstruct the history of SI. Genealogies also show that origination of new S-RNases rarely occurs within the lifetimes of species. Surprisingly, genealogies of F-box genes purported to provide pollen specificity often do not correspond to those of their cognate S-RNases, indicating we have much to learn about how this system works and evolves.

Abbreviations

cpDNA	DNA from plant plastids
GSI	Gametophytic SI
mya	Million years ago
SC	Self-compatibility
SCR/SP11	*S*-locus cysteine-rich protein (the pollen *S*-determinant in Brassica)

J.R. Kohn
Section of Ecology, Behavior and Evolution, Division of Biological Sciences, University of California San Diego, 9500 Gilman Drive, La Jolla CA, 92093–0116 USA, e-mail: jkohn@ucsd.edu

SFBB	*S*-locus F-box brothers
SI	Self-incompatibility
SLF/SFB	*S*-locus F-box (*SFB/SLF – S-Locus F-box/S-haplotype-specific F-box*-the pollen *S*-determinant in many GSI systems; *SLF* in *Antirrhinum* and *Petunia*; *SFB* in *Prunus*)
S-locus	Self-incompatibility locus
SRK	*S*-locus Receptor Kinase (the pistil *S*-determinant in Brassica)
S-RNase	S-ribonuclease

5.1 Introduction

The self-incompatibility (SI) locus of flowering plants has attracted the attention of population geneticists since extreme *S*-locus polymorphism was described by Emerson (1939) in natural populations of *Oenothera organensis*. Prior to molecular characterisations of the *S*-locus in several plant families, empirical studies focused on characterising the numbers of alleles in populations (reviewed in Lawrence 2000), while the main goal of theoretical treatments (Wright 1939, 1960, 1964; Fisher 1958, 1961; Moran 1962; Ewens and Ewens 1966) was to explain how selection could maintain the large numbers of alleles observed. However, it was appreciated very early on (Kingman 2000; W.J. Ewens personal communication) that the *S*-locus might also provide a unique historical perspective. The advent of *S*-locus sequence data from natural populations of self-incompatible plants (reviewed in Richman and Kohn 2000; Castric and Vekemans 2004) together with the extension of theoretical treatments of the *S*-locus to include coalescent approaches to loci under balancing selection (Takahata 1990, 1993; Takahata and Nei 1990; Clark 1993; Clark and Kao 1994; Vekemans and Slatkin 1994) has brought a resurgence of interest in the population genetics of the *S*-locus, including ways that it may be used to infer historical phenomena and to gain insight into longstanding evolutionary questions.

There are two primary reasons why the *S*-locus has attracted the attention of biologists. First, it is a system of self/non-self recognition and rejection. All such systems have inherent appeal in terms of their mechanisms of action and evolutionary properties. Second, the property that drives the evolution of the *S*-locus is negative frequency-dependent selection; alleles that are rare in a population have more potential mates, while those that are common have fewer. This is the force that Wright (1939) recognised could explain the extreme polymorphism discovered by Emerson (1939), and this is also the force that preserves historical information stretching much further back in time than it is possible to go using standard loci (Takahata 1990; Vekemans and Slatkin 1994). Selection that perpetually increases the frequency of rare alleles preserves polymorphism over very long periods of time. It is this property, extreme age of polymorphism, that is the focus of this chapter.

This chapter primarily uses examples from three angiosperm families, the Solanaceae, Plantaginaceae and Rosaceae, which use S-RNases as the stylar component of specificity in the gametophytic SI (GSI) reaction (reviewed in Takayama and Isogai 2005); see also Chaps. 9 and Chap. 10 McClure, this volume. What is known from S-RNase-based systems is compared to information from the sporophytic *SRK/SCR*-based incompatibility in the Brassicaceae. Another, non-homologous, GSI system is known in the Papaveraceae, but sequence information from only five pistil *S*-locus alleles derived from two species of *Papaver* are currently available (Kurup et al. 1998).

Two characteristics of the molecular variation found among S-RNases exemplify the extreme age of *S*-locus polymorphism. First, alleles from the same individual can be extremely divergent. In the Solanaceae, two stylar S-RNase alleles found in the same obligately heterozygous individual often differ at more than 50% in their amino acid residues (Richman et al. 1995). Second, phylogenies of *S*-alleles show abundant evidence of shared ancestral polymorphism (Ioerger et al. 1990; Richman et al. 1996b; Richman and Kohn 2000; Castric and Vekemans 2004). Alleles drawn from different species and genera cluster together in phylogenetic reconstructions, indicating that the allelic lineages they represent were present in the common ancestor of the species sampled. This pattern of shared ancestral polymorphism confirms theoretical expectations that negative frequency-dependent selection will preserve polymorphism for very long periods of time. In the Solanaceae, much of the observed *S*-locus polymorphism arose prior to the most recent common ancestor of all of the species whose *S*-alleles have been examined (Igic et al. 2004, 2006). This ancestor is thought to have occurred 35–45 million years ago (mya) (Ioerger et al. 1990; Paape et al. 2008). Similar ancient polymorphism exists at the sporophytic *S*-locus of the Brassicaceae (reviewed in Castric and Vekemans 2004).

Polymorphism that persists for tens of millions of years can be used to answer questions about the locus itself, the history of lineages that carry it and the consequences of its loss. Because the *S*-locus enforces outcrossing, its evolution and loss is intimately tied to breeding system transitions between outcrossing and partial or complete selfing. The historical information preserved within *S*-locus polymorphism can be used to trace, with an unusual degree of certainty, the history of such transitions across entire plant families. This provides a unique tool with which to address longstanding problems in biology such as whether selfing lineages are shorter lived than outcrossing ones (Stebbins 1957, 1974; Takabayashi and Morrell 2001), and whether complex biological traits such as self-incompatibility, once lost, can be regained in the same form.

5.2 Long-Term Demographic Information from the *S*-locus

The earliest studies comparing small samples of S-RNase sequences from different Solanaceae (e.g. species of *Petunia*, *Nicotiana* and *Solanum* sect. *Lycopersicon*) found strong evidence of ancestral polymorphism shared among the taxa sampled (Ioerger et al. 1990). However, studies that surveyed *S*-locus variation in natural populations (Richman et al. 1995, 1996a) also found variation in the level of ancestral polymorphism preserved in different genera (Richman et al. 1996b). In particular, members of the closely allied genera *Physalis* and *Witheringia* have *S*-alleles that represent only three lineages that predate their most recent common ancestor (Richman et al. 1996b; Richman and Kohn 1999; Lu 2001; Stone and Pierce 2005), while all other genera of Solanaceae sampled have *S*-alleles from many more ancient lineages (Richman 2000; Richman and Kohn 2000; Savage and Miller 2006). Richman et al. (1996b) compared the restricted *S*-allele lineage diversity in *Physalis crassifolia* to the more diverse S-allele assemblage sampled from *Solanum carolinense*. Using a coalescent approach, they showed that the long-term population size of *P. crassifolia* was one or two orders of magnitude smaller than that of *S. carolinense*. This contrasts with current estimates of population size derived from *S*-allele numbers. *P. crassifolia* populations currently harbour far more alleles at the *S*-locus than are found in populations of *S. carolinense*.

While the study of Richman et al. (1996a) used coalescence-based methods as a tool to explore deep historical demographic events, species-specific genealogies of *S*-alleles do not conform standard expectations of a birth–death process (Uyenoyama 1997). In particular, *S*-allele genealogies are more star-like than expected, exhibiting rapid diversification near the base of the genealogy with a subsequent apparent slowdown in the rate of allelic diversification. Explanations for this observation vary. Uyenoyama (1997) advanced the hypothesis that deleterious recessives hitchhiking on the obligately heterozygous *S*-locus retard the apparent diversification rate. Richman and Kohn (1999) found evidence that more divergent alleles were preserved when ecological factors reduce the effective population size and the number of *S*-alleles maintained. Whatever the cause, the more star-like than expected genealogies of S-alleles require that caution be exercised when applying coalescence-based methods to this locus.

Nevertheless, *S*-allele sequences from species of *Physalis* and *Witheringia* provide an extremely strong phylogenetic signal. To date, a total of 93 *S*-allele sequences from three species of *Physalis* (Richman et al. 1996a; Richman and Kohn 1999; Lu 2001) and two of *Witheringia* (Richman and Kohn 2000; Stone and Pierce 2005) have been published and all fall within only three lineages that predate the most recent common ancestor of these genera. A simple question to ask is, how old is this *S*-locus bottleneck? Clearly this event occurred prior to the most recent common ancestor of *Physalis* and *Witheringia*, but more recently than the divergence of these genera from any that do not show restriction of *S*-locus sequence variation. Paape et al. (2008) assayed *S*-alleles from several members of the subtribe Iochrominae (Solanaceae), a group found to be sister to the clade containing *Physalis* and *Witheringia* in a large molecular phylogenetic analysis of Solanaceae (Olmstead et al. in press). Even with a limited sample, *S*-alleles from the Iochrominae represent several lineages not observed in *Physalis* and *Witheringia* (Fig. 5.1). This means that the restriction of the *S*-locus had to have occurred after the most recent common ancestor of the group containing the Iochrominae, *Physalis* and *Witheringia*, but before the most recent common ancestor of *Physalis* and *Witheringia*. Using cpDNA sequence information calibrated with fossil data, Paape et al. (2008) estimated that the most recent common ancestor of *Physalis* and *Witheringia* occurred some 14 mya while their most recent common ancestor with the Iochrominae occurred

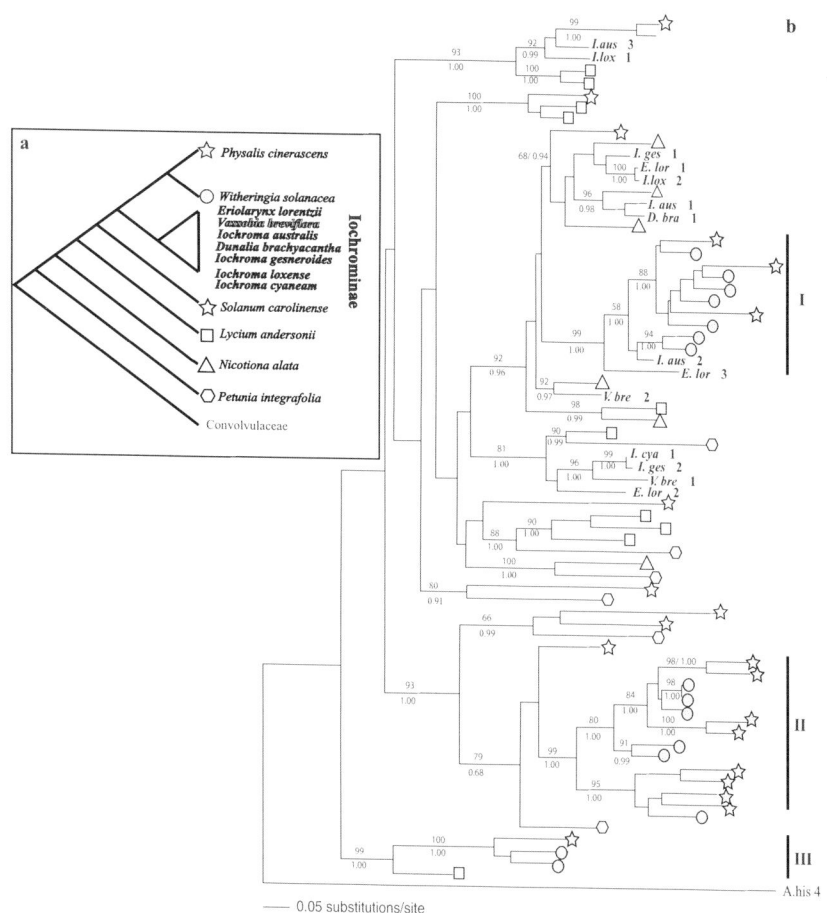

Fig. 5.1 (**a**) The phylogeny of Solanaceae species from which S-alleles were sampled. All nodes in the phylogeny have >90% bootstrap support in the large phylogenetic analysis of Olmstead et al. (in press). (**b**) Maximum-likelihood phylogeny of 72 S-alleles. Symbols correspond to alleles from taxa in the species phylogeny. All alleles from the genera *Physalis* and *Witheringia* are restricted to one of the three lineages indicated by Roman numerals. The 14 alleles sampled from various Iochrominae species (*boldface*) represent several lineages not found in *Physalis* or *Witheringia*, which predate the divergence of *Solanum* from the other genera sampled. The S-allele DNA phylogeny was constructed in PAUP* v4.0 (Swofford 2002). Bootstrap scores are indicated above branches and posterior probabilities >80% generated by Mr. Bayes v3.0 (Ronquist and Huelsenbeck 2003) are below branches. Only a subset of alleles known from *Physalis*, *Witheringia* and other Solanaceae were used to simplify the computation and presentation of the data. For additional details, see Paape et al. (2008) from which the figure is redrawn

approximately 18 mya. The restriction of *S*-locus variation had to have occurred between these two dates, exemplifying the great age of the historical information available from the *S*-locus.

A more difficult question than the age of the restriction in *S*-locus variation in *Physalis* and *Witheringia* is what could have caused it. Because of the power of negative frequency-dependent selection to preserve variation, the *S*-locus is particularly resistant to all but the most severe types of bottleneck events. For instance, a population with constant size of only 100 individuals nevertheless maintains six alleles at equilibrium (Wright 1939). A weak bottleneck followed by allelic turnover that further reduced the number of *S*-lineages appears unlikely, given the great strength of selection preserving alleles when their number is below equilibrium, and also given the slow pace of allelic turnover observed elsewhere in the Solanaceae. Extreme types of bottleneck events would apparently be required to explain the fact that only three ancient lineages are represented among the *S*-alleles of *Physalis* and *Witheringia*. A founder event involving perhaps as few as two individuals would seem the only likely way for such a restriction to arise, but also would require the lack of further gene flow between source and founder populations. Subsequent gene flow would almost certainly introduce additional lineages of *S*-alleles. Whatever its cause, information to date suggests that lineages rarely survive such restrictions and remain self-incompatible. Observations of species whose *S*-alleles represent many ancient lineages, the common finding, imply that no ancestor of those species suffered a similar restriction at the *S*-locus.

Only one study has found restricted *S*-locus variation outside of *Physalis* and *Witheringia*, and this bottleneck seems certain to have been caused by a founder event associated with long-distance dispersal. Miller et al. (2008) examined *S*-alleles from several self-incompatible species of *Lycium* from southern Africa. The genus *Lycium* is found primarily in the new world, and phylogenetic analyses (Levin and Miller 2005; Levin et al. 2007) strongly suggest that it originated in South America. However, *Lycium* species are also found on several oceanic islands as well as southern Africa. African species are monophyletic within the genus, suggesting a single colonisation event brought *Lycium* to the Old World. Fewer *S*-allele lineages that pre-date the genus *Lycium* are represented among African *Lycium* S-alleles than are found among samples of the same size from several New World species (Miller et al. 2008). In fact, current data are consistent with the idea that as few as three individuals could have made up the founding Old World population, though that estimate could rise with additional sampling. The restriction of *S*-locus variation in Old World *Lycium* is estimated to have occurred less than 10 mya. However, it pre-dates the diversification of the monophyletic assemblage of more than 30 African species of *Lycium,* providing another stunning example of the time-depth of information preserved at the *S*-locus.

5.3 Implications of Shared Ancestral Polymorphism

5.3.1 Tracing the History of Mating System Change

In addition to demographic information, ancestral *S*-locus polymorphism provides important historical implications about breeding systems that extend back in time to the most recent common ancestor of all SI Solanaceae. For instance, species of *Brugmansia, Lycium, Nicotiana, Petunia, Physalis, Solanum,* and *Witheringia* have been assayed for *S*-locus polymorphism (Igic et al. 2006 and references therein). Each one possesses *S*-alleles representing multiple lineages that were present in the common ancestor these genera. This implies a continuous history of self-incompatibility from the time of that common ancestor to the present (Igic et al. 2004, 2006).

What justifies the assertion that all ancestors of sampled SI species were themselves SI going all the way back to the most recent common ancestor of those species? When SI is lost, polymorphism at the *S*-locus is rendered selectively neutral and is expected to collapse in 4 N_e generations (Hudson 1990), or sooner if loss of SI is caused by a selective sweep of a non-functional S-allele. Once polymorphism is lost, the GSI system cannot be regained because, with fewer than three different alleles, all individuals are mutually incompatible (Wright 1939). Even if it were possible to regain the system following complete collapse of *S*-locus polymorphism, such an occurrence would leave an indelible mark on the *S*-locus; a lineage whose entire complement of *S*-alleles forms a monophyletic clade relative to *S*-alleles from other taxa. This has never been observed in the Solanaceae. In addition, once SI is lost, mutations in other genes whose products are required for self-incompatibility to operate are expected to accumulate. Little or no polymorphism at the *S*-locus, together with multiple loss of function mutations, are commonly observed in SC taxa recently derived from SI ones (Stone 2002; Igic et al. 2008).

Character states of ancestral taxa, and transition rates among character states, are usually reconstructed based solely upon the character state distribution among extant taxa (Pagel 1999). Compared to this situation, shared ancestral polymorphism provides an unparalleled degree of certainty regarding the character states of ancestors. Igic et al. (2004, 2006) used evidence from shared ancestral polymorphism as an aid in reconstructing the history of self-incompatibility throughout the Solanaceae. The Solanaceae comprises some 2,600 species, an estimated 40% of which are SI (Whalen and Anderson 1981; Igic and Kohn 2006). SI and SC taxa are broadly intermixed on the family phylogeny, as is common in large families with self-incompatibility (Heilbuth 2000; Ferrer and Good-Avila 2007; see also Chap. 4, this volume). When only the self-incompatibility status of extant taxa are considered, standard methods (Pagel 1999) of reconstructing character state transition rates find strong statistical support for multiple gains of SI within the Solanaceae, rejecting the hypothesis that SI has never been regained once lost. However, when inference from shared ancestral *S*-locus polymorphism is used to unite all Solanaceae whose *S*-alleles have been sampled with a continuous history of SI, the hypothesis that the rate of transition from SC to SI in the Solanaceae is zero cannot be rejected (Igic et al. 2006). Within the Solanaceae, self-incompatibility follows Dollo's law (Gould 1970) that a complex character, once lost, is never regained.

Dollo's law is subject to several interpretations (Bull and Charnov 1985). The meaning here is not that SI, once lost, is never regained by some new mechanism, but that RNase-based SI has not been regained once lost. Across the angiosperms, it is clear that multiple gains of various forms of incompatibility have occurred (Weller et al. 1995; Igic et al. 2008). Nevertheless, the rate of gain of incompatibility systems is very far exceeded by the frequency of loss.

A similar phylogenetic analysis concerning the history of sporophytic SI in the Asteraceae (Ferrer and Good-Avila 2007) also finds many cases of closely related extant taxa with alternative SI and SC character states; see also Chap. 4, this volume. No molecular information concerning the basis of SI is available for Asteraceae. Ferrer and Good-Avila (2007) used standard reconstruction procedures (Pagel 1999) to conclude that multiple gains of incompatibility occurred in the family, precisely the conclusion that is reached in the Solanaceae in the absence of information from shared ancestral polymorphism (Igic et al. 2004, 2006). Since molecular evidence of shared ancestral polymorphism in the Solanaceae controverts the conclusion of multiple origins of SI, the finding of multiple origins of SI within the Asteraceae, based solely on the distribution of character states among extant taxa, must be viewed with caution.

5.3.2 Diversification Rate Differences and Character State Reconstruction

Two questions arise from the observation that shared ancestral polymorphism provides evidence for irreversible loss of self-incompatibility at family level. First, if SI is frequently lost but never regained within the Solanaceae, is SI becoming less frequent through time? If not then SI species must, on average, have a higher diversification rate (defined as the speciation rate minus the extinction rate) than SC taxa. The second question is, why do methods of estimating character state transition rates and ancestral states (Pagel 1999) fail when only the character states of extant taxa are used? The answer to this second question is likely to be related to the answer to the first, as will be seen below.

Using simple macroevolutionary models that assume irreversible transitions from SI to SC, Igic et al. (2004, 2008) showed that the frequency of SI species will decline unless the diversification rate associated with self-incompatibility is greater than the sum of the diversification rate associated with self-compatibility plus the transition rate of species from SI to SC states. A remaining challenge is to calculate character state specific rates of diversification and the SI to SC transition rate from detailed phylogenetic data. Doing so will test whether SI is being maintained due to some macro-evolutionary advantage it provides. Stebbins (1974) argued that the transition from outcrossing enforced by self-incompatibility to predominant selfing was the most commonly repeated evolutionary pathway in flowering plants. In his view, this line of evolution was often a short-term solution to some ecological challenge but tended to be an evolutionary dead end (Stebbins 1957), an assertion that has proven difficult to test (Takabayashi and Morrell 2001). Macro-evolutionary approaches such as those outlined above will provide useful evidence in this regard.

If SI does provide an increased diversification rate relative to SC, it is unlikely to do so by increasing the speciation rate of taxa that possess it. Selfing taxa usually show greater levels of inter-population genetic differentiation (Hamrick and

Godt 1989), and would therefore be expected to speciate more readily than outcrossing taxa. However, enforced outcrossing might reduce the rate of extinction relative to selfing taxa, due to increased levels of genetic variation preserved within outcrossing populations (Hamrick and Godt 1989; Charlesworth and Charlesworth 1995; Glémin et al. 2006).

Why do commonly used methods of ancestral state and transition rate estimation fail? A little-appreciated assumption of the models underlying these methods is that the character states do not themselves affect the diversification rate (Igic et al. 2006; Maddison 2006). If diversification rates differ among character states, current methods will tend to overestimate the transition rate towards the state that provides the greater diversification rate (Maddison 2006). This could explain the substantial transition rates from SC to SI that are inferred in the Solanaceae and Asteraceae when evidence from shared ancestral polymorphism is not used in the analyses (Igic et al. 2004, 2006; Ferrer and Good-Avila 2007).

5.4 The Pace of New Allele Formation

Another area of S-locus evolution where genealogical approaches can provide insight concerns the evolution of new S-alleles. Different pollen and pistil expressed genes encode specificity in all incompatibility systems that have been characterised at the molecular level. This brings up the knotty problem of how new alleles arise, because a mutation changing the specificity of, for instance, the pollen would result in a haplotype that is self-compatible and potentially lost from the population if self-compatibility is selected against. Mutations in both the female and the male components of the system would apparently be required to form a new specificity (Charlesworth 2000). Several interesting proposals for how new alleles can arise have been suggested (Matton et al. 1999; Uyenoyama and Newbigin 2000; Uyenoyama et al. 2001, Chookajorn et al. 2004), but all require that polymorphism within an allele be maintained either within or between populations while the appropriate mutations are accumulating. Negative frequency-dependent selection, while tending to preserve extreme polymorphism among alleles, is expected to have the opposite effect on polymorphism within alleles, reducing polymorphism below levels found at standard loci. If n alleles are maintained in a population, the population effective size of each allele is $1/n$ times the effective size of the population (Clark 1993) and the expected level of polymorphism is reduced accordingly. In fact, the extreme difference in within- vs. among-allele polymorphism can be used to confirm that balancing selection is acting on a locus, as has been done for the mating-type loci of fungi (May et al. 1999).

Genealogical approaches can ask at least two questions relevant to the formation of new S-alleles. First, what is the level of polymorphism within functionally equivalent alleles either within or between populations? Since polymporphism is necessary for the transitional stages of new allele formation under current models, polymorphic alleles could represent opportunities to study how this process occurs.

Second, what is the tempo of new allele formation relative to species formation? If, species-specific monophyletic clades of alleles are common, then new alleles are frequently arising within the lifespan of species. If, on the other extreme, the closest relative of each allele is found in some other species, then new allele formation would appear to be extremely rare relative to the lifespan of species, and finding transitional stages in the process of new allele formation would require interspecific comparisons (Sato et al. 2006; Surbanovski et al. 2007).

Raspé and Kohn (2007) examined S-RNases from *Sorbus aucuparia* (Rosaceae, subfamily Maloideae) from the Pyrenees Mountains and compared them to sequences recovered previously from a population in Belgium (Raspé and Kohn 2002). Of the alleles recovered, 10 were found in both populations. Nine out of the ten alleles showed no nucleotide polymorphism, despite the 1,000 km separation of the two populations. The tenth showed a single nucleotide change, but this could not be confirmed as only a single sequence was recovered from the original Belgian population. These findings confirm the theoretical prediction that polymorphism within alleles will be rare (Clark 1993). In addition, phylogenetic analysis finds that nearly every allele from *Sorbus aucuparia* is more closely related to an allele from another genus (other Maloideae assayed are members of the genera *Crataegus*, *Malus* or *Pyrus*) than to other alleles in this species (Fig. 5.2). The same is true of the other Maloideae. Monophyletic clades of alleles drawn from single species are rare, represented by only two intra-specific sister pairs, neither of which is strongly supported by bootstrap analysis (Fig. 5.2; Raspé and Kohn 2007). This implies that stylar S-alleles have rarely diversified since the origination of the species in which they now occur, or that whenever a new allele does arise, it displaces its progenitor allele, leaving no apparent diversification (Uyenoyama et al. 2001).

The data from the Maloideae contrast with findings in the Brassicaceae, where within-specificity polymorphism has been documented in *Brassica oleracea* (Miege et al. 2001) and where studies of similar alleles in closely related species are beginning to unlock stages in the development of new specificities (Sato et al. 2006). In addition, we know from studies of Solanaceae that diversification rates of S-alleles increase following bottlenecks (Richman 2000; Paape et al. 2008; Miller et al. 2008), as is expected with greater selection when the population is below equilibrium allele number. The slow pace of recent diversification among the Maloideae may reflect demographic stasis and reduced selection for new alleles. To date both the small number of SI systems that have been characterised at the molecular level, and the limited sampling of natural populations that has been done within each of them, restrict our ability to draw firm conclusions about how and how rapidly new alleles arise.

5.5 Remaining Issues of S-RNase Evolution

Some major questions remain unanswered concerning the evolution of stylar S-RNases. Most pressing is the apparent disparity in patterns of diversification seen in the Solanaceae and Plantaginaceae relative to what is observed in the Rosaceae. The

5 What Genealogies of S-alleles Tell Us

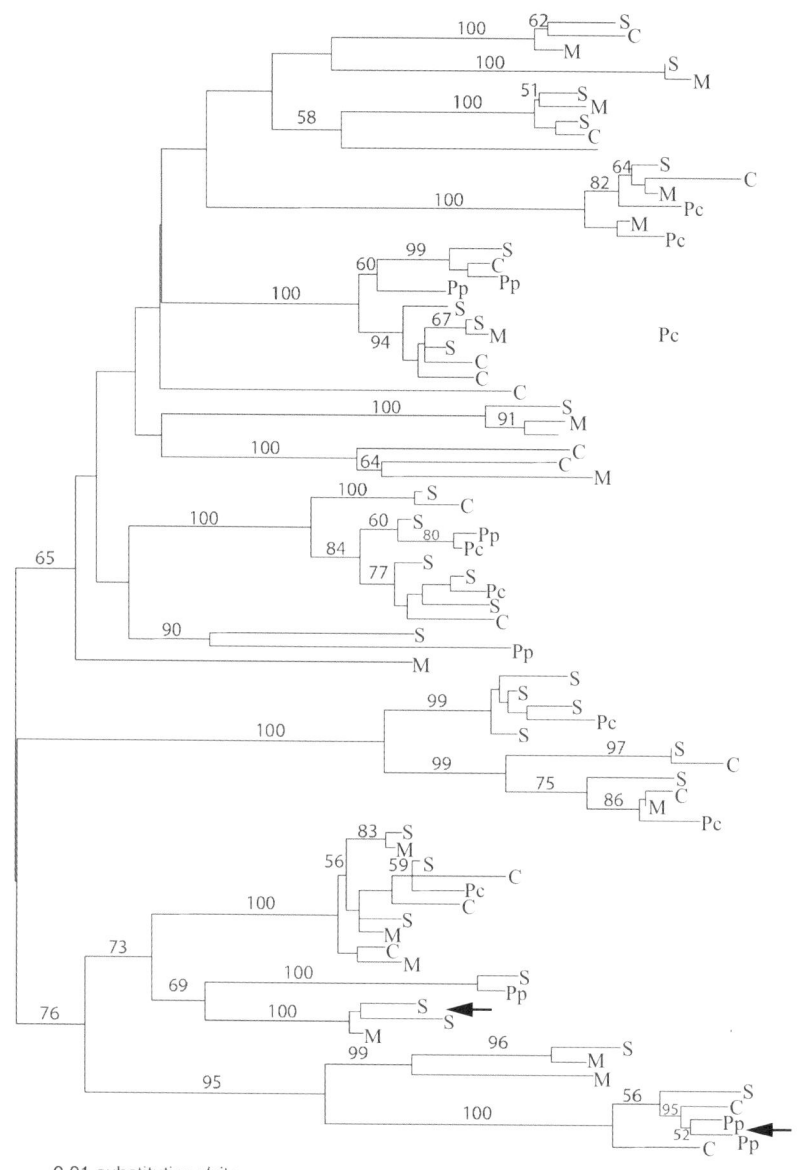

Fig. 5.2 Midpoint-rooted maximum-likelihood phylogeny of 80 S-alleles from the Maloideae (redrawn from Raspé and Kohn 2007). Arrows indicate the only two instances in which sister alleles are drawn from the same species. Neither sister relationship receives strong bootstrap support

Solanaceae and Plantaginaceae appear to harbour similar high levels of nucleotide diversity and shared ancestral polymorphism at the S-RNase locus, though the Plantaginaceae is currently less well-sampled (Xue et al. 1996; Vieira and Charlesworth 2002). Compared to these families, S-RNases from the Rosaceae have lower levels of nucleotide diversity and S-RNases from two subfamilies (Maloideae and Amygdaloideae (*Prunus*)) form reciprocally monophyletic clades (Igic and Kohn 2001; Ma and Oliviera 2002; Steinbachs and Holsinger 2002). Population allele numbers in species of Rosaceae and Solanaceae appear similar, and so smaller population sizes appear unlikely to have caused the increased turnover seen in Roasaceae S-allele genealogies. In addition, the Rosaceae appears to be no older than the Solanaceae (Wikström et al. 2001) so increased time is an unlikely cause of increased turnover.

The genealogies of Maloideae and Amygdaloideae S-RNases also differ from one another. Those from *Prunus* show very little phylogenetic structure. That is, S-RNase genealogies from species of *Prunus* are extremely star- or comb-like, with very few internal branches achieving statistical support from bootstrap analyses (Nunes et al. 2006). More structure is seen in genealogies of Maloid S-RNases where genealogical relationships among alleles are often strongly supported (Fig. 5.2; Raspé and Kohn 2007). These puzzling findings suggest that S-RNase evolution in the Rosaceae, particularly in *Prunus*, differs in some important respects from the Solanaceae and Plantaginaceae. One possibility is increased levels of recombination in the Rosaceae. Recombination reduces apparent coalescence time and could both increase the observed rate of turnover and reduce the internal structure of gene genealogies (Schierup et al. 2001). To date, however, evidence for recombination at the S-RNase locus is mixed (Vieira et al. 2003; Nunes et al. 2006) and no study has demonstrated increased rates of recombination in the Rosaceae relative to other families using S-RNase-based gametophytic SI.

5.6 Pollen Specificity Genes

While the pistil S-RNases have been studied now for more than two decades, only recently has the pollen specificity component of these systems begun to be elucidated (Entani et al. 2003; Ikeda et al. 2004; Qiao et al. 2004; Sijacic et al. 2004; Ushijima et al. 2003, 2004; Sassa et al. 2007). Expectations for the gene for the pollen specificity component of any self-incompatibility system are the same as those for the pistil component: tight linkage to the cognate locus, unusually high nucleotide polymorphism, long coalescence time, and evidence at the molecular level for positive selection. In addition, genealogies of female and male specificity genes should be concordant, reflecting long coevolutionary histories. Recombination events between pollen and style genes should result in self-compatible haplotypes, expected to be lost from the population if SI is selected for.

In the Brassicaceae, where *SRK* and *SCR* (also known as *SP11*) genes specify pistil and pollen specificity, respectively (reviewed in Takayama and Isogai 2005; see also Chaps. 6 and 7, this volume), these expectations are met. High levels of synonymous and non-synonymous nucleotide polymorphism are seen in both genes, though polymorphism may be somewhat higher in *SCR/SP11* than *SRK* (Sato et al. 2002). This could reflect a greater proportion of the pollen than the pistil gene under diversifying selection. Importantly, Sato et al. (2002) found that the genealogies of twelve pollen and twelve pistil alleles were largely congruent. The hypothesis of strict coevolution of linked pollen and pistil genes, expected for the *S*-locus, could not be rejected.

The recent discoveries implicating F-box proteins as the pollen specificity component of the *S*-locus in all three families that utilise S-RNases (Entani et al. 2003; Ikeda et al. 2004; Qiao et al. 2004; Sijacic et al. 2004; Ushijima et al. 2003, 2004; Sassa et al. 2007; see also Chaps. 9 and 10, this volume) held the promise of rapid gains in understanding the co-evolution of pistil and pollen components in this system. Phylogenetic clustering and structural similarity of S-RNases relative to other plant RNases led to the conclusion that *S*-RNase-based incompatibility is homologous among the Solanaceae, Plantaginaceae and Rosaceae (Igic and Kohn 2001; Steinbachs and Holsinger 2002). The fact that all three families are reported to use *S*-locus *F*-box genes as the pollen specificity component of the SI reaction (see Chap. 9, this volume) would appear to bolster the case for homology of these systems. The alternative, that convergent evolution led to the independent evolution of *S*-RNase/F-box incompatibility systems in different families and/or subfamilies is possible, but we so far lack good candidates for independent ancestral RNases (Igic and Kohn 2001; Steinbachs and Holsinger 2002).

However, several aspects of current findings regarding *S*-locus F-box genes differ very markedly from what would be expected if specificity determining F-Box and *S*-RNase genes had the same evolutionary history (see Chap. 10, this volume). Multiple F-box loci are found at the *S*-locus of each family or subfamily known to use *S*-RNase-based incompatibility (Sassa et al. 2007; Wheeler and Newbigin 2007). This is not surprising given that over 600 F-box loci found are found in the genome of *Arabidopsis thaliana* (Wang et al. 2004). What is surprising is that sequences of F-box genes from the *S*-loci of *Petunia* (Solanaceae), *Antirrhinum* (Plantaginaceae), *Malus* and *Pyrus* (Rosaceae:Maloideae) and *Prunus* (Rosaceae:Amygdaloideae) form family- or subfamily-specific monophyletic clades in phylogenetic reconstructions (Sassa et al. 2007; Wheeler and Newbigin 2007). This means that the locus implicated as encoding pollen specificity in a particular group is more closely related to the other F-box genes at its *S*-locus than to the F-box genes purported to specify pollen mating type in other groups. This contrasts sharply with S-RNases where, for instance, Plantaginaceae and Solanaceae S-RNases form a monophyletic clade relative to other plant RNases (Igic and Kohn 2001; Steinbachs and Holsinger 2002).

Second, levels of polymorphism of the F-box genes thought encode pollen specificity are surprisingly low in some taxa, while in others they seem to conform to expectations of long coalesence times. In *Prunus* a *S*-linked F-Box locus (*SFB*) has been identified that has similar levels of polymorphism as the S-RNase locus, and which shows evidence of positive selection at the molecular level (Ikeda et al. 2004;

Nunes et al. 2006). Loss-of-function mutations in the *SFB* gene are associated with pollen-part self-compatibility in several studies (Ushijima et al. 2004; Sonneveld et al. 2005, Hauck et al. 2006) However, while distances among *Prunus* S-RNase alleles and those among corresponding *SLF* alleles are correlated, genealogies of *Prunus* pistil and purported pollen alleles do not strictly correspond (Nunes et al. 2006), perhaps due to rare recombination events.

In the subfamily Maloideae (Rosaceae). F-box loci currently called *S*-locus F-box brothers (*SFBB*) occur. These loci have levels of polymorphism similar to one another and to S-RNases from Maloideae. It is unclear which locus is involved in pollen specificity and Sassa et al. (2007) suggest that these loci may act in concert to determine mating type. The species of Rosaceae used for studies of self-incompatibility are all trees, making confirmatory experimental molecular approaches difficult.

In contrast to the Rosaceae, the implicated F-Box genes of the Solanaceae and Plantaginaceae have reduced levels of nucleotide polymorphism. This occurs despite the fact that S-RNases of these two families have higher levels of nucleotide polymorphism than do those of *Prunus* or the Maloideae. The contrast in levels of polymorphism between pistil and putative pollen genes is particularly striking in *Antirrhinum* (Plantaginaceae). Levels of both amino acid and synonymous DNA site divergence (Ks) are at least an order of magnitude lower at the *S*-locus F-box (*SLF*) locus than the S-RNase locus (Xue et al. 1996; Zhou et al. 2003; Sassa et al. 2007; Wheeler and Newbigin 2007). Together these facts imply that the histories of the implicated F-Box genes and their S-RNase cognates are markedly different. These findings are so far from expectations that it is difficult to make sense of them at present, but several possibilities have been suggested.

First, the fact that each purported pollen-specifying *S*-locus F-box gene is most closely related to other F-Box genes in its own genome might suggest independent evolution of S-RNase-based self-incompatibility in different groups. However, because the pollen F-box genes within a family or subfamily often do not have evolutionary histories that correspond to their cognate S-RNases, it can hardly be expected that pollen specifying F-Box genes from different families would necessarily group together. Either some or all of the implicated F-box genes do not function as the pollen specificity component, or the way that these evolve is quite unexpected. Considerable experimental evidence implicates the *SLF* loci of *Antirrhinum* and *Petunia* as the pollen specificity components of the incompatibility reaction (Qiao et al. 2004; Sijacic et al. 2004; Hua et al. 2007; see Chap. 9, this volume). However, definitive experiments, such as transformation from one specificity to another, have yet to be done. If these loci specify incompatibility, then the challenge is to explain why they have different apparent histories than their cognate S-RNases. One possibility could be that inter-locus recombination among F-box genes homogenises sequences among loci, leaving only the variation that encodes specificity. Another possibility is that over time different F-box loci get recruited to operate in pollen specificity, so that F-box genes in different families and sub-families are unrelated.

5.7 Conclusions

Genealogical analysis of the pistil components of both the sporophytic self-incompatibility system of the Brassicaceae and the gametophytic S-RNase-based system in the Solanaceae, Plantaginaceae and Rosaceae has confirmed many of the predictions that arise from negative frequency-dependent selection. In particular, predictions of long coalescence times and abundant shared ancestral polymorphism are met, though the tempo of allele formation and turnover varies somewhat among groups. In the Solanaceae, S-RNase variation had been useful in studies of the ancient demographic histories of lineages, in answering longstanding questions of mating system evolution and in crystallising the realisation that current methods of estimating character evolution are prone to error, particularly when the character analysed affects the diversification rate. In the Brassicaceae, the pollen specificity component evolves in a similar fashion to the pistil gene and there is good correspondence between the pollen and pistil gene trees, indicating a long co-evolutionary history (see Chap. 6, this volume). In RNase-based incompatibility, F-box genes implicated as specifying mating type in pollen evolve differently than their stylar cognates in some or all cases. This unexpected finding should instigate a great deal of future research.

Acknowledgements The author thanks B. Igic, E. Newbigin, and T. Paape for useful discussions. Support was provided by NSF DEB-0639984 to J.R.K.

References

Bull JJ, Charnov EL (1985) On irreversrible evolution. Evolution 39:114–1155
Castric V, Vekemans X (2004) Plant self-incompatibility in natural populations: A critical assessment of recent theoretical and empirical advances. Mol Ecol 13:2873–2889
Charlesworth D (2000) How can two-gene models of self incompatibility generate new specificities? Plant Cell 12:309–310
Charlesworth D, Charlesworth B (1995) Quantitative genetics in plants: The effect of the breeding system on genetic variability. Evolution 49:911–920
Chookajorn T, Kachroo A, Ripoll DR, Clark AG, Nasrallah JB (2004) Specificity determinants and diversification of the *Brassica* self-incompatibility pollen ligand. Proc Natl Acad Sci USA. 101:911–917
Clark AG (1993) Evolutionary inferences from molecular characterization of self-incompatibility alleles. In: Takahata N, Clark AG (eds) Mechanisms of molecular evolution: Introduction to molecular paleopopulation biology. Sinauer, Sunderland, MA, pp 79–108
Clark AG, Kao Th (1994) Self-incompatibility: theoretical concepts and evolution. In: Williams EG, Clarke AE, Knox RB, (eds) Genetic control of self-incompatibility and reproductive development in flowering plants. Kluwer, Boston, MA, USA, pp 220–242
Emerson S (1939) A preliminary survey of the *Oenothera organensis* population. Genetics 24:524–537
Entani, T, Iwano M, Shiba H, Che FS, Isogai A, Takayama S (2003) Comparative analysis of the self-incompatibility *S*-locus region of *Prunus mume*: identification of a pollen-expressed F-box gene with allelic diversity. Gene Cell 8:203–213

Ewens WJ, Ewens PM (1966) The maintenance of alleles by mutation: Monte Carlo results for normal and self-sterility populations. Heredity 21:371–378

Ferrer MM, Good-Avila SV (2007) Macrophylogenetic analysis of the gain and loss of self-incompatibility in the Asteraceae. New Phytol 173:401–414

Fisher RA (1958) The genetical theory of natural selection, 2nd edn. Dover Publications, New York

Fisher RA (1961) Possible differentiation in the wild population of *Oenothera organensis*. Aust J Biol Sci 14:76–78

Glémin S, Bazin E, Charlesworth D (2006) Impact of mating systems on patterns of sequence polymorphism in flowering plants. Proc R Soc Biol Sci B 273:3011–3019

Gould, SJ (1970) Dollo on Dollo's law: Irreversibility and the status of evolutionary laws. J Hist Biol 3:189–212

Hamrick JL, Godt MJ (1989) Allozyme diversity in plant species. In: Brown AHD, Clegg MT, Kahler AL, Weir BS (eds) Plant population genetics, breeding and germplasm resources. Sinauer, Sunderland, MA, pp 43–63

Hauck NR, Yamane H, Tao R, Iezzoni A (2006) Accumulation of nonfunctional S-haplotypes results in the breakdown of gametophytic self-incompatibility in tetraploid *Prunus*. Genetics 172:1191–1198

Heilbuth JC (2000) Lower species richness in dioecious clades. Am Nat 156: 221–241

Hudson RR (1990) Gene genealogies and the coalescent process. Oxf Surv Evol Biol 7:1–44

Igic B, Bohs L, Kohn JR (2004) Historical inferences from the self-incompatibility locus. New Phytol 161:97–105

Igic B, Bohs L, Kohn JR (2006) Ancient polymorphism reveals unidirectional breeding system shifts. Proc Natl Acad Sci USA 103:1359–1363

Igic B, Lande R, Kohn JR (2008) Loss of self-incompatibility and its evolutionary consequences. Int J Plant Sci 169:93–104

Ikeda K, Igic B, Ushijima K, Yamane H, Hauck NR, Nakano R, Sassa H, Iezzoni AF, Kohn JR, Tao R (2004) Primary structural features of the S-haplotype-specific F-box protein, SFB, in *Prunus*. Sex Plant Reprod 16:235–243

Ioerger TR, Clark AG, Kao T-h (1990) Polymorphism at the self-incompatibility locus in Solanaceae predates speciation. Proc Natl Acad Sci USA 87:9732–9735

Kingman JFC (2000) Origins of the coalescent: 1974–1982. Genetics 156:1461–1463

Kurup S, Ride JP, Jordan N, Fletcher G, Franklin-Tong VE, Franklin FCH (1998) Identification and cloning of related self-incompatibility S-genes in *Papaver rhoeas* and *Papaver nudicaule*. Sex Plant Reprod 11:192–198

Lawrence MJ (2000) Population genetics of homomorphic self-incompatibility in flowering plants. Ann Bot 85 (suppl A):221–226

Levin RA, Miller JS (2005) Relationships within tribe Lycieae (Solanaceae): Paraphyly of *Lycium* and multiple origins of gender dimorphism. Am J Bot 92:2044–2053

Levin RA, Shak JR, Miller JS, Bernardello G, Venter AM (2007) Evolutionary relationships in tribe Lycieae (Solanaceae). Acta Hortic 745:225–239

Lu Y (2001) Roles of lineage sorting and phylogenetic relationship in the genetic diversity at the self-incompatibility locus of Solanaceae. Heredity 86:195–205

Ma RC, Oliviera MM (2002) Evolutionary analysis of S-RNase genes from Rosaceae species. Mol Genet Genom 267:71–78

Matton DP, Luu DT, Xike Q, Laublin G, O'Brien M, Maes O, Morse D, Cappadocia M (1999) Production of an S RNase with dual specificity suggests a novel hypothesis for the generation of new S alleles. Plant Cell 11:2087–2098

May G, Shaw F, Badrane H, Vekemans X (1999) The signature of balancing selection: Fungal mating compatibility gene evolution. Proc Natl Acad Sci USA 96:9172–9177

Miege C, Ruffio-Chable V, Schierup MH, Cabrillac D, Dumas C, Gaude T, Cock JM (2001) Intrahaplotype polymorphism at the *Brassica S*-locus. Genetics 159:811–822

Miller JS, Levin RA, Feliciano NM (2008) A tale of two continents: Baker's rule and the maintenance of self-incompatibility in *Lycium* (Solanaceae). Evolution Int J Org Evolution 62:1052–1065

Moran PAP (1962) The statistical processes of evolutionary theory. Clarendon Press, Oxford

Nunes MDS, Santos RAM, Ferreira SM, Viera J, Viera CP (2006) Variability patterns and positively selected sites at the gametophytic self incompatibility pollen *SFB* gene in a wild self-incompatible *Prunus spinosa* (Rosaceae) population. New Phytol 172:577–587

Olmstead RG, Bohs L, Abdel Magid H, Santiago-Valentin E, Collier SM, Garcia VF (in press) A molecular phylogeny of the Solanaceae. Taxon

Paape T, Igic B, Smith SD, Olmstead R, Bohs L, Kohn JR. (2008) A 15-million-year-old genetic bottleneck. Mol Biol Evol 25:655–663

Pagel M (1999) The maximum likelihood approach to reconstructing ancestral character states of discrete characters on phylogenies. Syst Biol 48:612–622

Qiao H, Wang F, Zhao L, Zhou J, Lai Z, Zhang Y, Robbins TP, Xue Y (2004) The F-box protein AhSLF-S2 controls the pollen function of S-RNase-based self-incompatibility. Plant Cell 16:2307–2322

Raspé O, Kohn JR (2002) The number of S-alleles in natural populations of *Sorbus acuparia* and *Crataegus monogyna* (Rosaceae). Heredity 88:458–465

Raspé O, Kohn JR (2007) Population structure at the *S*-locus of *Sorbus aucuparia*. Mol Ecol 16:1315–1325

Richman AD (2000) Evolution of balanced genetic polymorphism. Mol Ecol 9:1953–1963

Richman AD, Kohn JR (1999) Self-incompatibility alleles from *Physalis*: Implications for historical inference from balanced genetic polymorphisms. Proc Natl Acad Sci USA 96:168–172

Richman AD, Kohn JR (2000) Evolutionary genetics of self-incompatibility in the Solanaceae. Plant Mol Biol 42:169–179

Richman AD, Kao Th, Schaeffer SW, Uyenoyama MK (1995) S-allele sequence diversity in natural populations of *Solanum carolinense* (Horsenettle). Heredity 75:405–415

Richman AD, Uyenoyama MK, Kohn JR (1996a) *S*-allele diversity in a natural population of ground cherry *Physalis crassifolia* (Solanaceae) assessed by RT-PCR. Heredity 76:497–505

Richman AD, Uyenoyama MK, Kohn JR (1996b) Contrasting patterns of allelic diversity and gene genealogy at the self-incompatibility locus in two species of Solanaceae. Science 273:1212–1216

Sato K, Nishio T, Kimura R, Kusaba M, Suzuki T, Hatakeyama K, Okendon DJ, Satta Y (2002) Coevolution of the *S*-locus genes *SRK*, *SLG* and *SP11/SCR* in *Brassica oleraceae* and *B. rapa*. Genetics 162:931–940

Sato Y, Sato K, Nishio T (2006) Interspecific pairs of class II *S*-haplotypes having different recognition specificities between *Brassica oleraceae* and *Brassica rapa*. Plant Cell Physiol 47:340–345

Savage AE, Miller JS (2006) Gametophytic self-incompatibility in *Lycium parishii* (Solanaceae): Allelic diversity, genealogical structure, and patterns of molecular evolution. Heredity 96:434–444

Schierup MH, Mikkelsen AM, Hein J (2001) Recombination, balancing selection and phylogenies in MHC and self-incompatibility genes. Genetics 159:1833–1844

Sijacic P, Wang X, Skirpan AL, Wang Y, Dowd PE, McCubbin AG, Huang S, Kao T-h (2004) Identification of the pollen determinant of S-RNase mediated self-incompatibility. Nature 429:302–305

Sonneveld T, Tobutt KR, Vaughan SP, Robbins T (2005) Loss of pollen-S function in two self-compatible selections of *Prunus avium* is associated with deletion/mutation of an *S*-haplotype-specific F-box gene. Plant Cell 17:37–51

Stebbins GL (1957) Self-fertilization and population variability in the higher plants. Am Nat 91:337–354

Stebbins GL (1974) Flowering plants: Evolution above the species level. Belknap Press, Cambridge, MA

Steinbachs JE, Holsinger KE (2002) S-RNase-mediated gametophytic self-incompatibility is ancestral in eudicots. Mol Biol Evol 19:825–829
Stone JL (2002) Molecular mechanisms underlying the breakdown of gametophytic self-incompatibility. Quart Rev Biol 77:17–32
Stone JL, Pierce SE (2005) Rapid recent radiation of S-RNase lineages in *Witheringia solanacea* (Solanaceae). Heredity 94:547–555
Surbanovski N, Tobutt KR, Konstantinovic M, Maksimovic V, Sargent Dj, Stevanovic V, Ortega E, Boskovic RI (2007) Self incompatibility of *Prunus tenella* and evidence that reproductively isolated species of *Prunus* have different SFB alleles coupled with an identical S-RNase allele. Plant J 50:723–734
Takahata N (1990) A simple genealogical structure of strongly balanced allelic lines and trans-species evolution of polymorphism. Proc Natl Acad Sci USA 87:2419–2423
Takahata N (1993) Evolutionary genetics of human paleopopulations. In: Takahata N, Clarke AG (eds), Mechanisms of molecular evolution, Sinauer, Sunderland MA, pp 1–21
Takahata N, Nei M (1990) Allelic genealogy under overdominant and frequency-dependent selection and polymorphism of major histocompatibility complex loci. Genetics 124:967–978
Takayama S, Isogai A (2005) Self-incompatibility in plants. Annu Rev Plant Biol 58:467–489.
Takebayashi N, Morrell PL (2001) Is self-fertilization an evolutionary dead end? Revisiting an old hypothesis with genetic theories and a macroevolutionary approach. Am J Bot 88:1143–1150
Ushijima K, Sassa H, Dandekar AM, Gradziel TM, Tao R, Hirano H (2003) Structural and transcriptional analysis of the self-incompatibility locus of almond: identification of a pollen expressed F-box gene with haplotype-specific polymorphism. Plant Cell 15:771–781
Ushijima K, Yamane H, Watari A, Kakehi E, Ikeda K, Hauck NR, Iezzoni A, Tao R (2004) The S-haplotype-specific F-box protein gene, *SFB*, is defective in self-compatible haplotypes of *Prunus avium* and *P. mume*. Plant J 39:573–586
Uyenoyama MK (1997) Genealogical structure among alleles regulating self-incompatibility in natural populations of flowering plants. Genetics 147:1389–1400
Uyenoyama MK, Newbigin E (2000) Evolutionary dynamics of dual-specificity self-incompatibility alleles. Plant Cell 12:310–312
Uyenoyama MK, Zhang Y, Newbigin E (2001) On the origin of selfincompatibility haplotypes: Transition through self-compatible intermediates. Genetics 157:1805–1817
Vekemans X, Slatkin M (1994) Gene and allelic genealogies at a gametophytic self-incompatibility locus. Genetics 137:1157–1165
Vieira CP, Charlesworth D (2002) Molecular variation at the self-incompatibility locus in natural populations of the genera Antirrhinum and Misopates. Heredity 88:172–181
Vieira CP, Charlesworth D, Vieira J (2003) Evidence for rare recombination at the gametophytic self-incompatibility locus. Heredity 91:262–267
Wang L, Dong L, Zhang Y, Wu W, Deng X, Xue Y (2004) Genomewide analysis of S-locus F-box-like genes in *Arabidopsis thaliana*. Plant Mol Biol 56:929–945
Weller SG, Donoghue MJ, Charlesworth D (1995) The evolution of self-incompatibility in flowering plants: A phylogenetic approach. In: Hoch PC, Stephenson AG (eds) Experimental and molecular approaches to plant biosystematics, vol 53. Missouri Botanical Garden, Saint Louis, pp 355–382
Whalen MD, GJ Anderson (1981) Distribution of gametophytic self-incompatibility and infra-generic classification in *Solanum*. Taxon 30:761–767
Wheeler D, Newbigin E (2007) Expression of 10 S-class *SLF*-like sequences in Nicotiana alata pollen and its implications for understanding the pollen factor of the S-locus. Genetics 177:1–10
Wikström N, Savolainen V, Chase MW (2001) Evolution of the angiosperms: Calibrating the family tree. Proc R Soc Lond B 268:2211–2220
Wright S (1939) The distribution of self-sterility alleles in populations. Genetics 24:538–552
Wright S (1960) On the number of self-incompatibility alleles maintained in equilibrium by a given mutation rate in a population of a given size: A re-examination. Biometrics 16:61–85

Wright S (1964) The distribution of self-incompatibility alleles in populations. Evolution 18:609–619

Zhou J, Wang F, Ma W, Zhang Y, Han B, Xue Y (2003) Structural and transcriptional analysis of S-locus F-box genes in Antirrhinum. Sex Plant Reprod 16:165–177

Xue Y, Carpenter R, Dickinson HG, Coen ES (1996) Origin of allelic diversity in *Antirrhinum* S-locus RNases. Plant Cell 8:805–814

Chapter 6
Self-Incompatibility and Evolution of Mating Systems in the Brassicaceae

S. Sherman-Broyles and J.B. Nasrallah

Abstract Genetically determined self-incompatibility (SI) systems ensure high rates of out-crossing because they allow the pistil to recognise and reject genetically identical pollen. As such, SI systems are thought to be advantageous because populations with high levels of polymorphism have the genetic variability required for withstanding a wide range of environmental challenges. Nevertheless, SI has repeatedly been lost in plant lineages, and it has been noted that the most frequently travelled path in plant evolution is the path from out-crossing to self-fertility. In this chapter, we focus on the self-incompatibility system of the Brassicaceae and discuss results related to the diversification of the SI recognition repertoire and the various paths that underlie switches to self-fertility in the family.

Abbreviations

Col-0	Columbia accession of *A. thaliana*
Cvi-0	Cape Verdi Island accession of *A. thaliana*
CVR	C terminal variable region
EGF-like	Epidermal growth factor (EGF)-like
eSRK	A soluble form of the extracellular domain of SRK
HDI, HDII, HDIII	Hypervariable domains I, II, III
PAN_APPLE	A PAN module superfamily structural domain consensus sequence
PCR	Polymerase chain reaction
PUB8	A gene encoding an ARM repeat- and U box-containing protein

---

S. Sherman-Broyles and J.B. Nasrallah
Department of Plant Biology, Cornell University, Ithaca, NY 14853, USA, e-mail: sls98@cornell.edu and jbn2@cornell.edu

RLK	Receptor-like kinase
SCR/SP11	*S*-locus cysteine-rich protein (the pollen *S*-determinant in Brassica)
SI	Self-incompatibility
SLG	*S*-locus glycoprotein
S-locus	Self-incompatibility locus
SRK	*S*-locus receptor kinase (the pistil *S*-determinant in Brassica)
tSRK	A membrane-spanning truncated protein comprising the extracellular omain, the transmembrane domain and the juxtamembrane domain of SRK
ΨSCR	*SRK*-like sequences in *A. thaliana*

6.1 Introduction

Plant mating system evolution is characterised by two contrasting points of view. On the one hand, Darwin (1876) concluded that nature abhors self-fertilisation, as drastic reductions in fitness, or inbreeding depression, are frequently observed upon selfing of plant populations that are not under strong selective pressure to switch to self-fertilisation. On the other hand, Stebbins (1957) pointed out that the most frequently travelled path in plant evolution is the path from out-crossing to self-fertility. While self-fertility can initially have negative effects on fitness due to inbreeding depression, selfing populations may quickly become stable when deleterious or lethal load is purged (Byers and Waller 1999; Crnokrak and Barrett 2002; Lande and Schemske 1985). The type of mating system that a species or population displays is clearly important, as it largely determines the level of genetic polymorphism in the population (Hamrick and Godt 1996), and the level of polymorphism, together with population size, in turn determines the impact and effectiveness of natural selection. Populations with high levels of polymorphism have the genetic variability required for withstanding a wide range of environmental challenges.

Genetically determined self-incompatibility (SI) systems ensure high rates of out-crossing because they allow the pistil to recognise and reject genetically identical pollen. Early work suggested that SI was a basal characteristic in the angiosperms (Bateman 1952; Whitehouse 1950). Current phylogenetic analysis (see Chap. 4, this volume) suggests that the earliest angiosperms had some form of SI. It is now widely accepted that SI arose independently multiple times and that self-incompatible plants in different plant families utilise different recognition molecules and distinct pathways for arrest of self pollen tube development (Charlesworth et al. 2005; Nasrallah 2005; Takayama and Isogai 2005; see also chapters within Sect. II, this volume). The Brassicaceae family is characterised by the sporophytic form of SI, in which pollen SI phenotype is determined by the diploid genotype of the pollen-producing parent see Chap. 7, this volume), is thought to have arisen independently 9 or 10 times (Igic et al. 2008).

There is no doubt that the obligate out-crossing mode of mating conferred by SI is advantageous. In addition to producing high levels of polymorphisms at the population level, SI also confers a distinct advantage on individual plants that express a new SI specificity because the number of their potential mating partners is increased dramatically. Consequently, SI systems and their specificity-determining loci attain extraordinarily high levels of polymorphisms, and these polymorphisms typically persist for long periods of time, such that SI loci often exhibit trans-specific polymorphisms (i.e. the sharing of polymorphisms among closely related species) (Dwyer et al. 1991; Schierup et al. 2001; Bechsgaard et al. 2006; Kimura et al. 2002; Sato et al. 2003). Despite the advantages of SI, however, phylogenetic analyses have shown that SI was lost independently multiple times within SI lineages (Igic et al. 2006). For example, in the Brassicaceae almost half the taxa are self-compatible (Bateman 1955), including the model plant, *Arabidopsis thaliana*. How SI was lost to cause evolutionary switches to self-fertility is poorly understood. An answer to this and other critical evolutionary questions, including how different SI systems arose and how new self-recognition specificities were generated within each system, are important not only from an evolutionary perspective but also because they have the potential to reveal factors required for the operation of SI.

In this review, we focus on evolutionary studies of the sporophytic SI system of the Brassicaceae family. After presenting a brief overview of the SI specificity-determining *S*-locus and its 'self' recognition genes (see also Chaps. 7 and 8, this volume), we discuss what is known about diversification of the SI recognition repertoire and we outline the various paths that underlie switches to self-fertility in the family.

6.2 Structural and Sequence Diversification of *S*-locus Haplotypes and Their Recognition Genes

Molecular evolutionary studies of SI in the Brassicaceae gained momentum in recent years after the identification in *Brassica* species of the two highly polymorphic self recognition proteins encoded in the *S*-locus: the *S*-locus receptor kinase SRK, which is displayed at the surface of stigma epidermal cells, and the *S*-locus cysteine-rich protein SCR/SP11, which is localised in the pollen coat (see Fig. 6.1). SRK was first reported in 1991 (Stein et al. 1991), its plasma membrane localisation was established in 1996 (Stein et al. 1996) and proof that it determines SI specificity in the stigma was obtained in 2000 (Takasaki et al. 2000). SCR was first reported in 1999 by Schopfer et al. (1999) who provided the sequences for three SCR variants and also demonstrated, by analysis of loss-of-function induced mutants and by gain-of-function transgenic experiments, that SCR represented the pollen determinant of SI specificity. The same gene was described as an *S*-locus-linked sequence designated *SP-11* by Suzuki et al. (2000) and subsequently shown to represent the pollen determinant of SI using a pollination bioassay (Takayama

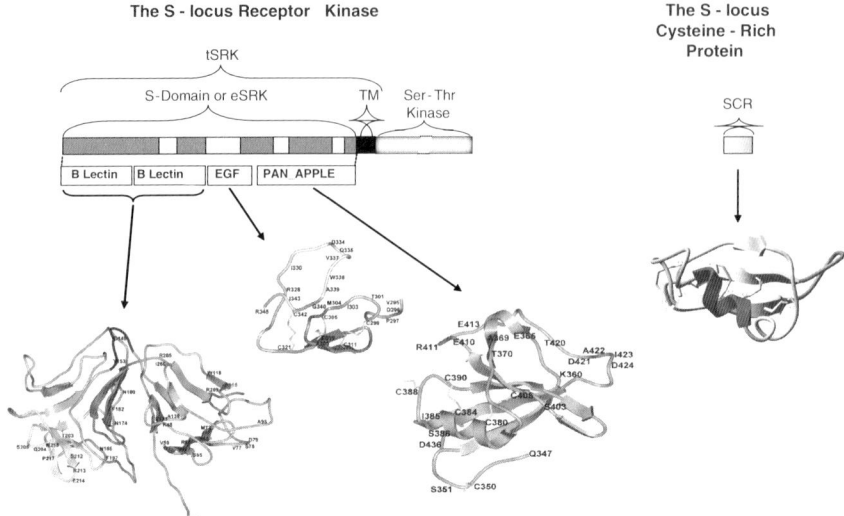

Fig. 6.1 The *S*-locus self-recognition proteins, SRK and SCR. The top diagrams are schematic representations of the mature forms of SRK (*to the left*) and SCR (*to the right*). SRK is shown as consisting of an extracellular S-domain or eSRK (*grey rectangle*), followed by a transmembrane domain (TM; *black square*), and a serine–threonine (Ser/Thr) kinase domain (*l grey rectangle*; not drawn to scale). Within the eSRK, the hypervariable regions, which contain candidate specificity-determining residues, are depicted as grey boxes and the conserved cysteines as white bars. Brackets indicate the two alternative forms of SRK detected in stigmas, tSRK and eSRK. Below the SRK diagram, the figure shows the location and 3D-models of the structural modules predicted in the eSRK (adapted from Naithani et al. 2007). To the right of the figure is a 3-D image of the CS$\alpha\beta$ structure of SCR (from Chookajorn et al. 2004). The regions between Cys3 and Cys4 (*light grey arrows*) and between Cys5 and Cys6 (*dark grey arrow*) are predicted to form surface-exposed loops that might contain specificity determinants

et al. 2001) and transgenic experiments (Shiba et al. 2001). Later studies demonstrated that SRK physically interacts with SCR via its extracellular domain *in vitro* and *in vivo,* that the SRK-SCR interaction is *S*-haplotype-specific and that it is this specificity that underlies recognition of self pollen (Chookajorn et al. 2004; Kachroo et al. 2001; Kemp and Doughty 2007; Shimosato et al. 2007; Takayama et al. 2001; see also Chap. 7, this volume, for a fuller account of the identification of *S*-locus components).

These early *Brassica* studies provided a framework for an ongoing mechanistic analysis of SI as described in detail in Chap. 8, this volume. They also allowed the isolation of *SRK* and *SCR* sequences from other species of the Brassicaceae, including *Raphanus*, *Arabidopsis lyrata,* and in *Capsella grandiflora* (Paetsch et al. 2006; Sakamoto et al. 1998; Kusaba et al. 2001; Nasrallah et al. 2007) and the transfer of the SI trait into the self-fertile model plant *A. thaliana* (Nasrallah et al. 2002, 2004). In turn, these developments set the stage for molecular evolutionary studies of the *S*-locus and its genes and for a molecular genetic investigation of switches in breeding system.

6.2.1 Conserved and Diverged Features of the S-locus

6.2.1.1 The SRK and SCR/SP11 Genes

An invariant feature of all *S*-locus variants or *S*-haplotypes examined to date in various Brassicaceae species is the presence of *SRK* and *SCR/SP11* genes. Variant forms of these genes are generally conserved in overall structure. *SRK*, like other members of the *S*-domain receptor-like kinase (RLK) family, which is defined on the basis of a unique extracellular 'S' (for *S*-locus) domain, consists of seven exons, with the first exon encoding the signal peptide and extracellular S domain the second exon encoding the transmembrane domain, and the remaining exons encoding the cytoplasmic kinase domain (Stein et al. 1991). This gene structure and the use of alternative polyA addition sites are associated with production of alternative *SRK* transcripts and accumulation of two major SRK isoforms: the full-length SRK receptor kinase (Stein et al. 1996) and a soluble version of the extracellular domain (designated eSRK) (Giranton et al. 2000). Additionally, a subset of SRK variants also exist as membrane spanning truncated protein (tSRK) consisting of the extracellular domain, the transmembrane domain and the juxtamembrane domain (Shimosato et al. 2007) (see Fig. 6.1 for a cartoon depicting these different domains). For their part, all *SCR/SP11* genes identified to date consist of one exon that encodes a signal peptide and a second exon that encodes the mature SCR/SP11 protein, a small protein of approximately 50 amino acids that shares structural similarities with defensins (Schopfer et al. 1999; Chookajorn et al. 2004; Takayama et al. 2001).

Beyond these similarities, the *S* loci of *Brassica*, *Raphanus*, *Arabidopsis* and *Capsella* species exhibit important inter-specific differences in genomic location and gene content. The first structural studies of the crucifer *S*-locus were carried out in *Brassica* species (Boyes et al. 1997), and comparative mapping of flanking markers placed the locus in a region that is co-linear with a region of *A. thaliana* chromosome 1 (Conner et al. 1998). However, the *S*-locus was mapped to a different chromosomal region in *Arabidopsis* species, namely the region of *A. thaliana* chromosome 4 between At4g21350 and At4g21380 (Kusaba et al. 2001). Thus, a translocation of the *S*-locus region must have occurred in either the *Brassica/Raphanus* or *Arabidopsis* lineages. Ascertaining which of the two chromosomal locations is ancestral will require currently unavailable data from additional crucifer genera.

6.2.1.2 The SLG Genes

Another feature of the *Brassica/Raphanus S*-locus that distinguishes it from the *S*-locus of *Arabidopsis* (Kusaba et al. 2001) and *Capsella* (Nasrallah et al. 2007) species is that the vast majority of its haplotypes contains a third gene in addition to *SRK* and *SCR* (see Fig. 6.2). This gene, the *S*-locus glycoprotein (*SLG*) gene, encodes an abundant glycoprotein expressed specifically in the stigma epidermal cell wall (Kandasamy et al. 1989). *SLG* was in fact the first *S*-locus gene identified by virtue of its high-level stigma-specific expression, polymorphism and linkage to the *S*-locus (Nasrallah et al. 1985; see also Chap. 7, this volume). The identification of *SLG* facilitated the subsequent identification of *SRK*, the extracellular domain of which shares extensive sequence similarity with *SLG* (Stein et al. 1991). Despite exhibiting several features expected of a stigma determinant of SI specificity, SLG does not function in this capacity in *Brassica* species. Its absence from a small number of *Brassica S*-haplotypes (Suzuki et al. 2000) and from the *S* loci of *Arabidopsis* (Kusaba et al. 2001) and *Capsella* species (Nasrallah et al. 2007) indicates that it is not required for SI and that the gene arose in the *Brassica/Raphanus* lineage, possibly by partial duplication of the *SRK* gene followed by acquisition of regulatory elements that causes accumulation of its products to $\sim$200-fold higher levels than *SRK* products. What then is the role of SLG and why has it persisted in the majority of *Brassica/Raphanus S*-haplotypes? A possibility is that SLG is required for the stabilisation or proper maturation of at least some variants of SRK. In the *B. rapa scf1* mutant, which exhibits breakdown of SI in the stigma and highly reduced *SLG* transcript and protein levels, SRK protein is undetectable despite normal levels of *SRK* transcripts (Nasrallah et al. 2002; Dixit et al. 2000). Furthermore, in tobacco leaves, SRK exhibits aberrant oligomerisation when expressed alone but not when expressed together with SLG (Dixit et al. 2000). Finally, co-expression of *SLG* with *SRK* in transgenic plants has been shown to enhance SRK-mediated SI in one case (Takasaki et al. 2000), although not in another (Silva et al. 2001). Why some SRK variants might require an SLG 'chaperone' while others do not is not known. It is possible that in plants lacking *SLG*, the cell wall-localised eSRK (Giranton et al. 2000) might fulfill a role similar to that proposed for SLG.

6.2.2 Intra-Specific Structural Heteromorphism and Sequence Polymorphism: Suppressed Recombination and Maintenance of SRK-SCR Linkage

The fact that crucifer SI is based on allele-specific interaction between SRK and *SCR/SP11*, combined with the fact that self-incompatible plants are typically heterozygous at the *S*-locus, means that, for SI to persist, absolute linkage of matched *SRK* and *SCR/SP11* alleles must be maintained. Indeed, despite extending over several hundred kilobases in some instances (Boyes et al. 1997; Casselman et al. 2000), the *S*-locus region is characterised by reduced recombination. The locus exhibits simple Mendelian inheritance as one genetic unit, and no recombinants between *SRK* and *SCR* were recovered even when large populations of plants segregating for *S*-haplotype were analysed (Casselman et al. 2000). Based on analysis of *A. lyrata S*-haplotypes, linkage disequilibrium in the region has been estimated to extend for 400–600 kb (Hagenblad et al. 2006; Kawabe et al. 2006; Kamau and Charlesworth 2005). Additionally, *S*-haplotypes exhibit trans-specific evolution over long periods of evolutionary time (Boyes et al. 1997; Casselman et al. 2000; Charlesworth and Awadalla 1998; Kimura et al. 2002; Sato et al. 2003; Charlesworth et al. 2006; Takuno et al. 2007). It should be noted, however, that recombination is thought to have occurred during the evolutionary history of *S*-locus genes (Awadalla and Charlesworth 1999; Kusaba et al. 1997; Kusaba and Nishio 1999; Takuno et al. 2007). Recombination must have also played an important role in the creation of the *S*-locus, as a recombination event must have brought *SRK* and *SCR* into close physical linkage.

Several features of the *S*-locus likely contribute to the suppressed recombination observed in the region. Reduced recombination in some *S*-haplotypes, such as *B. rapa S8, A. lyrata Sa* and *Capsella grandiflora S7,* may be ascribed to the very close physical proximity of *SRK* and *SCR,* with less than 2 kb of DNA separating the two genes (see Fig. 6.2). Given the wide variation in the size of *S*-haplotypes (Boyes

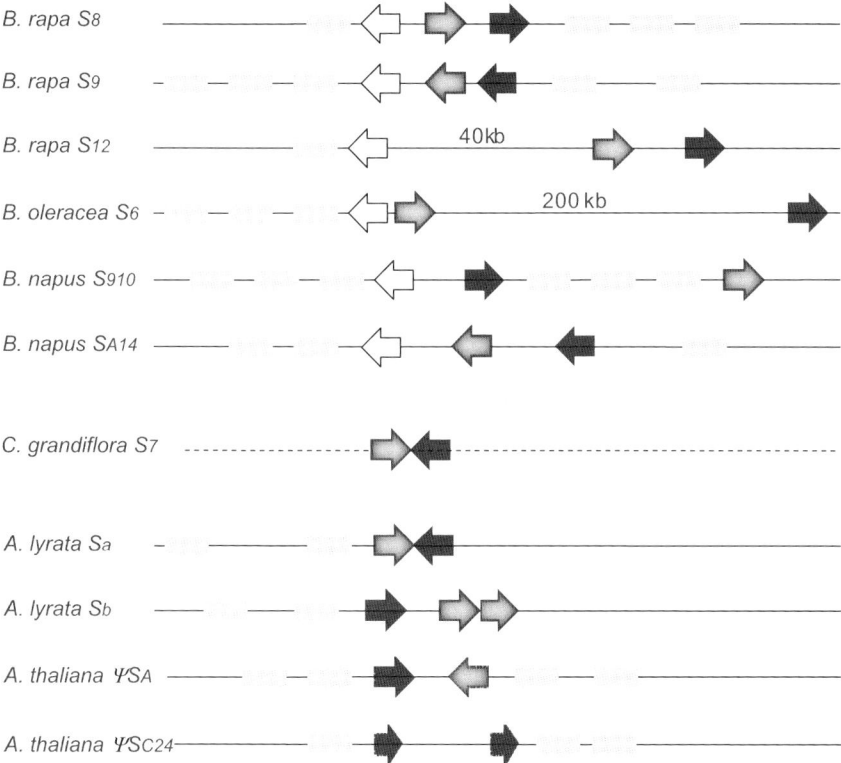

Fig. 6.2 Structural heteromorphism of *S*-haplotypes in three genera of the Brassicaceae. In each haplotype, the 5′ → 3′ gene orientation is depicted by black arrows for *SRK*, dark grey arrows for *SCR*, and light grey arrows for *SLG* (which is absent from the *S*-locus of *Arabidopsis* and *Capsella* species). *S*-locus-linked genes that do not function in SI are shown as grey boxes, except in *C. grandiflora,* for which no flanking genes have been identified (depicted by the *dashed line*). Note the intra-specific differences in the orientations of the *S*-locus genes relative to each other and to flanking genes, and in the distances separating the genes, all of which contribute to reduced recombination in the region. All *S*-haplotypes shown are functional except for the *A. thaliana* ΨS_A haplotype of the Col-0 accession, which contains a non-functional *SRK* allele and truncated *SCR* sequences, and the ΨS_{C24} haplotype of the C24 accession, which contains only truncated *SRK* sequences and no *SCR*

et al. 1997; Kusaba et al. 2001) (Fig. 6.2), additional features must contribute to suppression of recombination in the region. In particular, intra-specific comparisons of different *S*-haplotypes has revealed features that reduce co-linearity in the *S*-locus region. One such feature is the highly rearranged gene order, or structural heteromorphism, exhibited by different *S*-haplotypes, whereby *SRK* and *SCR* occupy different positions and orientations relative to each other and to flanking markers (Boyes et al. 1997; Nasrallah 2000) (see Fig. 6.2). Another prominent feature is the presence of haplotype-specific sequences, often related to transposable elements (Boyes et al. 1997; Fujimoto et al. 2006b; Uyenoyama 2005; Sherman-Broyles et al. 2007). These transposon-like sequences are generally younger than the *S*-haplotypes that harbour them, suggesting that the *S*-locus acts as a 'transposon trap', similar to other regions exhibiting suppressed recombination, which also accumulate transposons and other haplotype-specific sequences due to long-term independent evolution. As such, the *S*-locus of crucifers belongs in a relatively rare class of loci that fit the definition of a supergene, a term coined to refer to genes that are tightly linked and must co-evolve to maintain function.

6.2.3 Diversification of the S-locus Genes and the SI Recognition Repertoire

Balancing selection, in particular negative frequency-dependent selection, acts on the *S*-locus within populations. Because rare *S*-haplotypes have the advantage of increased opportunity for successful matings compared to more prevalent *S*-haplotypes (see Chap. 3 and Charlesworth 2006), they increase in frequency while common haplotypes decrease in frequency. Thus, *S*-haplotypes are long-lived and not lost due to genetic drift. Gene trees of *S*-locus genes have long-branch lengths, which is the signature of balancing selection. In contrast to gametophytic SI, in which allele frequencies approach equilibrium, in sporophytic SI, allele frequencies are not equal because two alleles can be expressed in each pollen grain and because of dominance effects and segregation distortion (transmission advantage). Dominant alleles can accumulate linked deleterious alleles contributing to their low frequency within populations. Recessive alleles have a transmission advantage, which results in higher frequencies and the occurrence of homozygous individuals within populations (Bechsgaard et al. 2004). Pollen-recessive *S*-haplotypes are not rejected by stigmas carrying the same allele; these types of matings can lead to homozygous individuals within the population and these alleles will reach higher frequencies.

While analysis of allele frequencies can shed light on the fate of *S*-haplotypes in populations, major unresolved questions relate to how new SI specificities were generated (see Chap. 5, this volume) and how the SRK receptor and its SCR/SP11 ligand co-evolve to maintain their interaction. Answers to these questions will require knowledge of which amino-acid residues determine specificity in the receptor (specifically its extracellular domain) and in its ligand. Progress on this front has been slow and largely based on statistical analysis of sequences, but some empirical results are beginning to emerge (Chookajorn et al. 2004; Kemp and Doughty 2007; Naithani et al. 2007).

6.2.3.1 SRK Polymorphisms

The SRK extracellular domain (designated S domain or eSRK) and SLG are the prototypic members of the *S* gene family (named for the S domain), which includes receptor-like kinases similar to SRK and receptor-like proteins similar to SLG (Dwyer et al. 1994; Shiu and Bleecker 2001). The S domain is a distinctive sequence that usually includes 12 conserved cysteine residues. The three-dimensional (3D) structure of this S domain has not been solved, but computer generated 3D protein models have been recently generated for 97% of the eSRK6 variant (Naithani et al. 2007) (see Fig. 6.1). These models, together with sequence similarities to domains of known structure, are consistent with a modular structure of the eSRK. At the N-terminus are two lectin-like domains, similar to the two lectin domains of the mannose-binding Scafet protein of bluebell. These lectin-like domains are followed by an epidermal growth factor (EGF)-like domain containing the first six conserved cysteines, and by a C-terminal PAN_APPLE domain containing the remaining six conserved cysteines. Lectin-like domains are typically associated with carbohydrate binding, while EGF and PAN_APPLE domains are protein interaction domains, and the presence of these domains suggests that the eSRK might engage in interactions, not only with proteins but also with carbohydrates. However, eSRK might not possess carbohydrate-binding activity. First, the consensus mannose-binding site QXDXNXVXY is missing in eSRK. Second, the presence of a lectin-like domain is not always indicative of mannose-binding activity, as illustrated by curculin (Barre et al. 1997), a protein from the fruits of *Curculigo latifolia* that exhibits appreciable sequence similarity to GNA (a mannose-binding lectin from the snowdrop *Galanthus nivalis*) but has no apparent capacity to bind mannose.

Little is known about the role of individual eSRK modules and which might interact with SCR. To date, only the PAN_APPLE domain has been identified by yeast two-hybrid interaction studies as a major pre-ligand association domain responsible for ligand-independent SRK self-association or self-dimerisation (Naithani et al. 2007). Solving the 3D structure of the eSRK, defining the role of individual modules, as well as identifying and mapping specificity-determining residues onto the 3D structure are all critical for understanding SRK function, how it interacts with its cognate SCR and how this interaction causes its activation. In the absence of this information, efforts have focused on sequence comparisons of SRK variants to predict which residues are likely to function as specificity determinants. Diversity among *SRK* alleles is as high as 35% within species of *Brassica* (Stein et al. 1991; Kusaba et al. 1997) and as high as 51% among alleles in *A. lyrata* (Schierup et al. 2001). As indicated earlier, *SRK* alleles from one species may share higher nucleotide identity with alleles from a different species than with alleles in the same species. This trans-specificity of alleles is a signature of diversifying selection and indicates that genes at the *S*-locus are quite ancient (Bechsgaard et al. 2006; Charlesworth et al. 2006; Dwyer et al. 1991).

Sequence alignment of allelic variants, first of SLG (Kusaba et al. 1997; Nasrallah et al. 1987) and subsequently of eSRK, demonstrated that polymorphic (i.e. non-synonymous) residues, although scattered over the length of the sequences, are concentrated in four regions, three of which are called hypervariable domains (HDI, HDII, and HDIII) (Kusaba et al. 1997; Nasrallah et al. 1987), and one called the C-terminal Variable Region (CVR). Three of these hypervariable regions are embedded entirely within a predicted structural module: HDI within the second lectin-like domain, HDIII within the EGF-like domain and the CVR within the PAN_APPLE domain (see Fig. 6.1). These hypervariable regions have been the focus of much attention as potential determinants of SI specificity. The possibility that these domains contain SI specificity determinants is supported by the observed ratio of non-synonymous to synonymous substitutions (K_a/K_s), which are often >1 in these regions (Sato et al. 2002; see also Chap. 5, this volume). Furthermore, statistical analysis of eSRK variants identified 40 residues having a high posterior probability of being under selective pressure to change in physico-chemical property, such as volume, polarity, or charge, the majority but not all of which were located within the HDs (Sainudiin et al. 2005). These results support the notion that hypervariability in the HDs of eSRK is due to diversifying selection, as might be expected for SI specificity determinants, rather than to relaxed constraint. It should be noted, however, that some allelic comparisons of the HDs produce K_a/K_s ratios that are only modestly elevated relative to the remainder of the molecule, possibly because the high level of overall sequence variation might obscure positively selected residues (Awadalla and Charlesworth 1999). Furthermore, it has been suggested that HDIII is unlikely to function in this capacity because it exhibits within-haplotype variation and contains polymorphisms that are shared with other specificities, while HDI and HDII are more likely involved because they are perfectly conserved in SRKs that exhibit the same SI specificity (Miege et al. 2001).

The task for the future is to test the predictions of these statistical studies by empirical assays of chimeric receptors generated by swapping domains between SRK variants and of mutant receptors produced by site-directed mutagenesis of specific residues. The expectation is that assaying these variants for modified affinity towards the corresponding *SCR/SP11* variants and for *in planta* specificity in recognition and inhibition of pollen will identify individual residues or cluster of residues that cause a change in SI specificity when swapped between different SRK variants. These experiments are critical for determining with confidence which residues are important for SRK function and for its specific interaction with its cognate SCR/SP11 ligand. The potential resolving power of domain swapping experiments is illustrated by recent studies in yeast, which identified the CVR in the self-interaction PAN_APPLE domain, and in particular a small cluster of 11 amino acids within the CVR, as being responsible for specificity in SRK self-interaction (Naithani et al. 2007). These results have uncovered a previously unknown facet of specificity in SRK function and demonstrated that at least some polymorphisms in the eSRK are relevant to interactions other than the interaction of the receptor with its ligand. The biological significance of allelic specificity in eSRK self-interaction remains to be established. The *S*-haplotypes of Brassicaceae are known to exhibit puzzling allelic interactions of dominance/recessiveness and mutual weakening in the stigma. These interactions are largely a function of SRK (Hatakeyama et al. 2001), but are not associated with significant changes in the expression of recessive or weakened alleles. It is tempting to speculate that these allelic interactions might be due to differences in the propensity of SRK variants to form heterodimers having modified affinity towards the SCR ligand (Naithani et al. 2007).

6.2.3.2 SCR/Sp11 Polymorphisms

SCR/SP11 alleles are even more polymorphic than *SRK*. In *Brassica*, alleles of *SCR/SP11* can share as little as 35% amino-acid sequence identity (Schopfer et al. 1999; Watanabe et al. 2000; Kusaba et al. 2001), although the group of three *SCR* alleles derived from the pollen-recessive class-II *S*-haplotypes are much more conserved and can differ by as little as five amino acids (Shiba et al. 2002). For other *SCR* alleles, including alleles from *Brassica* class-I *S*-haplotypes and alleles of *A. lyrata*, only a small number of residues are conserved in most SCRs, with some exceptions. These include eight cysteine residues (Cys1 through Cys8), the glycine residue in the GlyxCys2 motif and an aromatic residue in the Cys3xxxTyr/Phe motif. Between these conserved regions, both the number and identity of amino acids differ widely. Because of this extreme divergence, identification of new *SCR/SP11* alleles (especially in *A. lyrata*) can be impossible to achieve using PCR-based methods, but rather requires the more laborious strategy of chromosome walking from the less variable *SRK* or *S*-locus flanking genes (Schopfer et al. 1999; Kusaba et al. 2001). As a result, the number of cloned *SCR* variants is often much smaller than the number of cloned *SRKs*. For example, in *A. lyrata*, only a few *SCR* alleles are known (Kusaba et al. 2001; Dwyer et al. unpublished data), while more than 37 *SRK* alleles have been isolated (Bechsgaard et al. 2006). In *Capsella grandiflora*, only one *SRK-SCR* gene pair has been cloned by genomic library screening (Nasrallah et al. 2007), while six putative *eSRK* sequences were obtained by PCR (Paetsch et al. 2006).

Notably, *SCR/SP11* alleles can also differ in their expression pattern. As indicated earlier, all known *SCR* alleles in both *Brassica* and *A. lyrata* are expressed in the anther tapetum (Schopfer et al. 1999; Shiba et al. 2001; Kusaba et al. 2002), consistent with sporophytic control of pollen SI phenotype. However, while many *SCR/SP11* alleles also exhibit gametophytic expression in developing microspores, a subset of *SCR/SP11s*, specifically the pollen-recessive alleles, are expressed exclusively in the tapetum (Kusaba et al. 2002; Shiba et al. 2002; Kakizaki et al. 2003). And it is the silencing of these alleles in heterozygous combinations with pollen-dominant alleles that explains dominant-recessive interactions of *S*-haplotypes in pollen (Kusaba et al. 2002; Shiba et al. 2002). Curiously, the silencing of a recessive allele does not require that the 'suppressing' allele be transcribed (Fujimoto et al. 2006a). Furthermore, transformation of a *B. oleracea* plant homozygous for the pollen-recessive *S2* haplotype with an *SCR6* construct derived from the pollen dominant *S6* haplotype (Schopfer et al. 1999) does not cause silencing of the resident *SCR2* allele, suggesting that *SCR* transcripts are not themselves effectors of silencing. Silencing of a recessive *SCR/SP11* allele in heterozygotes was associated with increased promoter methylation (Shiba et al. 2006); see also Chap. 7 this volume, but it is not known what specific features in pollen-recessive *SCR/SP11* genes or in the larger context of the *S*-haplotypes in which they reside make these genes susceptible to silencing.

Despite their extensive sequence divergence, SCR variants are predicted to assume the same overall 3D structure. This structure, which was determined experimentally for the SCR8 variant of *B. rapa* (Mishima et al. 2003) and modelled for several other variants (Chookajorn et al. 2004), consists of a cysteine-stabilised $\alpha\beta$ fold similar to defensins (see Fig. 6.1). The buried hydrophobic core of this structure is formed by disulfide bridges linking Cys2 to Cys5, Cys3 to Cys6 and Cys4 to Cys7, while the N and C termini are held together by the Cys1-Cys8 disulfide bridge. The regions between Cys3 and Cys4, on the one hand, and between Cys5 and Cys6, on the other hand, are predicted to form surface-exposed loops (Fig. 6.1) that are likely involved in binding eSRK. Consistent with this prediction, analysis of *Brassica* SCR chimeras in which the Cys5-Cys6 region in *SCR6* was replaced with the corresponding region from *SCR13* identified four contiguous amino acids within this region that are responsible for specific interaction with eSRK13 and for specific activation of SRK13 in stigmas (Chookajorn et al. 2004). However, the reciprocal chimera (i.e. one in which the Cys5-Cys6 region in SCR13 was replaced with that from SCR6), as well as chimeras involving other regions of SCR6, did not acquire SCR6 specificity (Chookajorn et al. 2004). This result might be ascribed either to different amino-acid residues functioning as specificity determinants in different SCRs, or more trivially, to aberrant folding of the chimeras. In either case, this experiment illustrates the potential difficulties that will likely arise in structure-function studies of both SCR and SRK, and the results do not bode well for our future ability to derive general principles on the basis of analysis of only one or even a few variants.

6.2.3.3 Hypotheses for Generation of New SI Specificities

The little information available regarding SI specificity determinants also means that we are far from being able to explain how SRK and SCR co-evolve and how new SI specificities were generated. Clearly, extant SRK and SCR variants must represent an evolutionary compromise between the drive for novelty and pressure for maintaining function. As discussed in Chap. 5 (this volume), some hypotheses for the generation of new specificities in two-component SI systems (reviewed in Charlesworth et al. 2005) have invoked passage through intermediates having dual specificity in stigma or pollen (Matton et al. 1999; Kao and Tsukamoto, 2004) or through self-compatible intermediates, whereby a change in one gene that disrupts SI is followed by a compensatory mutation in the second gene that restores SI (Uyenoyama et al. 2001). However, such aberrant intermediates are unlikely to persist along with functional *S*-haplotypes (Uyenoyama and Newbigin 2000; Charlesworth, 2000; Uyenoyama et al. 2001). More recent hypotheses for the generation of novel specificities in various SI systems avoid this difficulty by proposing that all intermediates retain SI (Chookajorn et al. 2004; Uyenoyama and Newbigin, 2000; Charlesworth et al. 2005). For crucifers, one hypothesis accounts for empirical studies of the SRK-SCR interaction and the remarkable 'evolvability' of the SCR protein, which allows adoption of a wide range of specificities within the framework of the inherently flexible $CS\alpha\beta$ scaffold (Chookajorn et al. 2004). The SCR6/SCR13 mutagenesis and domain swapping experiments described earlier have revealed the unusual feature that SCR can tolerate significant changes in its sequence (Chookajorn et al. 2004), suggesting that a change in specificity can result from a small number of non-synonymous substitutions and that in nature, new specificities might arise by a similar process involving selection for increased-affinity SRK/SCR pairs (Chookajorn et al. 2004).

6.3 Evolutionary Switches from Self-Incompatibility to Self-Fertility

Self-compatibility (SC) is a derived characteristic frequently found within genera, families and groups of families that share the same type of SI, including the Brassicaceae (Goodwillie 1999; Igic et al. 2006; see also Chap. 4, this volume). While repeated loss of SI produces an excess of self-fertile taxa in phylogenetic trees, these taxa are thought to be less fit than self-incompatible taxa in the long term: SI taxa persist for long periods of time while self-fertile taxa are probably eliminated repeatedly due to their reduced capacity for diversification (Igic et al. 2006; Chap. 5, this volume). In any case, choice of mating system is a trade-off between competing forces; (see also Chap. 2, this volume). Self-incompatibility is maintained by a balance of inbreeding depression and heterosis. Inbreeding depression can be as high as 50% and most outcrossing species are heterozygous for one or more recessive lethal alleles (Barrett 1988), and this deleterious genetic load must be purged for self-fertile taxa to be established. Populations with large effective population sizes (N_e) may have higher levels of selfing than expected (Olmsted 1986). The relative fitness of selfed vs. outcrossed individuals will change as N_e decreases. In large populations, inbreeding depression reduces the advantage of reproductive assurance found in selfers. In small populations, the importance of reproductive assurance outweighs the reduction of fitness associated with inbreeding depression. The effects of inbreeding depression are diminished in time frames of 100 generations or less if cycles of pollinator limitation or population size reduction can decrease the genetic load in a gradual manner (Lande and Schemske 1985). Recent modelling of the evolution of selfing in a gametophytic SI system considered two modes of switching to SC, one involving mutation in the pollen *S* gene and the other involving a mutation in an unlinked modifier locus (Porcher and Lande 2005). Assuming equal frequencies of *S* alleles in the population, both scenarios show that under selection for reproductive assurance, inbreeding depression can be overcome and selfing alleles can spread, with mutations at modifier loci being less likely than mutations at pollen *S* genes.

In Brassicaceae, as in other families, self-fertile taxa are common. Modelling of the switch to self-fertility in this sporophytic SI system is complicated by the complex dominance relationships among *S*-haplotypes, which violate a major assumption typically made in modelling studies, namely that the frequency of *S* alleles would be equal in the population. Nevertheless, comparisons of self-fertile and self-incompatible populations within crucifer species are consistent with theoretical expectations. For example, self-fertile *Leavenworthia alabamica* populations show reduced inbreeding depression relative to self-incompatible populations (Busch 2005). Furthermore, the occurrence of small isolated self-fertile populations at the periphery of the *Leavenworthia* range argues for multiple origins of selfing in areas where sub-structure led to a decrease in *S*-haplotypes or to fewer pollinator visits.

Like other families (Goodwillie et al. 2005), self-incompatible species of the Brassicaceae can exhibit varying levels of partial self-compatibility caused by weakening of SI, as noted for *A. lyrata* (Mable 2003; Mable et al. 2005). This partial self-compatibility, also known as pseudo self-compatibility, is often considered to be a 'best-of-two worlds' mixed mating strategy, which combines the benefits of out-crossing with the ability to self when out-crossing is not possible, such as when few compatible mates are available (population bottlenecks) or when pollinators are scarce (Goodwillie et al. 2005; see also Chap. 2, this volume). Mutations at SI modifier loci are known which cause transient SI, whereby stigmas are initially self-incompatible but later lose the ability to reject self pollen (Liu et al. 2007; and see below). In addition, some *S*-locus haplotypes are inherently leaky and allow some degree of selfing, and such *S*-haplotypes can accumulate to high levels in populations (Prigoda et al. 2005).

6.3.1 Molecular Genetics of Switches to Self-Fertility

Analysis of spontaneously derived or induced self-fertile strains of predominantly self-incompatible species, of self-fertile close relatives of self-incompatible species and of inter-specific hybrids has indicated that similar types of genetic events underlie intra-specific and inter-specific switches from outcrossing to selfing. In all, mutations that lead to breakdown of SI may be grouped into two classes. One class includes mutations that disrupt the integrity of *SRK* or *SCR* and cause, respectively, stigma-specific and pollen-specific loss of SI (Goring et al. 1993; Nasrallah 1974, 2000, Nasrallah et al. 1992, 1994, 2007; Schopfer et al. 1999; Fujimoto et al. 2006, b; Okamoto et al. 2007). Indeed, the *S*-locus was repeatedly targeted by selection for self-fertility in nature, and disruption of its genes or of the linkage between its co-adapted *SRK* and *SCR* alleles is one of the primary routes to self-fertility. Indeed, analysis of the *S*-locus genes in self-compatible species or strains has provided evidence for the occurrence of multiple independent mutations of *SRK* and *SCR* in *A. thaliana* (Kusaba et al. 2001; Nasrallah et al. 2002) and *Brassica* species (Fujimoto et al. 2006a; Okamoto et al. 2007).

Another class of mutations consists of mutations at modifier loci, which in turn may be grouped into two sub-classes. Mutations in one sub-class disrupt SI by reducing the expression of *S*-locus genes. This sub-class is exemplified by a hypomorphic allele of *PUB8* in *A. thaliana* that is associated with reduced levels of *SRK* transcripts at later stages of stigma development (Liu et al. 2007; see below) and by the previously mentioned *scf1* mutation in *B. rapa*, which indirectly causes reductions in SRK levels by disrupting expression of the *SLG* gene (Nasrallah et al. 1992).

Other mutations at modifier loci disrupt the SI response without affecting expression of the *S*-locus recognition genes, and these mutations are expected to encode SI signalling intermediates or molecules required for their activity or expression. This class is defined by the *mod* mutation, which causes stigma-specific breakdown of SI without affecting SRK expression (Murase et al. 2004; see Chap. 8, this volume). Interestingly, this mutation was originally identified in the Yellow Sarson cultivar of *B. rapa*, indicating that the switch to self-fertility can be associated with multiple inactivating mutations at the *S*-locus and SI modifier loci. Presumably, disruption of the *S*-locus would relieve selective constraints on other genes that have SI-specific functions, causing accumulation of secondary mutations in these now-dispensible genes.

6.3.2 Breakdown of SI by Disruption of S-locus Gene Expression in Inter-Specific Hybrids

Inter-specific hybridisation has played a major role in plant evolution, producing novel plant species with or without change in ploidy. Hybrid speciation is promoted by selfing due to reduced probability of gene flow from parental species. When hybridisation involves a self-incompatible species, SI must be overcome to allow selfing. To understand how the normally dominant SI trait is lost in nascent hybrids, loss of SI was analysed in synthetic hybrids generated by crossing self-fertile and self-incompatible species of *Arabidopsis* and *Capsella* (Nasrallah et al. 2007). In *A. thaliana-lyrata* hybrids and their neo-allotetraploid derivatives, loss of SI occurred in stigmas, which exhibited aberrant processing of *SRK* mRNA and production of little or no fully spliced transcripts. Interestingly, normal processing was regained concomitant with restoration of SI in a backcross to *A. lyrata*, indicating that reduced levels of the SRK receptor resulting from aberrant transcript processing caused breakdown of SI in these plants. In *C. rubella-grandiflora* hybrids and their homoploid progenies, SI was lost in pollen due to suppression of *SCR/SP11* in heterozygous anthers, similar to the silencing of pollen-recessive *SCR* alleles in *S*-locus heterozygotes of *Brassica* and *A. lyrata*. Unlike the irreversible loss of SI caused by changes in *SRK* or *SCR/SP11* DNA sequence, the reversibility of these epigenetic mechanisms might provide nascent inter-specific hybrids in nature with flexibility in their selection of mating system, possibly contributing to reproductive success under uncertain ecological conditions.

6.3.3 The Case of Self-Fertility in A. thaliana

A. thaliana is a small weedy species whose native range is believed to center in western Asia and eastern Europe, from where it was spread by human activity to most continents (Hoffmann 2002). *A. thaliana* is an ephemeral plant typically found in short-lived disturbed habitats, and its populations likely go through recurrent cycles of colonisation and extinction events (Bergelson et al. 1998). And similar to other colonising species Baker's Law, (Baker 1955; Pannell and Barrett 2001), it is highly self-fertile, in contrast to its self-incompatible sister species *A. lyrata* and *A. halleri*. These three species are thought to have diverged from a common ancestor approximately 5 million years ago (Ramos-Onsins et al. 2004). Thus, the *A. thaliana* lineage must have lost SI and purged some of the genetic load associated with its out-crossing habit within the last 5 million years. However, despite its self-fertility, *A. thaliana* is not genetically homogenous. Outcrossing rates of $\sim 1\%$ have been estimated (Abbott and Gomes 1989), and recent genome-wide surveys of polymorphisms have shown that, while individual populations are homozygous, the species as a whole exhibits genetic variation at several loci, albeit at reduced levels relative to *A. lyrata* (Koornneef et al. 2004).

Because of its status as a model for molecular genetic studies, the evolutionary history of *A. thaliana* is the focus of much attention. Recent investigations into the genetic basis of the switch to self-fertility in this species have been spurred by the availability in stock centers of over 750 *A. thaliana* accessions collected from various geographical locations all around the globe and recent molecular analysis of SI in *A. thaliana*'s sister species, *A. lyrata*. A further impetus for these studies is the availability of a complete genome sequence (Arabidopsis Genome Initiative 2000) and of extensive genetic and molecular tools for gene identification and characterisation.

6.3.3.1 The *S*-locus of *A. thaliana* and Its SRK and SCR/SP11 Pseudogenes

The structure of the *A. thaliana S*-locus was first determined in the reference accession Columbia (Col-0). Analysis of a bacterial artificial clone derived from the *S*-locus region in this accession demonstrated the presence of *SRK*-like and *SCR/SP11*-like sequences (Kusaba et al. 2001). While the *SCR*-like sequences consist of three truncated and un-annotated sequences designated $\Psi SCR1$, $\Psi SCR2$ and $\Psi SCR3$, the *SRK*-like sequences correspond to At4g21370, an annotated gene having the exon/intron structure typical of *SRK* and predicted to encode a full-length

RLK. The Col-0 At4g21370 gene is expressed and its promoter is active in the stigma epidermis and transmitting tract similar to the *SRKs* of self-incompatible crucifers (Kusaba et al. 2001). However, cDNAs derived from this gene indicate the use of an un-annotated splice site that introduces a premature stop codon in exon 4, indicating that this gene is a pseudogene.

Subsequent investigations of various *A. thaliana* accessions revealed substantial *S*-locus polymorphisms (Nasrallah et al. 2004; Shimizu et al. 2004; Sherman-Broyles et al. 2007; Tang et al. 2007). These analyses demonstrated that the species has maintained three *S*-locus haplotypic classes, designated S_A, S_B and S_C (Sherman-Broyles et al. 2007) after the ΨSRK_A, ΨSRK_B and ΨSRK_C variants they contain (Shimizu et al. 2004). The S_A haplotypic class, which includes the *S*-haplotype of Col-0, is the most prevalent and occurs in at least 35 accessions; S_B is restricted to the Cape Verdi Island (Cvi-0) accession; and the S_C haplotypic class is found in at least ten accessions (Sherman-Broyles et al. 2007; Tang et al. 2007). A low level of within-haplotype sequence variation is observed, with values of $\pi = 0.0008$ for ΨSRK_A and $\pi = 0.0011$ for ΨSRK_C (Sherman-Broyles et al. 2007). In contrast, the $\pi = 0.1382$ estimate obtained by combining data from ΨSRK_A, ΨSRK_B and ΨSRK_C sequences (Shimizu et al. 2004) approaches the high level of $\pi = 0.36$ calculated for *A. lyrata SRK* sequences (Charlesworth et al. 2003).

Importantly, no functional *S*-haplotype has been identified in *A. thaliana*. Indeed, accessions belonging to each of the three *S* haplotypic classes can express some level of SI when transformed with an *A. lyrata SRK-SCR* gene pair, indicating that their resident *S*-haplotypes are non-functional. The nature of mutations that disrupt the *S*-locus genes differs among haplotypes, however. In the case of S_A haplotypes, mutations are observed in both *SRK* and *SCR,* with the ΨSRK_A allele either exhibiting the same premature stop codon as Col-0, in a few cases with an additional 1-base pair frameshift mutation in the *S* domain (Shimizu et al. 2004), or having a 5-base pair frameshift mutation in exon 5 (Shimizu et al. 2004). The S_B haplotype of Cvi-0 also exhibits a non-functional ΨSRK_B gene that contains a splice site mutation in intron 2, leading to a frameshift (Shimizu et al. 2004), but its SCR_B gene does not contain obvious disruptive mutations and might be functional (Tang et al. 2007). By contrast, the mutations that might have caused inactivation of the S_C haplotype have not been defined: the SRK_C allele does not exhibit mutations that disrupt the open reading frame (Shimizu et al. 2004) and SCR_C sequences have not been isolated. Whether inactivation of the S_C haplotype is associated with unrecognised mutations in SRK_C (e.g. regulatory mutations or mutations that abolish binding to the SCR ligand) or with mutations in SCR_C remains to be determined.

Relaxation of selective pressures that once preserved the integrity of *SRK* and *SCR* and maintained their linkage also resulted in extensive restructuring of the *S*-locus, as observed in some accessions. A case in point is the *S*-locus of the C24 accession, which shows evidence of major deletions and inter-haplotypic recombination events (Sherman-Broyles et al. 2007). This locus contains a truncated version of ΨSRK_A and lacks ΨSCR sequences due to a large deletion that eliminated 18 kb of sequence found in the Col-0 *S*-haplotype, and it also contains a truncated ΨSRK_C

allele, suggesting that it was produced by recombination between *S*-haplotypes in an S_AS_C heterozygous individual.

6.3.3.2 Cryptic Modifiers of SI in *A. thaliana*

While inactivation of the *S*-locus was clearly a major event in the switch to self-fertility, *A. thaliana* also harbours mutations at other loci required for SI. Indeed, introduction of the *A. lyrata SRKb-SCRb* construct into different accessions demonstrated differences in the ability of the stigma to mount and sustain a robust SI response over the course of its development (Nasrallah et al. 2002, 2004). *SRKb-SCRb* transformants of the C24 accession exhibited a robust and stable SI phenotype at all flower stages, similar to SI in naturally self-incompatible species (Nasrallah et al. 2004). In contrast, *SRKb-SCRb* transformants of Col-0, RLD and several other accessions exhibited transient SI, whereby a robust SI response is observed in stigmas of mature floral buds and just-opened flowers, followed by breakdown of SI at later stages of development (Nasrallah et al. 2002; Liu et al. 2007). In addition, a few accessions exhibited transient and weak SI, with only partial inhibition of self pollen at early stages of stigma development. Crosses between *SRKb-SCRb* plants of different accessions showed dominance of the stable SI phenotype and revealed the existence of cryptic loss-of-function mutations at several loci required for SI.

The transient SI phenotype observed in *SRKb-SCRb* transformants of several *A. thaliana* accessions is reminiscent of pseudo self-compatibility, an advantageous reproductive assurance strategy that allows selfing after opportunities for out-crossing have been exhausted. The gene responsible for transient SI was identified as At4g21350/*PUB8* (*PLANT U-BOX8*) by map-based cloning (Liu et al. 2007). *PUB8* is a gene located at one boundary of the *S*-locus, which encodes a previously uncharacterised ARM repeat- and U box-containing protein that differs from *Brassica* ARC1 (see Chap. 8, this volume) in structure and expression pattern, indicating that it is a novel ARM/U-box protein required for SI. In the stigmas of accessions exhibiting transient SI, a hypomorphic *PUB8* allele exhibits reduced expression relative to the *PUB8* allele of C24, and this reduction is associated with reduced *SRKb* transcript levels. This result indicates that a function of PUB8 is to regulate *SRKb* transcript levels, and that transient SI is due largely to reduced *SRKb* mRNA and suboptimal SRKb protein levels in late-stage stigmas.

PUB8 is a major modifier of SI in *A. thaliana:* the hypomorphic *PUB8* allele associated with transient SI is found in a third of 86 accessions, including Col-0. It is possible that *PUB8*-mediated transient SI might have produced a transitional phase of mixed mating in the switch to selfing in *A. thaliana*, at least in some populations. In view of reduced recombination in the *S*-locus region, a mutation at a modifier locus in linkage disequilibrium with *SRK/SCR* would have a significant impact on the distribution of *S*-haplotypes in nature. Thus, linkage of *PUB8* to *SRK/SCR* might explain why the Col-0 S_A haplotype is the most prevalent in the species. In any case, the cloning of *PUB8* underscores the value of using cryptic natural variation unveiled in the *A. thaliana SRK-SCR* transgenic model to identify factors required for SI.

6.3.3.3 Hypotheses for the Switch to Self-Fertility in *A. thaliana*

Shimizu et al. (2004) proposed the hypothesis that the switch to self-fertility in *A. thaliana* occurred as a result of a very recent selective sweep of the Col-0 $\Psi SCR1$ allele. This hypothesis was based on their amplification of *SCR* sequences essentially identical to the Col-0 $\Psi SCR1$ sequence from all 21 accessions analysed, irrespective of whether they harboured the S_A, S_B or S_C haplotype. However, these results were inconsistent with DNA gel blot analyses showing absence of $\Psi SCR1$-related sequences from S_B- and S_C-containing accessions as well as from several S_A-containing accessions (Nasrallah et al. 2004) and did not hold up to further scrutiny. Indeed, a larger survey of accessions by DNA amplification and gel blot analysis (Sherman-Broyles et al. 2007) and by whole genome re-sequencing with oligonucleotide arrays (Tang et al. 2007) confirmed that the Col-0 $\Psi SCR1$ sequence was detectable only in accessions carrying the $\Psi SRKA$ allele. Importantly, a highly diverged *SCR* sequence was isolated from the S_B haplotype of Cvi-0 (Tang et al. 2007). Furthermore, the Shimizu et al. (2004) hypothesis, and indeed the notion of a selective sweep of an *SCR* or *SRK* allele, is difficult to reconcile with the high levels of variation maintained at *S*-locus flanking genes, which would require the unlikely occurrence of two recombination events for insertion of $\Psi SCR1$ into S_B and S_C haplotypes (Charlesworth and Vekemans 2005).

The occurrence of independent *S*-locus mutations and of cryptic natural variation at SI modifier loci suggests that the switch to self-fertility in *A. thaliana* involved multiple loss-of-SI events rather than being caused by a single event. The switch to selfing might have occurred independently in distinct small populations, where a gene that promotes selfing would have strong selective advantage, consistent with theoretical expectations for an ephemeral species like *A. thaliana*. Functional *S*-haplotypes would have been lost because *SRK/SCR* function specifically in reproduction, while functional SI modifiers might persist in some populations either because they are required under certain conditions for SI-unrelated processes or because loss of SI occurred very recently.

As to the question of when *A. thaliana* became self-fertile, dates ranging from less than 500,000 to over a million years have been suggested. In an investigation of transpecific evolution of *S*-haplotypes among *A. thaliana, A. lyrata* and *A. halleri*, comparisons of *S*-domain sequences produced no evidence for a change in selection in *A. thaliana* relative to self-incompatible *A. lyrata* and *A. halleri* (Bechsgaard et al. 2006). These results indicate that inactivation of *SRK* occurred recently, at most 413,000 years ago. In contrast, genome-wide linkage disequilibrium patterns suggest that the switch to selfing in *A. thaliana* occurred by an as-yet-unidentified mutation on the order of a million years ago rather than within the last 500,000 years (Tang et al. 2007). A challenge for the future is to determine what mutational event(s) caused this switch.

6.4 Future Prospects

The study of self-incompatibility in the Brassicaceae and of mating system evolution in this family has entered an exciting phase. While much progress has been made in recent years towards a mechanistic understanding of SI, glaring gaps still exist (see also Chap. 8, this volume). Much work is required to explain exactly how recognition of self pollen by epidermal cells of the stigma translates into inhibition of pollen tube growth, with forward genetics and analysis of intra-specific natural variation likely to be the most rewarding approaches. Likewise, the basis of SI specificity in pistil and pollen determinants remains largely a mystery. Identifying the amino acid residues that determine specificity will require sifting through the many polymorphisms of SRK and SCR, not all of which are necessarily important for specificity. This work is difficult, yet critical, for addressing the challenging question of how SRK and SCR co-evolve to maintain their interaction and their competence for self-recognition, and to solve the puzzle of how the large repertoire of SI specificities expressed by self-incompatible species is generated.

Equally challenging questions relate to the evolutionary dynamics of SI in the Brassicaceae. The SRK-SCR recognition and signalling system could not have evolved de novo; rather, it must have been recruited from a pre-existing signalling system, possibly one unrelated to reproduction. Therefore, an understanding of how SI evolved in the first place may emerge from comparisons to other plant signalling pathways, as well as from elucidating the function of the *SRK*- and *SCR*-related genes contained in crucifer genomes (Dwyer et al. 1994; Shiu and Bleecker, 2001; Vanoosthuyse et al. 2001). In addition, genome-wide cross-species comparisons made possible by ongoing genome sequencing projects in *Brassica, A. lyrata* and *Capsella* will no doubt provide important data to test current hypotheses and possibly formulate new models for the origin of SI, its maintenance over evolutionary time and its repeated loss in various lineages of the family.

Acknowledgements Research in the authors' laboratory is supported by grants from the United States National Science Foundation and Department of Agriculture.

References

Abbott RJ, Gomes MF (1989) Population genetic structure and outcrossing rate of *Arabidopsis thaliana*. Heredity 42:411–418

Arabidopsis Genome Iniative (2000) Analysis of the genome sequence of the flowering plant *Arabidopsis thaliana*. Nature 408:796–815

Awadala P, Charlesworth D (1999) Recombination and selection at *Brassica* self-incompatibility loci. Genetics 152:413–425

Baker HG (1955) Self-compatibility and establishment after "long distance" dispersal. Evolution 9:347–349

Barre A, Van Damme EJM, Peumans WJ, Rougé P (1997) Curculin, a sweet-tasting and taste-modifying protein, is a non-functional mannose-binding lectin. Plant Mol Biol 33:691–698

Barrett SCH (1988) The evolution, maintenance, and loss of self-incompatibility systems. In Lovett Doust J, Lovett Doust L (eds) Plant reproductive ecology: Patterns and strategies, Oxford University Press, Oxford, pp 98–124

Bateman AJ (1952) Self-incompatibility systems in angiosperms. I. Theory. Heredity 6:285–310

Bateman AJ (1955) Self-incompatibility systems in angiosperms III. Cruciferae. Heredity 9:53–68

Bechsgaard J, Bataillon T, Schierup MH (2004) Uneven segregation of sporophytic self-incompatibility slleles in *Arabidopsis lyrata*. J Evol Biol 17:554–561

Bechsgaard J, Castric V, Charlesworth D, Vekemans V, Schierup MH (2006) The transition to self-compatibility in *Arabidopsis thaliana* and evolution within *S*-haplotypes over 10 million years. Mol Biol Evol 23:1741–1750

Bergelson J, Stahl EA, Dudek S, Kreitman M (1998) Genetic variation within and among populations of *Arabidopsis thaliana*. Genetics 148:1311–1323

Boyes DC, Nasrallah ME, Vrebalov J, Nasrallah JB (1997) The self-incompatibility *S*-haplotypes of *Brassica* contain highly divergent and rearranged sequences of ancient origin. Plant Cell 9:237–247

Busch JW (2005) The evolution of self-compatibility in geographically peripheral populations of *Leavenworthia alabamica* (Brassicaceae). Am J Bot 92:1503–1512

Byers DL, Waller DM (1999) Do plant populations purge their genetic load? Effects of population size and mating history on inbreeding depression. Annu Rev Ecol Syst 30:479–513

Casselman AL, Vrebalov J, Conner JA, Singhal A, Giovannoni J, Nasrallah ME, Nasrallah JB (2000) Determining the physical limits of the *Brassica S*-locus by recombinational analysis. Plant Cell 12:23–33

Charlesworth D (2000) How can two-gene models of self-incompatibility generate new specificities? Plant Cell 12:309–310

Charlesworth D (2006) Balancing selection and its effects on nearby genome regions. PLoS Genet 2:379–384

Charlesworth D, Awadalla P (1998) *Flowering* plant self-incompatibility: The molecular population genetics of *Brassica S*-loci. Heredity 81:1–9

Charlesworth D, Vekemans X (2005) How and when did Arabidopsis become highly self-fertilising? BioEssays 27:472–475

Charlesworth D, Bartolome C, Schierup MH, Mable BK (2003) Haplotype structure of the stigmatic self-incompatibility Gene in Natural Populations of *Arabidopsis lyrata*. Mol Biol Evol 20:1741–1753

Charlesworth D, Vekemans X, Castric V, Glemin S (2005) Plant self-incompatibility systems: A molecular evolutionary perspective. New Phytol 68:61–69

Charlesworth D, Kamau E, Hagenblad J, Tang C (2006) *Trans*-specificity at loci near the self-incompatibility loci in *Arabidopsis*. Genetics 172:2699–2704

Chookajorn T, Kachroo A, Ripoll DR, Clark AG, Nasrallah JB (2004) Specificity determinants and diversification of the *Brassica* self-incompatibility pollen ligand. Proc Natl Acad Sci USA 101:911–917

Conner JA, Conner P, Nasrallah ME, Nasrallah JB (1998) Comparative mapping of the *Brassica S*-locus region and its homeolog in Arabidopsis: Implications for the evolution of mating systems in the Brassicaceae. Plant Cell 10:801–812

Crnokrak P, Barrett SCH (2002) Purging the Genetic Load: A review of experimental evidence. Evolution 56:2347–2358

Darwin C (1876) The effects of cross- and self-fertilization in the vegetable kingdom. Murray, London

Dixit R, Nasrallah ME, Nasrallah JB (2000) Post-transcriptional maturation of the S receptor kinase of *Brassica* correlates with co-expression of the *S*-locus glycoprotein in the stigmas of two Brassica strains and in transgenic tobacco plants. Plant Physiol 124:297–311

Dwyer KG, Balent MA, Nasrallah JB, Nasrallah ME (1991) DNA sequences of self-incompatibility genes from *Brassica campestris* and *B. oleracea*: Polymorphism predating speciation. Plant Mol Biol 16:481–486

Dwyer KG, Kandasamy MK, Mahosky DI, Acciai J, Kudish BI, Miller JE, Nasrallah ME, Nasrallah JB (1994) A superfamily of *S*-locus-related sequences in *Arabidopsis*: Diverse structures and expression patterns. Plant Cell 6:1829–1843

Fujimoto R, Sugimura T, Fukai E, Nishio T (2006a) Suppression of gene expression of a recessive SP11/SCR allele by an untranscribed SP11/SCR allele in Brassica self-incompatibility. Plant Mol Biol 61:577–587

Fujimoto R, Okazaki K, Fukai E, Kusaba M, Nishio T (2006b) Comparison of the genome structure of the self-incompatibility *S*-locus in interspecific pairs of *S*-haplotypes. Genetics 173:1157–1167

Giranton J, Dumas C, Cock JM, Gaude T (2000) The integral membrane *S*-locus receptor kinase of Brassica has serine/threonine kinase activity in a membranous environment and spontaneously forms oligomers in planta. Proc Natl Acad Sci U S A 97:3759–3764

Goodwillie C (1999) Multiple origins of self-compatibility in *Linanthus* Section *Leptosiphon* (Polemoniaceae): Phylogenetic evidence from internal-transcribed-spacer sequence data. Evolution 53:1387–1395

Goodwillie C, Kalisz S, Eckert CG (2005) The evolutionary enigma of mixed mating in plants: occurrence, theoretical explanations, and empirical evidence. Annu Rev Evol Ecol Syst 36: 47–79

Goring DR, Glavin TL, Schafer U, Rothstein SJ (1993) An S receptor kinase gene in self-compatible *Brassica napus* has a 1-bp deletion. Plant Cell 5:531–539

Hagenblad J, Bechsgaard J, Charlesworth D (2006) Linkage disequilibrium between Incompatibility locus region genes in the plant *Arabidopsis lyrata*. Genetics 173:1057–1073

Hamrick JL, Godt MJW (1996) Effects of life history traits on genetic diversity in plant species. Phil Trans R Soc London Series B Biol Sci 351:1291–1298

Hatakeyama K, Takasaki T, Suzuki G, Nishio T, Watanabe M, Isogai A, Hinata K (2001) The S receptor kinase gene determines dominance relationships in stigma expression of self-incompatibility in *Brassica*. Plant J 26:69–76

Hoffmann MH (2002) Biogeography of *Arabidopsis thaliana* (L.) Heynh. (Brassicaceae). J Biogeogr 29:125–134

Igic B, Bohs L, Kohn JR (2006) Ancient polymorphism reveals unidirectional breeding system transitions. Proc Natl Acad Sci U S A 103:1359–1363

Igic B, Lande R, Kohn JR (2008) Loss of self-incompatibility and its evolutionary consequences. Int J Plant Sci 169:93–104

Kachroo A, Schopfer CR, Nasrallah ME, Nasrallah JB (2001) Allele-specific receptor-ligand interactions in Brassica self-incompatibility. Science 293:1824–1826

Kakizaki T, Takada Y, Ito A, Suzuki G, Shiba H, Takayama S, Isogai A, Watanabe M (2003) Linear dominance relationship among four class-II *S*-haplotypes in pollen is determined by the expression of SP11 in *Brassica* self-incompatibility. Plant Cell Physiol 44:70–75

Kamau E, Charlesworth D (2005) Balancing selection and low recombination affect diversity near the self-incompatibility loci of the plant *Arabidopsis lyrata*. Curr Biol 15:1773–1778

Kandasamy MK, Paolillo DJ, Faraday CD, Nasrallah JB, Nasrallah ME (1989) The *S*-locus specific glycoproteins of *Brassica* accumulate in the cell wall of developing stigma papillae. Dev Biol 134:462–472

Kao T, Tsukamoto T (2004) The molecular and genetic bases of S-RNase-based self-incompatibility. Plant Cell 16:S72–83

Kawabe A, Hansson B, Forrest A, Hagenblad J, Charlesworth D (2006) Comparative gene mapping in *Arabidopsis lyrata* chromosome 6 and 7 and *A. thaliana* chromosome IV: Evolutionary history, rearrangements and local recombination rates. Genet Res 88:45–46

Kemp BP, Doughty J (2007) *S* cysteine-rich (SCR) binding domain analysis of the *Brassica* self-incompatibility *S*-locus receptor kinase. New Phytol 175:619–629

Kimura R, Sato K, Fujimoto R, Nishio T (2002) Recognition specificity of self-incompatibility maintained after the divergence of *Brassica oleracea* and *Brassica rapa*. Plant J 29:215–223

Koornneef M, Alonso-Blanco C, Vreugdenhil D (2004) Naturally occurring genetic variation in *Arabidopsis thaliana*. Annu Rev Plant Biol 55:141–172

Kusaba M, Nishio T (1999) Comparative analysis of *S*-haplotypes with very similar *SLG* alleles in *Brassica rapa* and *Brassica oleracea*. Plant J 17:83–91

Kusaba M, Nishio T, Satta Y, Hinata K, Ockendon D (1997) Striking sequence similarity in inter- and intra-specific comparisons of class I *SLG* alleles from *Brassica oleracea* and *Brassica campestris*: Implications for the evolution and recognition mechanism. Proc Natl Acad Sci U S A 94:7673–7678

Kusaba M, Dwyer KG, Hendershot J, Vrebalov J, Nasrallah JB, Nasrallah ME (2001) Self-incompatibility in the genus *Arabidopsis*: Characterization of the *S*-locus in the outcrossing A. *lyrata* and its autogamous relative A. *thaliana*. Plant Cell 13:627–643

Kusaba M, Tung CW, Nasrallah ME, Nasrallah JB (2002) Monoallelic expression and dominance interactions in anthers of self-incompatible *Arabidopsis lyrata*. Plant Physiol 128:17–20

Lande R, Schemske DW (1985) The evolution of self-fertilization and inbreeding depression in plants I. Genetic models. Evolution 39:24–40

Liu P, Sherman-Broyles S, Nasrallah ME, Nasrallah JB (2007) A cryptic modifier causing transient self-incompatibility in *Arabidopsis thaliana*. Curr Biol 17:734–740

Mable BK (2003) Estimating the number, frequency, and dominance of S-alleles in a natural population of *Arabidopsis lyrata* (Brassicaceae) with sporophytic control of self-incompatibility. Heredity 90:422–431

Mable BK, Robertson AV, Dart S, DiBerardo C, Witham L (2005) Breakdown of Self-incompatibility in the PerenNial *Arabidopsis lyrata* (Brassicaceae) and its Consequences. Evolution 59:1437–1448

Matton DP, Luu DT, Xike Q, Laublin G, O'Brien M, Maes O, Morse D, Cappadocia M (1999) Production of an S RNase with dual specificity suggests a novel hypothesis for the generation of new S alleles. Plant Cell 11:2087–2098

Miege C, Ruffio-Chable V, Schierup MH, Cabrillac D, Dumas C, Gaude T, Cock JM (2001) Intrahaplotype polymorphism at the *Brassica S*-locus. Genetics 159:811–822

Mishima M, Takayama S, Sasaki K, Jee J, Kojima C, Isogai A, Shirakawa M (2003) Structure of the male determinant factor for *Brassica* self-incompatibility. J Biol Chem 278:36389–36395

Murase K, Shiba H, Iwano M, Che F, Watanabe M, Isogai A, Takayama S (2004) A membrane-anchored protein kinase involved in *Brassica* self-incompatibility signaling. Science 303:1516–1519

Naithani S, Chookajorn T, Ripoll DR, Nasrallah JB (2007) Structural modules for receptor dimerization in the *S*-locus receptor kinase extracellular domain. Proc Natl Acad Sci U S A 104:12211–12216

Nasrallah JB (2000) Cell-cell signaling in the self-incompatibility response. Curr Opin Plant Biol 3:368–73

Nasrallah JB (2005) Recognition and rejection of self in plant self-incompatibility: comparisons to animal histocompatibility. Trends Immunol 26:412–418

Nasrallah JB, Kao TH, Goldberg ML, Nasrallah JB (1985) A cDNA clone encoding an *S*-locus specific glycoprotein from *Brassica oleracea*. Nature 318:263–267

Nasrallah JB, Kao TH, Chen CH, Goldberg ML, Nasrallah ME (1987) Amino-acid sequence of glycoproteins encoded by three alleles of the *S*-locus of *Brassica oleracea*. Nature 326:617–619

Nasrallah JB, Rundle SJ, Nasrallah ME (1994) Genetic evidence for the requirement of *Brassica S*-locus receptor kinase gene in the self-incompatibility response. Plant J 5:373–384

Nasrallah JB, Liu P, Sherman-Broyles S, Schmidt R, Nasrallah ME (2007) Epigenetic mechanisms for breakdown of self-incompatibility in interspecific hybrids. Genetics 175:1965–1973

Nasrallah ME (1974) Genetic control of quantitative variation of self incompatibility proteins detected by immunodiffusion. Genetics 76:45–50

Nasrallah ME, Kandasamy MK, Nasrallah JB (1992) A genetically defined trans-acting locus regulates *S*-locus function in *Brassica*. Plant J 2:497–506

Nasrallah ME, Liu P, Nasrallah JB (2002) Generation of self-incompatible *Arabidopsis thaliana* by transfer of two *S*-locus genes from A. *lyrata*. Science 297:247–249

Nasrallah ME, Liu P, Sherman-Broyles S, Boggs N, Nasrallah JB (2004) Natural variation in expression of self-incompatibility in *Arabidopsis thaliana*: Implications for the evolution of selfing. Proc Natl Acad Sci U S A 101:16070–16074

Okamoto S, Odashima M, Fujimoto R, Sato Y, Kitashiba H, Nishio T (2007) Self-compatibility in *Brassica napus* is caused by independent mutations in *S*-locus genes. Plant J 50:391–400

Olmstead RG (1986) Self-incompatibility in light of population structure and inbreeding. IN: Biotechnology and Ecology of Pollen. Mulcahy, D., Mulcahy, G., and Ottaviano, E. (eds), pp 239–245. Springer-Verlag, NY

Paetsch M, Mayland-Quellhorst S, Neuffer B (2006) Evolution of the self-incompatibility system in the Brassicaceae: Identification of *S*-locus receptor kinase (SRK) in self-incompatible *Capsella grandiflora*. Heredity 97:283–290

Pannell JR, Barrett SCH (2001) Effects of population size and metapopulation dynamics on a mating-system polymorphism. Theor Popul Genetics 59:145–155

Porcher E, Lande R (2005) Loss of gametophytic self-incompatibility with inbreeding depression. Evolution 59:46–60

Prigoda NL, Nassuth A, Mable BK (2005) Phenotypic and genotypic expression of self-incompatibility haplotypes in *Arabidopsis lyrata* suggests unique origins of alleles in different dominance classes. Mol Biol Evol 22:1609–1620

Ramos-Onsins SE, Stranger BE, Mitchell-Olds T, Aguade M (2004) Multilocus analysis of variation and speciation in closely related species *Arabidopsis halleri* and *A. lyrata*. Genetics 166:373–388

Sainudiin R, Shuk WWW, Yogeeswaran K, Nasrallah JB, Yang Z, Nielsen R (2005) Detecting site-specific shysiochemical selective sressures: applications to the class I HLA of the human major histocompatibility complex and the SRK of the plant sporophytic self-incompatibility system. J Mol Evol 60:315–326

Sakamoto K, Kusaba M, Nishio T (1998) Polymorphism of the *S*-locus glycoprotein gene (*SLG*) and the *S*-locus related gene (*SLR1*) in *Raphanus sativus* L. and self-incompatible ornamental plants in the Brassicaceae Mol. Gen. Genet. 258:397–403

Sato K, Nishio T, Kimura R, Kusaba M, Suzuki T, Hatakeyama K, Ockendon D, Satta Y (2002) Coevolution of the *S*-locus genes, *SRK, SLG* and *SP11/SCR* in*Brassica oleracea and B. rapa*. Genetics 162:931–940

Sato Y, Fujimoto R, Toriyama K, Nishio T (2003) Commonality of self-recognition specificity of S-haplotypes between *Brassica oleracea* and *Brassica rapa*. Plant Mol Biol 52:617–626

Schierup MH, Mable BK, Awadalla P, Charlesworth D (2001) Identificaiton and characterization of a polymorphic receptor kinase gene linked to the self-incompatibility locus of *Arabidopsis lyrata*. Genetics 158:387–399

Schopfer CR, Nasrallah ME, Nasrallah JB (1999) The male determinant of self-incompatibility in *Brassica*. Science 286:1697–700.

Sherman-Broyles S, Boggs N, Farkas A, Liu P, Vrebalov J, Nasrallah ME, Nasrallah JB (2007) *S*-locus genes and the evolution of self-fertility in *Arabidopsis thaliana*. Plant Cell 19:94–106

Shiba H, Takayama S, Iwano M, Shimosato H, Funato M, Nakagawa T, Che F, Suzuki G, Watanabe M, Hinata K, Isogai A (2001) A pollen coat protein, SP11/SCR, determines the pollen *S*-specificity in the self-sncompatibility of *Brassica* species. Plant Physiol 125:2095–2103

Shiba H, Iwano M, Entani T, Ishimoto K, Shimosato H, Che F, Satta Y, Ito A, Takada Y, Watanabe M, Isogai A, Takayama S (2002) The dominance of alleles controlling self-incompatibility in *Brassica* pollen is regulated at the RNA level. Plant Cell 14:491–504

Shiba H, Kakizaki T, Iwano M, Tarutani Y, Watanabe M, Isogai A, Takayama S (2006) Dominance relationships between self-incompatibility alleles controlled by DNA methylation. Nat Genet 38:297–299

Shimizu KK, Cork JM, Caicedo AL, Mays CA, Moore RC, Olsen KM, Ruzsa S, Coop G, Bustamante CD, Awadalla P, Purugganan MD (2004) Darwinian selection on a selfing locus. Science 306:2081–2084

Shimosato H, Yokota N, Shiba H, Iwano M, Entani T, Che F, Watanabe M, Isogai A, Takayama S (2007) Characterization of the SP11/SCR High affinity binding site involved in self/nonself recognition in Brassica self incompatibility. Plant Cell 19:109–117

Shiu S, Bleecker AB (2001) Receptor-like kinases from Arabidopsis form a monophyletic gene family related to animal receptor kinases. Proc Natl Acad Sci U S A 98:10763–10768

Silva NF, Stone SL, Christe LN, Sulaman W, Nazarain KAP, Burnett M, Arnoldo MA, Rothstein SJ, Goring DR (2001) Expression of the S receptor kinase in self-compatible *Brassica napus* cv. Westar leads to the allele-specific rejection of self-incompatible *Brassica napus* pollen. Mol Genet Gen 265:552–559

Stebbins GL (1957) Self-fertilization and population variability in the higher plants. Am Nat 91:337–354

Stein J, Howlett B, Boyes DC, Nasrallah ME, Nasrallah JB (1991) Molecular cloning of a putative receptor protein kinase gene encoded at the self-incompatibility locus of *Brassica oleracea*. Proc Natl Acad Sci U S A 88:8816–8820

Stein J, Dixit R, Nasrallah ME, Nasrallah JB (1996) SRK, the stigma-specific *S*-locus receptor kinase of *Brassica*, is targeted to the plasma membrane in transgenic tobacco. Plant Cell 8:429–445

Suzuki T, Kusaba M, Matsushita M, Okazaki K, Nishio T (2000) Characterization of Brassica *S*-haplotypes lacking *S*-locus glycoprotein. FEBS Lett 482:102–108

Takasaki T, Hatakeyama K, Suzuki G, Watanabe M, Isogai A, Hinata K (2000) The S receptor kinase determines self-incompatibility in *Brassica* stigma. Nature 403:913–916

Takayama S, Isogai A (2005) Self-incompatibility in plants. Annu Rev Plant Biol 56:467–489

Takayama S, Shimosato H, Shiba H, Funato M, Che F, Watanabe M, Iwano M, Isogai A (2001) Direct ligand-receptor complex interaction controls *Brassica* self-incompatibility. Nature 413:534–538

Takuno S, Fujimoto R, Sugimura T, Sato K, Okamoto S, Zhang S, Nishio T (2007) Effects of recombination on hitchhiking diversity in the *Brassica* self-incompatibility locus complex. Genetics 177:949–958

Tang C, Toomajian C, Sherman-Broyles S, Plagnol V, Guo Y, Hu TT, Clark RM, Nasrallah JB, Weigel D, Nordborg M (2007) The evolution of selfing in *Arabidopsis thaliana*. Science 317:1070–1072

Uyenoyama, M.K. and E. Newbigin. 2000. Evolutionary dynamics of dual-specificity self-incompatibility alleles. Plant Cell 12: 310–311

Uyenoyama MK (2005) Evolution of tight linkage to mating type. New Phytol 165:63–70

Uyenoyama MK, Zhang Y, Newbigin E (2001) On the origin of self-incompatibility haplotypes: transition through self-compatible intermediates. Genetics 157:1805–1817

Vanoosthuyse V, Miege C, Dumas C, Cock JM (2001) Two large *Arabidopsis* gene families are homologous to the *Brassica* gene superfamily that encodes pollen coat proteins and the male component of the self-incompatibility response. Plant Mol Biol 16:17–34

Watanabe M, Ito Y, Takada Y, Ninomiya C, Kakizaki T, Takahata Y, Hatakeyama K, Hinata K, Suzuki G, Takasaki T, Satta Y, Shiba H, Takayama S, Isogai A (2000) Highly divergent sequences of the pollen self-incompatiblity (S) gene in class-I *S*-haplotypes of *Brassica campestris* (syn. *rapa*) L. FEBS Lett 473:139–144

Whitehouse HLK (1950) Multiple-allelomorph incompatibility of pollen and style in the evolution of the angiosperms. Ann Bot 54:199–216

Part II
Molecular and Cell Biology of Self-Incompatibility Systems

Chapter 7
Milestones Identifying Self-Incompatibility Genes in *Brassica* Species: From Old Stories to New Findings

M. Watanabe, G. Suzuki, and S. Takayama

Abstract Self-incompatibility (SI) in *Brassica* species is an elaborate system preventing self-fertilisation. It is important for high-quality seed production in *Brassica* vegetables, such as cabbage, turnip and Chinese cabbage. In classical genetics, *Brassica* SI is sporophytically controlled by a single *S*-locus with multiple alleles. This concept of the *S*-locus enabled us to begin molecular studies to determine how the SI system is regulated. Through vigorous molecular analysis by research groups worldwide, female and male *S* determinants were identified as a receptor-type protein kinase (SRK) and its ligand (SP11/SCR). In this chapter, we highlight six milestones, which are historically important findings in *Brassica* SI research, especially with respect to identification of genes regulating the recognition of self-pollen.

Abbreviations

ARC1	Arm repeat containing; an E3 ubiquitin ligase
EGF-like	Epidermal growth factor (EGF)-like
eSRK	A soluble form of the extracellular domain of SRK
FDD	Fluorescent differential display
HV	Hypervariable

M. Watanabe
Faculty of Science, Tohoku University, Sendai 980-8587, Japan;
The 21st Century Center of Excellence Program, Iwate University, Morioka 020-8550, Japan;
Laboratory of Plant Reproductive Genetics, Graduate School of Life Sciences, Tohoku University, 2-1-1, Katahira, Aoba-ku, Sendai 980-8577, Japan, e-mail: nabe@ige.tohoku.ac.jp

G. Suzuki
Division of Natural Science, Osaka Kyoiku University, Kashiwara 582-8582, Japan

S. Takayama
Graduate School of Biological Sciences, Nara Institute of Science and Technology, Ikoma 630-0192, Japan

IEF	Iso-electric focusing
KAPP	Kinase-associated protein phosphatase
MLPK	*M* locus protein kinase
PAC	P1-derived artificial chromosome
PCPs	Pollen coat proteins
RLK	Receptor-like kinase
RT-PCR	Reverse transcription polymerase chain reaction
SDS-PAGE	Sodium dodecyl sulfate polyacrylamide gel electrophoresis
SI	Self-incompatibility
SLG	*S*-locus glycoprotein
S-locus	Self-incompatibility locus
SP11/SCR	*S*-locus cysteine-rich protein (the pollen *S*-determinant in *Brassica*)
SRK	*S*-locus receptor kinase (the pistil *S*-determinant in *Brassica*)
SSI	Sporophytic self-incompatibility
THL1 and THL2	Thioredoxin *h* proteins
THL1 and THL2	Thioredoxin h-like proteins
tSRK	A membrane-spanning truncated protein comprising the extracellular domain, the transmembrane domain and the juxtamembrane domain of SRK
ΨSCR	*SRK*-like sequences in *A. thaliana*

7.1 Self-Incompatibility as an Agriculturally Important Trait

The oldest record of hand-pollination in plants was in a relief made during the Assyrian dynasty about 800 B.C., which is preserved in the Metropolitan Museum of Art in New York, USA. On the relief, two genies fertilise a sacred tree by hand-pollination. This represents a very early example of manual fertilisation of date palms, showing that ancient humans differentiated male and female flowers (Relief: sacred tree attended by winged beings; http://www.metmuseum.org/). Hand-pollination has been used for fruit and seed set, and in addition, is important as a driving force to create new varieties of crops, through plant breeding. It has contributed greatly to the production of the many present varieties of vegetables, fruits and cereals.

In modern plant breeding, F_1-hybrid breeding is the most superior and popular system because of its uniformity and hybrid vigour, heterosis. The first F_1-hybrid breeding system was established in maize, as 'hybrid corn' in the USA in 1920. One problem of this system was its laborious emasculation of the male flower, tassel. In Japan, the F_1-hybrid variety of eggplants was first produced in 1924. After this success, F_1-hybrid varieties of cucurbitaceous plants (cucumber, water melon, etc.) were introduced. These solanaceous and cucurbitaceous plants have an advantage over the F_1 hybrid because, due to the fruit structure, they can produce many hundreds of F_1-hybrid seeds by hand-pollination with one flower.

In contrast, the seed quantity per flower for cruciferous plants (cabbage, Chinese cabbage, turnip, etc.) is only 10–20, making it difficult to achieve an efficient F_1-hybrid system by using hand-pollination. Therefore, it was necessary to establish an efficient F_1-hybrid breeding system in Brassicaceae. In 1940, a Japanese seed company, Sakata Seed Co., succeeded in producing an F_1-hybrid variety of cabbage (cv. Suteki Kanran) by using the self-incompatibility (SI) trait. Because SI is a reproductive system preventing self-fertilisation and promoting out-crossing, the F_1-hybrid seeds can be produced automatically from two parental, mix-planted SI lines with assistance of bee pollination. In 1950, another Japanese seed company, Takii & Co. Ltd., also succeeded in producing F_1-hybrid varieties of cabbage (cv. Choko-1c) and Chinese cabbage (cv. Choko-1cc). These *Brassica* F_1 hybrids could not have been produced without resolving two ambivalent problems in the SI trait. A rigid SI trait is necessary to remove contamination of the self-pollinated seeds during F_1-hybrid-seed production; however, it is disadvantageous for establishment of two parental lines, which require a large amount of seed propagation. To resolve these ambivalent problems, breeders selected the rigid SI lines from a large number of seed stocks, and used *bud pollination* to produce parental lines. By using bud pollination, in which young flower buds are used as a maternal parent, SI can be overcome, and self-pollinated seeds can be produced. However, production of a large number of parental seeds of the F_1 hybrid by bud pollination remains difficult and inefficient.

To establish an economical F_1-hybrid breeding system in cruciferous crops, discovery of breakdown of SI by 5% CO_2 treatment was especially significant (Nakanishi et al. 1969; Nakanishi and Hinata 1973, 1975). For large-scale propagation of the parental seeds of F_1 hybrid, bud pollination has been replaced by pollination with honey bees, using CO_2 treatment in a glasshouse. The timing of the CO_2 treatment, when just pollen grains are germinated on the stigma papilla cells, is important, indicating that overcoming the SI by CO_2 treatment might be related to the metabolism of pollen tube penetration (Nakanishi and Hinata 1973). There is a genetic variation in the level of breakdown of SI by CO_2 gas treatment in *Brassica* and *Raphanus* (Nakanishi and Hinata 1973; Niikura and Matsuura 2000). By using these different lines of sensitivity to CO_2, genetic analysis was performed in radish, which showed that the genetic locus regulating the CO_2 sensitivity was independent of the self-incompatibility *S*-locus, and functioned in the stigma side as a recessive gene (Niikura and Matsuura 2000). These genetic data are informative and contribute to a better understanding of the mechanism of the CO_2-treated breakdown of SI in the future. Thus, the economical F_1-hybrid breeding system in cruciferous crops was established in Japan, and high quality hybrid varieties have been developed. Now, Japanese F_1-hybrid seeds are exported to countries all over the world.

As we described in the historical background of F_1-hybrid breeding in this introductory chapter, there is a close relationship between SI and the breeding of *Brassica* vegetables, resulting in *Brassica* SI being an agriculturally important and attractive research subject in plant science. The molecular players of the SI recognition reaction in *Brassica* are now well known, and *Brassica* SI is the most extensively studied SI system. This review article spotlights six milestones, which are historically important findings in *Brassica* SI research, especially with respect to the identification of genes regulating the recognition of self-pollen. A large number of review articles published recently provide general understanding of *Brassica* SI (de Nettancourt 2001; Isogai and Hinata 2002; Kachoo et al. 2002; Takayama and Isogai 2003, 2005; Watanabe et al. 2006).

7.2 The First Milestone: Bateman's Idea for Sporophytic Control of the *S*-locus

SI research in cruciferous plants started with genetic analysis. The SI trait of cruciferous plants was first surveyed in the crucifer, bitter cress, *Cardamine pratensis* (Correns 1912). This genetic experiment had several problems owing to the high polyploidy of the plant material. In a second genetic experiment, Kakizaki (1930) suggested that SI of cabbage was controlled by a gametophytic model with two loci. In Kakizaki's model, *S* alleles functioned as an oppositional gene, *T* alleles functioned as a sympathetic gene and the *S*-locus was epistatic over the *T* locus (Kakizaki 1930). However, these two models could not show an accurate genetic model of SI in cruciferous plants. Bateman (1952, 1954, 1955) proposed an elegant genetic model in which SI was regulated by a sporophytically controlled single locus with multiple alleles in most cruciferous plants containing genus *Iberis*, *Capsella*, *Brassica*, *Raphanus*, etc. In other words, the behaviour of the pollen tube (gamete) is determined by the SI phenotype of the other tissues of a paternal plant (sporophyte). It is commonly called sporophtyic SI (SSI) today. The establishment of this sporophytic model was significant. After that, a number of experiments supported his idea, and his model was applied to other plant species of Brassicaceae (Thompson 1957; Sampson 1964; Thompson and Taylor 1966) and other families (Convolvulaceae and Asteraceae; Martin 1968; Crowe 1954). Bateman's sporophytic idea was a great milestone for establishing the genetic concept of SI research.

This genetic model when the pollen-producing plant (anther) shared the same *S* alleles with the stigma, pollen germination and pollen tube penetration are generally disturbed at the stigma surface on the papilla cell, rendering the combination incompatible. In contrast, when the pollen-producing plant had different *S* alleles from the stigma, pollen could germinate and penetrate into the papilla cells, resulting in a compatible combination (Fig. 7.1). It is worth noting that there are dominance relationships between *S* alleles in both pollen and stigma in the sporophytic model (Bateman 1954, 1955; Thompson and Taylor 1966; Hatakeyama et al. 1998a). About a half of century after the observation of dominance relationships between *S* alleles, their molecular mechanisms were revealed as described below.

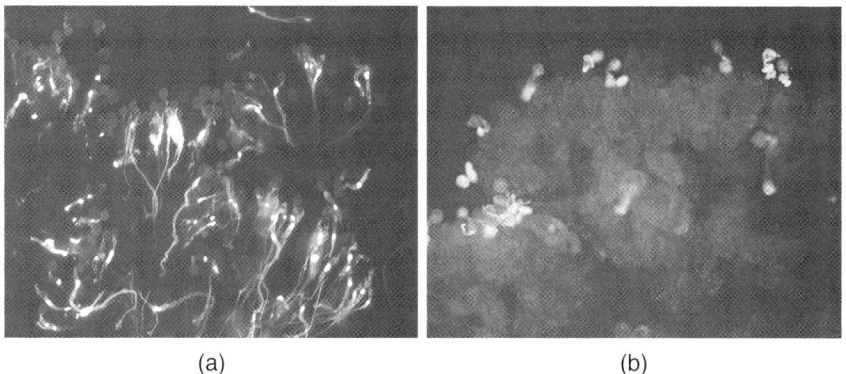

Fig. 7.1 Representative SI reaction in *Brassica rapa* (syn. *campestris*) L. Photographs were obtained by UV fluorescence microscopy. (**a**) Cross-pollination (compatibility). (**b**) Self-pollination (incompatibility) (Courtesy of Yoshinobu Takada and Sachiyo Isokawa (unpublished data))

7.3 The Second Milestone: Identification of SLG by Using IEF

Molecular studies of the *Brassica* SI system have stemmed from the finding of the *S*-locus protein, whose behaviour is consistent with the genetic model of the SI trait. The presence of *S*-specific products in stigmas was first detected by immunological methods (Nasrallah and Wallace 1967; Hinata et al. 1982). Isoelectric focusing (IEF) electrophoresis, which could separate proteins according to their pI values, was applied to analysis of the *S*-specific products in stigmas (Nishio and Hinata 1977). This IEF methodology had a great advantage in detecting allelic differences in the *S*-specific products, because significant allelic differences could be shown by fractionation of proteins according to their pI values, which are affected by the composition of amino acids. It is noteworthy that the IEF technique was also applied in the finding of the *S*-specific proteins in the plant species Solanaceae, Rosaceae, Papaverceae, with gametophytic SI (Singh and Kao 1992; Nakanishi et al. 1992; Sassa et al. 1992, 1993).

It has been established that the *S*-specific proteins in *Brassica* had different pI values corresponding to the respective *S* alleles, and co-segregated with the *S* alleles without exception. They were highly expressed in the mature stigma, which was coincident with the expression of SI (Nishio and Hinata 1977; Hinata and Nishio 1978; Nou et al. 1991, 1993a, b). Partial purification of the *S*-specific proteins revealed their polymorphic nature among *S* alleles, as estimated in the IEF analysis (Nishio and Hinata 1982). In addition, the complete purification and the detailed molecular characteristics of the *S*-specific proteins revealed that they consist of conserved and highly polymorphic amino acid regions and of conserved carbohydrate chains (Takayama et al. 1986a, b, 1987; Isogai et al. 1987). As a result of these findings, corresponding cDNA and genomic DNA sequences were determined, and the *S*-specific proteins were designated as SLG (*S*-locus glycoprotein; Nasrallah et al. 1987, 1988; Lalonde et al. 1989).

The SLG proteins were easily detected because they are abundant in the stigma. It is now known that SRK (*S* receptor kinase) (a kinase having SLG-like receptor domain; see Sect. 7.4) is the female determinant of SI, and SLG is not fundamentally essential to the SI recognition reaction (Suzuki et al. 2000b; Takasaki et al. 2000). However, it is clear that identifying SRK would have been more laborious and time-consuming without the finding of *SLG*; positional cloning with conventional mapping might have been necessary to isolate *SRK*, because its expression level was very low (Watanabe et al. 1994). Thus, the identification of SLG led to the successful discovery of the SI factors, and activated SI studies in *Brassica* at a molecular level. Because of this, the molecular analysis of the control mechanism of the *Brassica* SI system is now the most well understood system among the many other SI systems in the plant kingdom. From this viewpoint, identification of SLG represents an important milestone in *Brassica* SI research.

The function of SLG has not been clarified to date, although SLG is highly expressed in the stigma and exists in most of the *S* alleles (Watanabe et al. 1994; Kusaba et al. 2000; Suzuki et al. 2000b; Takasaki et al. 2000; Sato et al. 2002; Suzuki et al. 2003; Shimosato et al. 2007). There are several other SLG-like proteins in the stigma of *Brassica* (SLR1, SLR2, SLR3; Isogai et al. 1988; Lalonde et al. 1989; Scutt et al. 1990; Boyes et al. 1991; Watanabe et al. 1992; Cock et al. 1995), though their functions are also unclear. These common characteristics of SLG and other SLG-like proteins indicate that they have some significant functions in pollen–stigma interaction, which hopefully will be understood in the near future.

7.4 The Third Milestone: Identification of *SRK*, the Female *S* Determinant Gene

At the end of the 1980s, the functional relationship between SLG and the SI recognition reaction was not understood. Although identification of SLG was an important breakthrough for *Brassica* SI research, SLG was the only molecule known as an *S*-locus protein at that time, and did not show any sequence similarity to that of other proteins. The finding of maize *ZmPK1*, the first plant receptor-like protein kinase gene identified, was a turning point, because its extracellular receptor domain of ZmPK1 was similar to SLG (Walker and Zhang 1990). Simultaneously, Chen and Nasrallah (1990) reported two *SLG*-like genes on the *S*-locus in an S^2 allele of *B. oleracea* in the same year. These two reports in 1990 suggested that the ZmPK1-like receptor kinase might be involved in the *Brassica* SI. The second *S*-linked gene, the *S* receptor kinase (*SRK*) gene, was identified in *B. oleracea* the following year (Stein et al. 1991). Because the *S*-locus contains multiple genes (*SLG*, *SRK*, etc.) as one genetic segregational unit, an '*S* allele' is now referred to as an '*S*-haplotype' (Nasrallah and Nasrallah 1993). The extracellular domain (S domain) of SRK showed a high similarity to SLG (about 90%) and was connected via a single-pass transmembrane domain to a serine/threonine-type protein kinase catalytic domain (Stein et al. 1991). The expression of *SRK* was specifically detected in the stigma tissues, like the *SLG* gene (Watanabe et al. 1994). Based on the structural characteristics of SRK, the interaction between the S domain of SRK and the male *S* determinant was expected to trigger a phosphorylation cascade in the stigmatic papilla cell, leading to the rejection of self-pollen.

Thus, identification of the *ZmPK1* gene and the consequent identification of the *SRK* gene, which made it possible to propose a model involving receptor-kinase-mediated signal transduction as a control mechanism for *Brassica* SI, constitutes the third milestone in the history of the *Brassica* SI research (see also Chap. 8 for an account of advances in understanding the mechanisms involved in the *Brassica* SI response).

In the 1990s, Nasrallah's group in the USA led the molecular analysis of *SRK* and related genes (Boyes and Nasrallah 1993; Tantikanjana et al. 1993; Rundle et al. 1993; Boyes and Nasrallah 1995; Stein et al. 1996; Conner et al. 1998), and other SI research groups in Canada, England, France and Japan followed (Doughty et al. 1993, 1998; Goring et al. 1993; Watanabe et al. 1994; Delorme et al. 1995; Suzuki et al. 1995, 1997; Yu et al. 1996; Kusaba et al. 1997; Gu et al. 1998; Hatakeyama et al. 1998b; Stahl et al. 1998). Through these international studies, SI researchers realised that the phosphorylation cascade triggered by stigma-localised SRK might be important in SI, and the extracellular S domain of SRK, which is highly polymorphic between *S*-haplotypes, might have a role as a receptor of a pollen-coat-localised ligand. See also Chap. 6 for a discussion of SRK polymorphism.

7.5 The Fourth Milestone: Functional Evidences of *SRK* in SI

After molecular identification of the two *S*-linked genes (*SLG* and *SRK*), which were identified in the 1990s, determining the function of these genes in SI recognition reaction was required. Circumstantial evidence for the involvement of SRK and SLG in SI was obtained by the identification of self-compatible lines of *Brassica* species carrying non-functional *SRK*s or exhibiting reduced levels of *SLG* expression (Nasrallah et al. 1992, 1994; Goring et al. 1993; Watanabe et al. 1997). To directly show the function of SRK or SLG, a large number of transformation experiments have been performed. All 'gain-of-function' approaches, however, have resulted in the breakdown of SI in the transformants, most likely as a result of co-suppression between the endogenous *SLG* and/or *SRK* and the transgenes (Conner et al. 1997; Stahl et al. 1998; Takasaki et al. 1999). Our Japanese research group conducted 'loss-of-function' experiments using an antisense *SLG* gene, and demonstrated that the transformants became self-compatible (Shiba et al. 1995, 2000; Takasaki et al. 2001). The expression of *SRK*, as well as that of *SLG*, was suppressed in these transformants, suggesting that SLG and/or SRK are necessary for the SI recognition reaction.

In the case of the gain-of-function experiments, we tried to avoid co-suppression by introducing a class-I S^9 haplotype *SRK* (which is identical to an S^{28} haplotype) into *B. rapa* cultivar Osome, which carries a class-II S^{60} haplotype (Takasaki et al. 2000). High sequence diversity between the class-I SRK^9 transgene and the endogenous class-II SRK^{60} enabled us to apply a transgenic approach without co-suppression. In addition, a highly efficient transformation method developed for *B. rapa* cultivar Osome (Takasaki et al. 1997) contributed to the production of transgenic plants showing the gain-of-function phenotype. As a result, we clearly showed that the transgenic plants expressing the SRK^9-transgene acquired the S^9 haplotype specificity in the stigma, but not in the pollen, and we concluded by showing that SRK is the sole determinant of the *S*-haplotype specificity of the stigma (Takasaki et al. 2000). At the same time, independently, Canadian SI researchers also succeeded in demonstrating the gain-of-function experiments using *SRK* and *SLG* in *B. napus* (Cui et al. 2000; Silva et al. 2001). An important factor in the successful production of gain-of-function transformants may be their use of self-compatible amphidiploid *B. napus* cultivar Westar as a recipient, although the self-compatible phenotype of Westar is the result of a mutation on the pollen-side, and its *SRK* is functional (Okamoto et al. 2007). Although Silva et al. (2001) did not find that *SLG* had an enhancing effect on the SI response as Takasaki et al. (2000) did, both studies concluded that *SRK* is involved in the SI recognition. Because of their technical difficulty, it took almost a decade from identification of the *SRK* gene until its functionality was demonstrated. Thus, finally these gain-of-function experiments directly demonstrated the function of *SRK*. This undoubtedly was an extremely important step forward in SI research on *Brassica*.

7.6 The Fifth Milestone: Establishment of Bioassay System

After finding the *SLG* and *SRK* genes, *Brassica* SI researchers focused on identification of 'a pollen *S* gene', whose product was thought to interact with SRK in an *S*-allele-specific manner. Under the circumstances, it was necessary to look for some clues to detect biologically active (i.e. causing an SI reaction) molecules in pollen. Establishment of the bioassay system for *Brassica* SI reaction may have been an important milestone at this stage. Bioassay systems are generally important for identification of biologically active molecules. For example, in the course of isolating gibberellin (GA), establishing a bioassay system with α-amylase activity was key (Jones and Varner 1967). In *Papaver rhoeas*, which has a gametophytic SI, an in vitro bioassay system was established as an ideal system for identification of the female *S* determinant (Foote et al. 1994). In this system, where pollen was grown in vitro in an artificial germination medium, *S*-specific inhibition of pollen tube growth was observed when the *S* allele of the pollen grain was identical to that of the stigma extract in the medium (Franklin-Tong et al. 1988). Unfortunately, such an in vitro system was not applicable to *Brassica* SI, and development of another bioassay method was needed for *Brassica* SI research.

Dickinson's group in England led the way in establishing a novel bioassay system using *Brassica* pollen coat proteins (PCPs). Most of the PCPs on the pollen exine are thought to be derived from the anther tapetum, which performs essential roles in pollen development. The tapetum is known to supply nutrients to the developing microspores and is a precursor for the synthesis of pollen exine. The tapetum degrades at the later stage of pollen development, and the remnants of tapetum are also deposited on the pollen exine (Heslop-Harrison 1975). Several types of PCPs have been isolated and characterised (Doughty et al. 1993, 1998; Hiscok et al. 1995; Stanchev et al. 1996; Toriyama et al. 1998). In the course of analysing PCPs, Dickinson and his colleagues developed a bioassay system by pre-treating the stigma surface with purified PCPs using micromanipulation. When the stigma was pre-treated with PCPs derived from self-pollen, cross (compatible) pollen grains failed to hydrate and germinate on the stigma surface. In contrast, the pre-treatment of stigma with PCPs derived from cross-pollen did not affect cross pollen hydration or germination. By using this system, they provided evidence that a <10 kDa molecular size fraction enriched with small basic cysteine-rich proteins contains the male *S* determinant of *B. oleracea* (Stephenson et al. 1997). Actually, a pollen *S* gene, *SP11/SCR*, isolated by the genomic approach (Suzuki et al. 1999; Schopfer et al. 1999; see Sect. 7.7) was found to encode a small cysteine-rich protein as suggested by Stephenson et al. (1997). It is noteworthy that the bioassay system was used to demonstrate that SP11/SCR is the male *S* determinant (Takayama et al. 2000, 2001), and to assess the SI reaction (Murase et al. 2004). Thus, this represented an important milestone, as although it did not identify the male *S* determinant, it provided a strong hint to its identity.

7.7 The Sixth Milestone: Identification of *SP11/SCR*, the Male *S* Determinant Gene

Because the *Brassica* SI reaction is genetically regulated by the *S*-locus as one segregation unit, an unidentified pollen *S* gene must be located at the *S*-locus, in the vicinity of *SLG* and *SRK*. However, when genomic analyses around *SLG* and *SRK* were being conducted independently in several countries, the physical size of the *S*-locus was estimated at maximum 220 kb in the S_6 haplotype and 350 kb in the S_2 haplotype of *B. oleracea* (Boyes and Nasrallah 1993). This size was too large to conduct chromosome walking using a lambda phage vector, which is limited to 20-kb cloning. Therefore, our group in Japan did not try to conduct chromosome walking in the *S*-locus when genomic fragments of *SLG* and *SRK* were independently cloned into the lambda phage vector (Suzuki et al. 1995). Instead, we directly cloned (chromosome landing) the *S*-locus using a P1-derived artificial chromosome (PAC) vector in which a large genome fragment ($>\sim100$ kb) can be cloned (Ioannou et al. 1994). A 76-kb genomic fragment containing both *SLG* and *SRK* genes in the S^9 haplotype of *B. rapa* was successfully cloned into a PAC vector (Suzuki et al. 1997). Using *B. rapa* for this *S*-locus cloning was very advantageous because physical distances between *SLG* and *SRK* in *B. rapa* are generally smaller than those in *B. oleracea* (Suzuki et al. 2000a). By using this cloned *S*-locus, Suzuki et al. (1999) determined the nucleotide sequence of the 76 kb genomic fragment containing *SLG* and *SRK*, screened the cDNA libraries of immature anthers and finally found the male *S* determinant gene, *SP11* (*S*-locus protein 11). Conventional cDNA screening was also important for successful identification of *SP11*, because *SP11* could not be found from the nucleotide sequence of the *S*-locus; *SP11* is a small gene whose G + C content is lower than other genes, which is why it was not identified as a 'coding region' by the GENSCAN gene-finding program (Suzuki et al. 1999). In our study to identify the 'pollen *S* determinant gene', another genetic approach using a fluorescent differential display (FDD) technique was also effective (Takayama et al. 2000). In parallel with the genomic analysis on the *S*-locus region of the S^9 haplotype that resulted in the identification of S^9-*SP11*, we compared the cDNAs from S^8 and S^{12} haplotypes by using FDD and succeeded in identifying the allelic S^8-*SP11* gene as an S^8 haplotype-specific anther expressed gene. The identification of allelic *SP11* genes allowed us to design a common primer that could amplify most allelic *SP11* genes from other *S*-haplotypes (Watanabe et al. 2000). The S^8-*SP11* gene identified by FDD analysis was identical to *SCR*^{8} (*S*-locus cysteine-rich), which was independently identified by Nasrallah's group in the USA by the genomic analysis of the S^8 haplotype (Schopfer et al. 1999); see also Chap. 6 for a discussion on analysis of *SP11/SCR* polymorphisms. The function of SP11/SCR as a male *S* determinant was confirmed by transformation experiments and the pollination bioassay (Schopfer et al. 1999; Takayama et al. 2000; Shiba et al. 2001). Thus, the finding of the male *S* determinant, the last milestone, was accomplished by our Japanese and USA groups at the same time. The worldwide competition to identify both the pollen and pistil SI genes responsible for the recognition of self-pollen was now complete.

7.8 After Identifying the SI Genes and Future Milestones

At the end of the twentieth century, the long-anticipated pollen *S* gene was identified in *Brassica*, and all the SI researchers of the world were pleased that the first set of female and male *S* determinant in SI species was identified (see Franklin-Tong and Franklin 2000). After this finding, *Brassica* SI researchers focused on more veiled subjects such as *physical interaction between SRK and SP11/SCR*, *downstream signalling molecules after interaction between SRK and SP11/SCR*, *molecular mechanisms of dominance relationships* and *evolution of SI genes*. Here we provide a brief overview of these subjects.

7.8.1 Demonstrating Physical Interaction Between SRK and SP11/SCR

The demonstration of physical interaction between SRK (receptor) and SP11/SCR (S ligand) was independently achieved by Nasrallah's group and by our Japanese group (Kachoo et al. 2001; Takayama et al. 2001). Kachroo et al. demonstrated a direct interaction between SRK and SCR using recombinant proteins. The extracellular domain of SRK (eSRK) was shown to interact with SCR in an S-haplotype-specific manner through an immunoprecipitation approach. Takayama et al. (2001) used a different approach, labelling SP11 by iodination and then demonstrating its S-haplotype-specific binding to stigmatic microsomal membranes. Scatchard analysis revealed its high-affinity binding ($K_d = 0.7$ nM) (Takayama and Isogai 2003). The chemical cross-linking experiment suggested that SP11 specifically binds to SRK and a 60-kDa protein detectable by anti-SRK antibody. This 60-kDa protein was later identified as a membrane-anchored truncated SRK (tSRK), which must be post-translationally generated from SRK (Shimosato et al. 2007). Interestingly, the alternatively spliced soluble form of SRK (eSRK) as well as SLG exhibited no high-affinity binding, while artificial dimerised form of eSRK exhibited high-affinity binding to SP11. These findings suggest that SRK dimerisation, supported by membranous environment, is crucial for the formation of stabilised SP11-SRK complex. This is in agreement with the finding that SRK formed oligomer in the absence of ligand in microsome of stigma, while eSRK as well as SLG are mostly monomeric (Giranton et al. 2000). Using a surface-display system in yeast, the ligand-independent SRK dimerisation was suggested to depend on the two regions within the extracellular domain of SRK: the C-terminal PAN_APPLE domain and the EGF-like domain (Naithani et al. 2007; see Chap. 6, this volume, for a more in-depth discussion of the different forms of SRK, eSRK, tSRK and the PAN_APPLE and EGF-like domains). An important point is that, in supporting to its purposed functional role in SI, SRK was found to be autophosphorylated after interacting with SP11 of the same S-haplotype (Takayama et al. 2001). Although the molecular basis of selectivity in SP11-SRK interactions (Kemp and Doughty 2007) and of SRK activation remains to be clarified, it is now clear that the S-haplotype-specific interaction between SP11 and SRK triggers the autophosphorylation of SRK, activating the signalling cascade that results in the rejection of self-pollen.

Although the precise interaction sites of SP11 and SRK were not determined, the tertiary conformation of S^8-SP11 was established by NMR spectroscopy. Based on the structure-based sequence alignment of allelic SP11 with a different S-haplotype, Mishima et al. suggested that the 3D-structure of SP11 is highly similar among allelic genes, and that the hyper-variable (HV) C3-C4 and C5-C6 regions should be important for allelic specificity (Mishima et al. 2003). From the bioassay with chimeric SP11 proteins and transformation of chimeric *SP11* genes, C2-C3 and C5-C6 regions are suggested to be important for determining the recognition specificity (Sato et al. 2003, 2004; Chookajorn et al. 2004). As for the SP11/SCR-binding site of SRK, a recent report by an English group proposed an intriguing model consisting of a two-stage recognition process between SP11/SCR and the HV region of the S-domain of SRK (Kemp and Doughty 2007). In the future, after determining the actual tertiary structure of SRK, the precise interaction sites of SP11 and SRK will be clarified.

7.8.2 Downstream of the Interaction: Identifying Components and Mechanisms Involved in Mediating the Rejection of Self Pollen

The finding of SRK as the female determinant gave an important clue to the possible mechanism of *Brassica* SI. SRK belongs to the family of large plant receptor-like kinases (RLK), and different types of genes encoding RLK have been identified and analysed (reviewed in Torii 2004; Morillo and Tax 2006); RLKs are involved in many physiological phenomena, for example epidermal patterning, nodulation, hormone signalling, pollen-tube guidance and pathogen response. Various phosphorylation cascades downstream of RLKs might make a complex network in the cell to respond to the extracellular signals. The pathway downstream of SRK activation by interaction with SP11/SCR currently provides one of the few model cascades of such RLK-mediated signal pathways in plant cells. Therefore, this will form a crucial milestone for the future, as it will provide important information on events downstream of SRK signalling, which will form a key aspect of *Brassica* SI research as well as for the general understanding of the RLK-mediated signal transduction.

Today, there is relatively little information on the downstream of SRK signalling. However, see Chap. 8 for a more detailed account of what is known about this topic. Briefly, it is known that ARC1 (arm repeat containing) and MLPK (*M* locus protein kinase) are the only candidate molecules playing roles in the downstream of SRK. Although some other molecules, such as thioredoxin h-like proteins (THL1 and THL2), kinase-associated protein phosphatase (KAPP), calmodulin and sorting nexin, have been shown to interact with SRK, all these molecules are expressed as negative regulators of SRK signalling (Cabrillac et al. 2001; Vanoosthuyse et al. 2003). The ARC1 has an E3 ubiquitin ligase activity, which can interact with the SRK-kinase domain in a phosphorylation-dependent manner, is a putative positive effector of the *Brassica* SI response (Stone et al. 1999, 2003). The MLPK is a novel membrane-anchored cytoplasmic serine/threonine protein kinase whose transient expression restores the ability to reject self-pollen on a single recessive self-compatible (SC) mutant, *m* (Murase et al. 2004). Further, biochemical and molecular biological analyses suggested that MLPK localises to the papilla cell membrane and interacts directly with SRK to transduce SI signalling (Kakita et al. 2007).

As the entire SI signalling cannot be explained only by SRK, SP11/SCR and the two positive downstream-players, ARC1 and MLPK, many other molecules may be involved in the signal transduction of SI, from the autophosphorylation of SRK to the pollen rejection (see Fig. 7.2). To identify additional molecular players in the downstream of the SRK signalling, more SC mutants are necessary in Brassicaceae in the future. Recently identified genetically modified SI *A. thaliana* (Nasrallah et al. 2002, 2004) is a potential resource for SC-mutant analysis, because *A. thaliana* is the most useful model plant whose genome sequence has been completely determined (The Arabidopsis Genome Initiative 2000). Detailed analysis of the many SC mutants will reveal a complex cascade with the interaction of many cytoplasmic

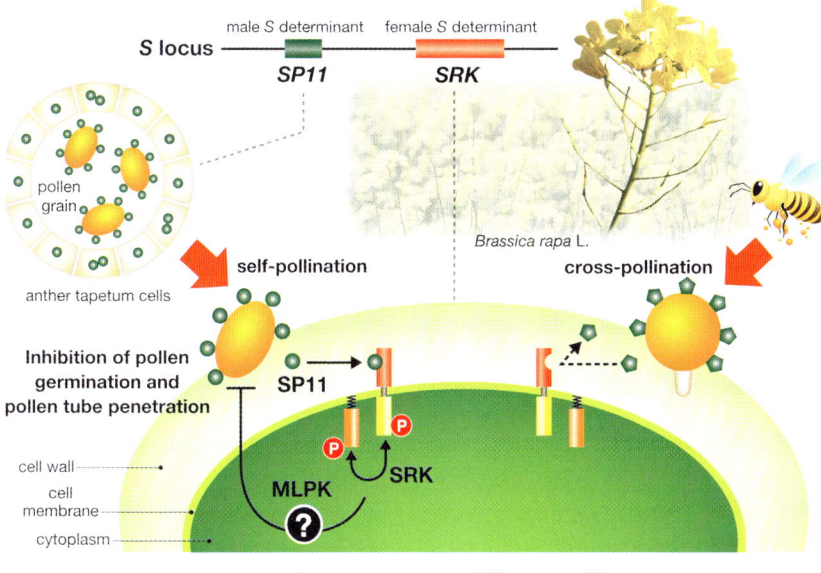

Fig. 7.2 Schematic model for self-pollen recognition in *Brassica* species

proteins. It is hoped that additional SI signalling research will lead to the study of inter-cellular signal transduction in plants, as in the case of the intensive research of EGF receptor-mediated signal transduction in mammalian cells.

7.8.3 Molecular Mechanisms of Dominance Relationships

A key characteristic of SSI in Brassica is the presence of a complex set of dominance relationships (Bateman 1954, 1955; Thompson and Taylor 1966; Hatakeyama et al. 1998a). Because the Brassicaceae has a sporophytic SI system and the pollen *S* determinant *SP11/SCR* as well as the female *S* determinant *SRK* is sporophytically expressed, the SI phenotypes of stigma and pollen are determined by the relationships between the two *S*-haplotypes they carry, as described earlier. From the genetic analysis, the characteristics of the dominance relationships in Brassicaceae are as follows: (i) co-dominance occurs more frequently than dominance or recessiveness; (ii) dominance or recessiveness is observed more frequently in the pollen than in the stigma; (iii) dominant or recessive is independent between the pollen and stigma; (iv) non-linear dominance relationships are also observed, and occur more frequently in the stigma than in the pollen (Bateman 1954, 1955, Thompson and Taylor 1966, Hatakeyama et al. 1998a). Based on this genetic data, the molecular mechanisms of the dominance relationships between *S*-haplotypes were studied both in the stigma and pollen sides. In the stigma, the dominance relationships may be post-transcriptionally regulated by the female S determinant, SRK, and be unrelated to the transcript level of *SRK* (Hatakeyama et al. 2001). On the other hand, dominance relationships on the pollen side are determined by the expression level of *SP11/SCR*. Shiba et al. found that the *SP11* transcript from a recessive S-haplotype, but not from a dominant S-haplotype, is greatly diminished in the anther tapetum of the S heterozygotes (Shiba et al. 2002). Some S-haplotypes show a liner dominance relationship among them, such as $S^9 > S^{44} > S^{60} > S^{40} > S^{29}$ (Hatakeyama et al. 1998a). Of these S-haplotypes, S^{44}, S^{60}, S^{40} are dominant in some cases but recessive in others, and the dominance relationships are thus affected by the allelic partner in S heterozygotes. Kakizaki et al. analysed 10 S heterozygotes derived from these five S-haplotypes, and found that the expression of the recessive *SP11* allele was always suppressed in the presence of the dominant *SP11* allele (Kakizaki et al. 2003).

The most surprising finding of our study was that the $5'$-promoter region of the recessive *SP11* allele was specifically methylated in the anther tapetum of heterozygote (Shiba et al. 2006). This promoter region is essential for *SP11* transcription at the tapetum, and its methylation occurs specifically in the anther immediately prior to the initiation of *SP11* transcription. These results suggest that tissue-specific monoallelic de novo DNA methylation is involved in determination of the dominance relationships at the pollen side. This is the first finding suggesting involvement of epigenetics (DNA methylation) in classical Mendelism (dominance relationships), and the occurrence of mammalian-like tissue- and developmental stage-specific de novo methylation in plants.

7.8.4 Evolution of SI Genes

In examining the evolution of SI, as the fourth subject, the finding of the pollen S gene led to an intensive analysis of the sequence-divergence of a set of highly polymorphic S genes; see also Chap. 6 for a detailed account of this topic. The allelic genes of *SP11* were isolated from *B. rapa*, *B. oleracea* and *Raphanus sativus* by RT-PCR method, and the sequence comparison of *SP11* and *SRK* genes between S-haplotypes showed the co-evolution of ligands and receptors and the trans-specific evolution of *SP11* and *SRK* genes (Watanabe et al. 2000, Kimura et al. 2002, Sato et al. 2002, Okamoto et al. 2007). In the case of the female side, several phylogenetic analyses of *SLG* and *SRK* revealed an important HV in SLG and SRK (Hinata et al. 1995, Sato et al. 2002). Further detailed analysis of the HV of SP11 and SRK may lead to an explanation of the extremely high diversity of S-haplotypes, which is one of the most mysterious questions remaining in the evolution of SI.

When considering the evolutionary aspect of SI, another approach using *Brassica* relative, *Arabidopsis*, has been taken. The origin of self-fertility of *Arabidopsis thaliana* was estimated from a comparison with the allelic variation of accessions. From the comparative genome between self-compatible *A. thaliana* and self-incompatible *A. lyrata*, the mutated S-locus containing ψ*SCR* and ψ*SRK* was identified on an *ARK3* region in chromosome IV of *A. thaliana* (Kusaba et al. 2001). Based on the lack of ψ*SCR* variation, Shimizu et al. proposed that the transition to self-compatibility in *A. thaliana* occurred very recently, and pseudogenisation of ψ*SCR* led to the loss of SI (Shimizu et al. 2004), and that the self-compatibility arose multiple times with different genetic bases (Shimizu et al. 2008). However, there are highly debated arguments for a critical step in the evolution of selfing and its timing (Bechsgaard et al. 2006; Sherman-Broyles et al. 2007; Tang et al. 2007). Tang et al. indicated that a selective sweep of ψ*SCR* seemed unlikely because of the extremely high sequence diversity of the *S*-locus region (Tang et al. 2007). See also Chap. 6, this volume, for an account of advances in understanding the evolution of mating systems in the Brassicaceae using information from *SRK* and *SP11/SCR* sequences.

Thus, the finding of a set of *Brassica S* genes has accelerated molecular-based analyses of SI-related mechanisms, and generated novel research themes in the last decade. Furthermore, recent *Brassica* SI research has successfully taken on the exciting topics of biological science, such as an *epigenetics* (Shiba et al. 2006). These subjects in *Brassica* SI constitute ongoing research projects that will clarify these interesting questions in the future.

7.9 Prospects

Thus far we have provided an overview of important milestones in *Brassica* SI research in identifying the SI genes, *SRK* and *SP11/SCR*. As described earlier, SRK is a membrane-bound receptor kinase, whose receptor domain can interact with its ligand, SP11/SCR. It took almost 50 years of international effort to identify a set of the male and female *S* determinants from Bateman's idea related to sporophytic control of the *S*-locus. Currently other studies on Brassica SI are focusing on the interesting topic of the evolution of SI in the crucifers (see Chap. 6) and a major ongoing effort is underway to attempt to elucidate exactly how the mechanism of pollen inhibition is achieved in *Brassica*. Although much is known about the components at the *S*-locus and also the involvement of other unlinked components (see Chap. 8 for further details), rather little is known about the signalling cascades involved and how this feeds back to signal to incompatible pollen. Moreover, virtually nothing is known about what exactly is responsible for causing pollen inhibition in the *Brassica* SI system. This is in contrast to the gametophytic SI systems, where (although much remains to be elucidated with respect to other aspects) something is known about how pollen is inhibited. For example, in the GSI *S*-RNase system, release of *S*-RNases will undoubtedly result inhibition of pollen tube growth (see Chap. 10 for further details) and in the *Papaver* SI system, a Ca^{2+}-based signalling system triggers a number of intracellular responses resulting in first pollen tube inhibition through actin depolymerisation and inhibition of soluble inorganic pyrophosphatases and then later programmed cell death (see Chap. 11 for further details). It will be extremely interesting to see what the cell biology reveals about the components that are involved in pollen inhibition in the *Brassica* SI system, as it is clear that the control of pollen germination and growth is complex. Doubtless, in the next 50 years, with advanced concepts and technology, we will be pursuing our goal while reaching new milestones in *Brassica* SI research. This will hopefully eventually allow our knowledge about this important SI system to be applied in a plant breeding context, which was the original aim of initiating studies into the molecular basis of SI.

Acknowledgements We are grateful to Yoshinobu Takada and Keita Suwabe (Tohoku University) for their helpful discussion. We also thank Sachiyo Isokawa (Tohoku University) for providing photographs. This work was supported in part by Grants-in-Aid for Scientific Research on Priority Areas (Nos. 18075003, 18075008, 18075011, and 18075012) and Grants-in-Aid for the 21st Century COE Program (to NAIST and Iwate University) and for the Global COE Program (to NAIST) from the Ministry of Education, Culture, Sports, Science, and Technology of Japan (MEXT), Grants-in-Aid for Creative Scientific Research (No. 16G3016) and a Grants-in-Aid for Scientific Research (No. 14360002, 17380001, 18380069, 20380002, and 20678001) from the Japan Society for Promotion of Science (JSPS), and a grant from the Intelligent Cosmos Academic Foundation. We apologise to all those whose work in this field could not be cited because of limited space.

References

The Arabidopsis Genome Initiative (2000) Analysis of the genome sequence of the flowering plant *Arabidopsis thaliana*. Nature 408:796–815
Bateman AJ (1952) Self-incompatibility systems in angiosperms. I. Theory. Heredity 6:285–310
Bateman AJ (1954) Self-incompatibility systems in angiosperms. II. *Iberis amara*. Heredity 8: 305–332
Bateman AJ (1955) Self-incompatibility systems in angiosperms. III. Cruciferae. Heredity 9:53–68
Bechsgaard JS, Castric V, Charlesworth D, Vekemans X, Schierup MH (2006) The transition to self-compatibility in *Arabidopsis thaliana* and evolution within *S*-haplotypes over 10 Myr. Mol Biol Evol 23:1741–1750
Boyes DC, Nasrallah JB (1993) Physical linkage of the *SLG* and *SRK* genes at the self-incompatibility locus of *Brassica oleracea*. Mol Gen Genet 236:369–373
Boyes DC, Nasrallah JB (1995) An anther-specific gene encoded by an *S*-locus haplotype of *Brassica* produces complementary and differentially regulated transcripts. Plant Cell 7:1283–1294
Boyes DC, Chen C-H, Tantikanjana T, Esch JJ, Nasrallah JB (1991) Isolation of a second *S*-locus-related cDNA from *Brassica oleracea*: Genetic relationships between the *S*-locus and two related loci. Genetics 127:221–228
Cabrillac D, Cock JM, Dumas C, Gaude T (2001) The *S*-locus receptor kinase is inhibited by thioredoxins and activated by pollen coat proteins. Nature 410:220–223
Chen C-H, Nasrallah JB (1990) A new class of *S* sequences defined by a pollen recessive self-incompatibility allele of *Brassica oleracea*. Mol Gen Genet 222:241–248
Chookajorn T, Kachroo A, Ripoll DR, Clark AG, Nasrallah JB (2004) Specificity determinants and diversification of the *Brassica* self-incompatibility pollen ligand. Proc Natl Acad Sci USA 101:911–917
Cock JM, Stanchv B, Delorme V, Croy RRD, Dumas C (1995) *SLR3*: A modified receptor kinase gene that has been adapted to encode a putative secreted glycoprotein similar to the *S*-locus glycoprotein. Mol Gen Genet 248:151–161
Conner JA, Tantikanjana T, Stein JC, Kandasamy MK, Nasrallah JB, Nasrallah ME (1997) Transgene-induced silencing of *S*-locus genes and related genes in *Brassica*. Plant J 11:809–823

Conner JA, Conner P, Nasrallah ME, Nasrallah JB (1998) Comparative mapping of the *Brassica* S-locus region and its homolog in *Arabidopsis*: Implications for the evolution of mating system in the Brassicaceae. Plant Cell 10:801–812

Correns C (1912) Selbstasterilitat und individualstoffe. Festschw. Med-Nasturwiss. Ges. Munster 84:186–217

Crowe LK (1954) Incompatibility in *Cosmos bipinnatus*. Heredity 8:1–11

Cui Y, Bai Y-M, Brugiere N, Arnoldo MA, Rothstein SJ (2000) The *S*-locus glycoprotein and the *S* receptor kinase are sufficient for self-pollen rejection in *Brassica*. Proc Natl Acad Sci USA 97:3713–3717

de Nettancourt D (2001) Incompatibility and incongruity in wild and cultivated plants, 2nd edn. Springer, Berlin Heidelberg New York

Delorme V, Giranton J-L, Hatzfeld Y, Friry A, Heizmann P, Ariza MJ, Dumas C, Gaude T, Cock JM (1995) Characterization of the *S*-locus genes, *SLG* and *SRK*, of the *Brassica* S^3 haplotype: Identification of a membrane-localized protein encoded by the *S*-locus receptor kinase. Plant J 7:429–440

Doughty J, Hedderson F, McCubbin A, Dickinson HG (1993) Interaction between a coating-borne peptide of the *Brassica* pollen grain and stigmatic *S* (self-incompatibility)-locus-specific glycoproteins. Proc Natl Acad Sci USA 90:467–471

Doughty J, Dixon S, Hiscok SJ, Willis AC, Parkin IAP, Dickinson HG (1998) PCP-A1, a defensin-like *Brassica* pollen coat protein that binds the *S*-locus glycoprotein, is the product of gametophytic gene expression. Plant Cell 10:1333–1347

Dwyer KG, Balent MA, Nasrallah JB, Nasrallah ME (1991) DNA sequences of self-incompatibility genes from *Brassica campestris* and *Brassica oleracea*: Polymorphism pre-dating speciation. Plant Mol Biol 16:481–486

Foote HC, Ride JP, Franklin-Tong VE, Walker EA, Lawrence MJ, Franklin FCH (1994) Cloning and expression of a distinctive class of self-incompatibility (*S*) gene from *Papaver rhoeas* L. Proc Natl Acad Sci USA 91:2265–2269

Franklin-Tong VE, Franklin FCH (2000) Self-incompatibility in *Brassica*: The elusive pollen *S* gene is identified! Plant Cell 12:305–308

Franklin-Tong VE, Lawrence MJ, Franklin FCH (1988) An in vitro bioassay for the stigmatic product of the self-incompatibility gene in *Papaver rhoeas* L. New Phytol 110:109–118

Goring DR, Glavin, TL, Schafer U, Rothstein SJ (1993) An *S* receptor kinase gene in a self-compatible *Brassica napus* has 1-bp deletion. Plant Cell 5:531–539

Gu T, Mazzurco M, Sulaman W, Matias DD, Goring DR (1998) Binding of an arm repeat protein to the kinase domain of the *S*-locus receptor kinase. Proc Natl Acad Sci USA 95:382–387

Hatakeyama K, Watanabe M, Takasaki T, Ojima K, Hinata K (1998a) Dominance relationships between *S*-alleles in self-incompatible *Brassica campestris* L. Heredity 79:241–247

Hatakeyama K, Takasaki T, Watanabe M, Hinata K (1998b) Molecular characterization of *S*-locus genes, *SLG* and *SRK* in a pollen-recessive self-incompatibility haplotype of *Brassica rapa* L. Genetics 149:1587–1597

Hatakeyama K, Takasaki T, Suzuki G, Nishio T, Watanabe M, Isogai A, Hinata K (2001) The *S* receptor kinase gene determines dominance relationships in stigma expression of self-incompatibility in *Brassica*. Plant J 26:69–76

Heslop-Harrison J (1975) Incompatibility and the pollen-stigma interaction. Annu Rev Plant Physiol 26:403–425

Hinata K, Nishio T (1978) *S*-allele specificity of stigma proteins in *Brassica oleracea* and *Brassica campestris*. Heredity 41:93–100

Hinata K, Nishio T, Kimura J (1982) Comparative studies on S-glycoproteins purified from different *S*-genotypes in self-incompatible *Brassica* species. II. Immunological specificities. Genetics 100:649–657

Hinata K, Watanabe M, Yamakawa S, Satta Y, Isogai A (1995) Evolutionary aspects of the *S*-related genes of the *Brassica* self-incompatibility system: synonymous and nonsynonymous base substitutions. Genetics 140:1099–1104

Hiscok SJ, Doughty J, Willis AC, Dickinson HG (1995) A 7-kDa pollen coating-borne peptide from *Brassica napus* interacts with *S*-locus glycoprotein and *S*-locus related glycoprotein. Planta 196:367–374

Ioannou PA, Amemiya CT, Garnes J, Kroisel PM, Shizuya H, Chen C, Batzer MA, de Jong PJ (1994) A new bacteriophage P1-derived vector for the propagation of large human DNA fragments. Nat Genet 6:84–89

Isogai A, Hinata K (2002) Molecular mechanism for the recognition reaction in the self-incompatibility of *Brassica* species. Proc Jpn Acad 78B:241–249

Isogai A, Takayama S, Tsukamoto C, Ueda Y, Shiozawa H, Hinata K, Okazaki K, Suzuki A (1987) *S*-locus-specific glycoproteins associated with self-incompatibility in *Brassica campestris*. Plant Cell Physiol 28:1279–1291

Isogai A, Takayama S, Shiozawa H, Tsukamoto C, Kanbara T, Hinata K, Okazaki K, Suzuki A (1988) Existence of a common glycoprotein homologous to S-glycoproteins in two self-incompatible homozygotes of *Brassica campestris*. Plant Cell Physiol 29:1331–1336

Jones RL, Varner JE (1967) The bioassay of gibberellins. Planta 72:155–161

Kachoo A, Schopfer CR, Nasrallah ME, Nasrallah JB (2001) Allelic-specific receptor-ligand interactions in *Brassica* self-incompatibility. Science 293:1824–1826

Kachoo A, Nasrallah ME, Nasrallah JB (2002) Self-incompatibility in the Brassicaceae: receptor-ligand signaling and cell-to-cell communication. Plant Cell 14:S227–S238

Kakita M, Murase K, Iwano M, Matsumoto T, Watanabe M, Shiba H, Isogai A, Takayama S (2007) Two distinct forms of MLPK localize to the plasma membrane and interact directly with SRK to transducer self-incompatibility signaling. Plant Cell 19:3961–3973

Kakizaki T, Takada Y, Ito A, Suzuki G, Shiba H, Takayama S, Isogai A, Watanabe M (2003) Linear dominance relationship among four class-II *S*-haplotypes in pollen side determined by the expression of *SP11* in *Brassica* self-incompatibility. Plant Cell Physiol 44:70–75

Kakizaki Y (1930) Studies on the genetics and physiology of self- and cross-incompatibility in the common cabbage (*Brassica oleracea* L. var. *capitata* L.). Jpn J Bot 5:133–208

Kemp BP, Doughty J (2007) *S* cysteine-rich (SCR) biding domain analysis of the *Brassica* self-incompatibility *S*-locus receptor kinase. New Phytol 175:619–629

Kimura R, Sato K, Fujimoto R, Nishio T (2002) Recognition specificity of self-incompatibility maintained after the divergence of *Brassica oleracea* and *Brassica rapa*. Plant J 29:215–223

Kumar V, Trick M (1994) Expression of the *S*-locus receptor kinase multigene family in *Brassica oleracea*. Plant J 6:807–813

Kusaba M, Nishio T, Hinata K, Ockendon DJ (1997) Sequences of class I *SLG*s of *Brassica oleracea* and *B. campestris*: Evolution and recognition. Proc Natl Acad Sci USA 94:7673–7678

Kusaba M, Dwyer K, Hendershot J, Vrebalov J, Nasrallah JB, Nasrallah ME (2001) Self-incompatibility in the genus *Arabidopsis*: Characterization of the *S*-locus in the outcrossing *A. lyrata* and its autogamous relative *A. thaliana*. Plant Cell 13:627–643

Lalonde BA, Nasrallah ME, Dwyer KG, Chen C-H, Barlow B, Nasrallah JB (1989) A highly conserved *Brassica* gene with homology to the *S*-locus-specific glycoprotein structural gene. Plant Cell 1:249–258

Martin FW (1968) The system of self-incompatibility in *Ipomoea*. J Hered 59:263–267

Mishima M, Takayama S, Sasaki K, Jee J-G, Kojima C, Isogai A, Shirakawa M (2003) Structure of the male determinant factor for *Brassica* self-incompatibility. J Biol Chem 278:36389–36395

Morillo SA, Tax FE (2006) Functional analysis of receptor-like kinases in monocots and dicots. Curr Opin Plant Biol 9:460–469

Murase K, Shiba H, Iwano M, Che F-C, Watanabe M, Isogai A, Takayama S (2004) A membrane-anchored protein kinase involved in *Brassica* self-incompatibility signaling. Science 303:1516–1519

Nakanishi T, Hinata K (1973) An effective time for CO_2 gas treatment in overcoming self-incompatibility in *Brassica*. Plant Cell Physiol 14:873–879

Nakanishi T, Hinata K (1975) Self-seed production by CO_2 gas treatment in self-incompatible cabbage. Euphytica 24:117–120

Nakanishi T, Esashi Y, Hinata K (1969) Control of self-incompatibility by CO_2 gas in *Brassica*. Plant Cell Physiol 10:925–927

Nakanishi TM, Yamazaki T, Funadera K, Tomonaga H, Ozaki T, Kawai Y, Ichii T, Sato Y, Kurihara A (1992) Isoelectric focusing analysis of stylar proteins associated with self-incompatibility alleles in Japanese pear. J Jpn Soc Hort Sci 61:239–248

Naithani S, Chookajorn T, Ripoll DR, Nasrallah JB (2007) Structural modules for receptor dimerization in the *S*-locus receptor kinase extracellular domain. Proc Natl Acad Sci USA 104:12211–12216

Nasrallah JB, Nasrallah ME (1993) Pollen-stigma signaling in the sporophytic self-incompatibility response. Plant Cell 5:1325–1335

Nasrallah JB, Kao T-h, Chen C-H, Goldberg ML, Nasrallah ME (1987) Amino-acid sequence of glycoproteins encoded by three alleles of the *S*-locus of *Brassica oleracea*. Nature 326:617–619

Nasrallah JB, Yu S-M, Nasrallah ME (1988) Self-incompatibility genes of *Brassica oleracea*: Expression, isolation and structure. Proc Natl Acad Sci USA 85:5551–5555

Nasrallah JB, Rundle SJ, Nasrallah ME (1994) Genetic evidence for the requirement of the *Brassica S*-locus receptor kinase gene in the self-incompatibility response. Plant J 5:373–384

Nasrallah ME, Wallace DH (1967) Immunogenetics of self-incompatibility in *Brassica oleracea*. Heredity 22:519–527

Nasrallah ME, Kandasamy MK, Nasrallah JB (1992) A genetically defined trans-acting locus regulates *S*-locus function in *Brassica*. Plant J 2:497–506

Nasrallah ME, Liu P, Nasrallah JB (2002) Generation of self-incompatible *Arabidopsis thaliana* by transfer of two *S*-locus genes from *A. lyrata*. Science 297:247–249

Nasrallah ME, Liu P, Sherman-Broyles S, Boggs NA, Nasrallah JB (2004) Natural variation in expression of self-incompatibility in *Arabidopsis thaliana*: Implications for the evolution of selfing. Proc Natl Acad Sci USA 101:16070–16074

Niikura S, Matsuura S (2000) Genetic analysis of the reaction of self-incompatibility to a 4% CO_2 gas treatment in the radish (*Raphanus sativus* L.). Theor Appl Genet 101:11789–1193

Nishio T, Hinata K (1977) Analysis of S-specific proteins in stigmas of *Brassica* L. by isoelectric focusing. Heredity 38:391–396

Nishio T, Hinata K (1982) Comparative studies on S-glycoproteins purified from different *S*-genotypes in self-incompatible *Brassica* species. I. Purification and chemical properties. Genetics 100:641–647

Nou IS, Watanabe M, Isogai A, Shiozawa H, Suzuki A, Hinata K (1991) Variation of *S*-alleles and *S*-glycoproteins in a naturalized population of self-incompatible *Brassica campestris* L. Jpn J Genet 66:227–239

Nou IS, Watanabe M, Isuzugawa K, Isogai A, Hinata K (1993a) Isolation of *S*-alleles from a wild population of *Brassica campestris* L. at Balcesme, Turkey and their characterization by S-glycoproteins. Sex Plant Reprod 6:71–78

Nou IS, Watanabe M, Isogai A, Hinata K (1993b) Comparison of *S*-alleles and S-glycoproteins between two wild populations of *Brassica campestris* in Turkey and Japan. Sex Plant Reprod 6:79–86

Okamoto S, Odashima M, Fujimoto R, Sato Y, Kitashiba H, Nishio T (2007) Self-compatibility in *Brassica napus* is caused by independent mutations in *S*-locus genes. Plant J 50:391–400

Okamoto S, Sato Y, Sakamoto K, Nishio T (2004) Distribution of similar self-incompatibility *S*-haplotypes in different genera, *Raphanus* and *Brassica*. Sex Plant Reprod 17:33–39

Rundle SJ, Nasrallah ME, Nasrallah JB (1993) Effects of inhibitors of protein serine/threonine phosphatases on pollination in *Brassica*. Plant Physiol 103:1165–1171

Sampson DR (1964) A one-locus self-incompatibility system in *Raphanus raphanistrum*. Can J Genet Cytol 6:435–445

Sato K, Nishio T, Kimura R, Kusaba M, Suzuki T, Hatakeyama K, Ockendon DJ, Satta Y (2002) Coevolution of the *S*-locus genes *SRK*, *SLG* and *SP11/SCR* in *Brassica oleracea* and *B. rapa*. Genetics 162:931–940

Sato Y, Fujimoto R, Toriyama K, Nishio T (2003) Commonality of self-recognition specificity of *S*-haplotypes between *Brassica oleracea* and *Brassica rapa*. Plant Mol Biol 52:617–626

Sato Y, Okamoto S, Nishio T (2004) Diversification and alteration of recognition specificity of the pollen ligand SP11/SCR in self-incompatibility of *Brassica* and *Raphanus*. Plant Cell 16:3230–3241

Scutt CP, Gates PJ, Gatehouse JA, Boulter D, Croy RRD (1990) A cDNA encoding an *S*-locus specific glycoprotein from *Brassica oleracea* plants containing the S^5 self-incompatibility allele. Mol Gen Genet 220:409–413

Sassa H, Hirano H, Ikehashi H (1992) Self-incompatibility-related RNase in styles of Japanese pear (*Pyrus serotina* Rehd.). Plant Cell Physiol 33:811–814

Sassa H, Hirano H, Ikehashi H (1993) Identification and characterization of stylar glycoproteins associated with self-incompatibility genes of Japanese pear (*Pyrus serotina* Rehd.). Mol Gen Genet 241:17–25

Schopfer CR, Nasrallah ME, Nasrallah JB (1999) The male determinant of self-incompatibility in *Brassica*. Science 286:1697–1700

Sherman-Broyles S, Boggs N, Farkas A, Liu P, Vrebalv J, Nasrallah ME, Nasrallah JB (2007) *S*-locus genes and the evolution of self-fertility in *Arabidopsis thaliana*. Plant Cell 19:94–106

Shiba H, Hinata K, Suzuki A, Isogai A (1995) Breakdown of self-incompatibility in *Brassica* by the antisense RNA of the *SLG* gene. Proc Jpn Acad Ser B 71:81–83

Shiba H, Kimura N, Takayama S, Hinata K, Isogai A (2000) Alteration of the self-incompatibility phenotype in *Brassica* by transformation of the antisense *SLG* gene. Biosci Biotech Biochem 64:1016–1024

Shiba H, Takayama S, Iwano M, Shimosato H, Funato M, Nakagawa T, Che F-S, Suzuki G, Watanabe M, Hinata K, Isogai A (2001) A pollen coat protein, SP11/SCR, determines the pollen *S*-specificity in the self-incompatibility of *Brassica* species. Plant Physiol 125:2095–2103

Shiba H, Iwano M, Entani T, Ishimoto K, Shimosato H, Che F-S, Satta Y, Ito A, Takada Y, Watanabe M, Isogai A, Takayama S (2002) The dominance of alleles controlling self-incompatibility in *Brassica* pollen is regulated at the RNA level. Plant Cell 14:491–504

Shiba H, Kakizaki T, Iwano M, Tarutani Y, Watanabe M, Isogai A, Takayama S (2006) Dominance relationships between self-incompatibility alleles controlled by DNA methylation. Nature Genet 38:63–67

Shimizu KK, Cork JM, Caicedo AL, Mays CA, Moore RC, Olsen KM, Ruzsa S, Coop G, Bustamante CD, Awadalla P, Purugganan MD (2004) Darwinian selection on a selfing locus. Science 306:2081–2084

Shimizu KK, Shimizu-Inatsugi R, Tsuchimatsu T, Purugganan MD (2008) Multiple origins of self-compatibility in *Arabidopsis thaliana*. Mol Ecol 17:704–714

Shimosato H, Yokota N, Shiba H, Iwano M, Entani T, Che F-S, Watanabe M, Isogai A, Takayama S (2007) Characterization of the SP11/SCR high-affinity binding site involved in self/nonself recognition in *Brassica* self-incompatibility. Plant Cell 19:107–117

Silva NF, Stone SL, Christie LN, Sulaman W, Nazarian KAP, Burnett LA, Arnoldo MA, Rothstein SJ, Goring DR (2001) Expression of the *S* receptor kinase in self-incompatible *Brassica napus* cv. Westar leads to the allele-specific rejection of self-incompatible *Brassica napus* pollen. Mol Genet Genomics 265:552–559

Singh A, Kao T-H (1992) Gametophytic self-incompatibility: Biological, molecular genetics, and evolutionary aspects. Inter Rev Cytol 140:449–483

Stahl RJ, Arnoldo MA, Glavin TL, Goring DR, Rothstein SJ (1998) The self-incompatibility phenotype in *Brassica* altered by the transformation of a mutant *S*-locus receptor kinase. Plant Cell 10:209–218

Stanchev BS, Doughty J, Scutt CP, Dickinson HG, Croy RRD (1996) Cloning of *PCP1*, a member of a family of pollen coat protein (PCP) genes from *Brassica oleracea* encoding novel cysteine-rich proteins involved in pollen-stigma interaction. Plant J 10:303–313

Stein JC, Howlett, B, Boyes, DC, Nasrallah ME, Nasrallah JB (1991) Molecular cloning of a putative receptor protein kinase gene encoded at the self-incompatibility locus of *Brassica oleracea*. Proc Natl Acad Sci USA 88:8816–8820

Stein JC, Dixit R, Nasrallah ME, Nasrallah JB (1996) SRK, the stigma-specific S-locus receptor kinase of *Brassica*, is targeted to the plasma membrane in transgenic tobacco. Plant Cell 8: 429–445

Stephenson AG, Doughty J, Dixon S, Elleman C, Hiscock S, Dickinson HG (1997) The male determinant of self-incompatibility in *Brassica oleracea* is located in the pollen coating. Plant J 12:1351–1359

Stone SL, Arnold M, Goring DR (1999) A breakdown of *Brassica* self-incompatibility in *ARC1* antisense transgenic plants. Science 286:1729–1731

Stone SL, Anderson EM, Mullen RT, Goring DR (2003) ARC1 is an E3 ubiquitin ligase and promotes the ubiquitination of proteins during the rejection of self-incompatible *Brassica* pollen. Plant Cell 15:885–898

Suzuki G, Watanabe M, Toriyama K, Isogai A, Hinata K (1995) Molecular cloning of members of the S-multigene family in self-incompatible *Brassica campestris* L. Plant Cell Physiol 36:1273–1280

Suzuki G, Watanabe M, Toriyama K, Isogai A, Hinata K (1997) Direct cloning of the *Brassica* S-locus by a P1-derived artificial chromosome (PAC) vector. Gene 199:133–137

Suzuki G, Kai N, Hirose T, Fukui K, Nishio T, Takayama S, Isogai A, Watanabe M, Hinata K (1999) Genomic organization of the S-locus: Identification and characterization of genes in *SLG/SRK* region of S^9 haplotype of *Brassica campestris* (syn. *rapa*). Genetics 153:391–400

Suzuki G, Watanabe M, Nishio T (2000a) Physical distances between S-locus genes in various S-haplotypes of *Brassica rapa* and *B. oleracea*. Theor Appl Genet 101:80–85

Suzuki G, Kakizaki T, Takada Y, Shiba H, Takayama S, Isogai A, Watanabe M (2003) The S-haplotypes lacking *SLG* in the genome of *Brassica rapa*. Plant Cell Rep 21:911–915

Suzuki T, Kusaba M, Matsushita M, Okazaki K, Nishio T (2000b) Characterization of *Brassica* S-haplotypes lacking S-locus glycoprotein. FEBS Lett 482:102–108

Tang C, Toomajian C, Sherman-Broyles S, Plagnol V, Guo Y-L, Hu TT, Clark RM, Nasrallah JB, Weigel D, Nordborg M (2007) The evolution of selfing in *Arabidopsis thaliana*. Science 317:1070–1072

Takasaki T, Hatakeyama K, Ojima K, Watanabe M, Toriyama K, Hinata K (1997) Factors influencing *Agrobacterium*-mediated transformation of *Brassica rapa* L. Breed Sci 47:127–134

Takasaki T, Hatakeyama K, Suzuki G, Watanabe M, Isogai A, Hinata K (2000) The S receptor kinase determines self-incompatibility in *Brassica* stigma. Nature 403:913–916

Takasaki T, Hatakeyama K, Watanabe M, Toriyama K, Hinata K (2001) Homology-dependent suppression of stigma phenotype by an antisense S-locus glycoprotein (*SLG*) gene in *Brassica rapa* L. Breed Sci 51:89–94

Takayama S, Isogai A (2003) Molecular mechanism of self-recognition in Brassica self-incompatibility. J Exp Bot 54:149–156

Takayama S, Isogai A (2005) Self-incompatibility in plants. Annu Rev Plant Biol 56:467–489

Takayama S, Isogai A, Tsukamoto C, Ueda Y, Hinata K, Okazaki K, Suzuki A (1986a) Isolation and some characterization of S-locus-specific glycoproteins in *Brassica campestris* associated with self-incompatibility. Agric Biol Chem 50:1365–1367

Takayama S, Isogai A, Tsukamoto C, Ueda Y, Hinata K, Okazaki K, Koseki K, Suzuki A (1986b) Structure of carbohydrate chains of S-glycoproteins in *Brassica campestris* associated with self-incompatibility. Agric Biol Chem 50:1673–1676

Takayama S, Isogai A, Tsukamoto C, Ueda Y, Hinata K, Okazaki K, Suzuki A (1987) Sequences of S-glycoproteins, products of the *Brassica campestris* self-incompatibility locus. Nature 326:102–104

Takayama S, Shiba H, Iwano M, Shimosato H, Che F-S, Kai N, Watanabe M, Suzuki G, Hinata K, Isogai A (2000) The pollen determinant of self-incompatibility in *Brassica campestris*. Proc Natl Acad Sci USA 97:1920–1925

Takayama S, Shimosato H, Shiba H, Funato M, Che F-S, Watanabe M, Iwano M, Isogai A (2001) Direct ligand-receptor complex interaction controls *Brassica* self-incompatibility. Nature 413:534–538

Tantikanjana T, Nasrallah ME, Stein JC, Chen C-H, Nasrallah JB (1993) An alternative transcript of the S-locus glycoprotein gene in a class II pollen-recessive self-incompatibility haplotype of *Brassica oleracea* encodes a membrane- anchored protein. Plant Cell 5:657–666

Thompson KF (1957) Self-incompatibility in marrow-stem kale, *Brassica oleracea* var. *acephala*. I. Demonstration of a sporophytic system. J Genet 55:45–60

Thompson KF, Taylor JP (1966) Non-linear dominance relationships between *S* alleles. Heredity 21:345–362

Torii KU (2004) Leucine-rich repeat receptor kinases in plants: Structure, function, and signal transduction pathways. Inter Rev Cytol 234:1–46

Toriyama K, Hanaoka K, Okada T, Watanabe M (1998) Molecular cloning of a cDNA encoding a pollen extracellular protein as a potential source of a pollen allergen in *Brassica rapa*. FEBS Lett 424:234–238

Vanoosthuyse V, Tichtinsky G, Dumas C, Gaude T, Cock JM (2003) Interaction of calmodulin, a sorting nexin and kinase-associated protein phosphatase with the *Brassica oleracea* S-locus receptor kinase. Plant Physiol 133:919–929

Walker JC, Zhang R (1990) Relationship of a putative receptor protein kinase from maize to the S-locus glycoproteins of *Brassica*. Nature 345:743–746

Watanabe M, Nou IS, Takayama S, Yamakawa S, Isogai A, Suzuki A, Takeuchi T, Hinata K (1992) Variations in and inheritance of NS-glycoprotein in self-incompatible *Brassica campestris* L. Plant Cell Physiol 33:343–351

Watanabe M, Takasaki T, Toriyama K, Yamakawa S, Isogai A, Suzuki A, Hinata K (1994) A high degree of homology exists between the protein encoded by *SLG* and the *S* receptor domain encoded by *SRK* in self-incompatible *Brassica campestris* L. Plant Cell Physiol 35:1221–1229

Watanabe M, Ono T, Hatakeyama K, Takayama S, Isogai A, Hinata K (1997) Molecular characterization of *SLG* and *S*-related genes in a self-compatible *Brassica campestris* L. var. *yellow sarson*. Sex Plant Reprod 10:332–340

Watanabe M, Ito A, Takada Y, Ninomiya C, Kakizaki T, Takahata Y, Hatakeyama K, Hinata K, Suzuki G, Takasaki T, Satta Y, Shiba H, Takayama S, Isogai A (2000) Highly divergent sequences of the pollen self-incompatibility (*S*) gene in class-I *S*-haplotypes of *Brassica campestris* (syn. *rapa*) L. FEBS Lett 473:139–144

Watanabe M, Suzuki G, Shiba H, Takayama S (2006) Molecular mechanisms of self-incompatibility in Brassicaceae. In: da Silva JAT (ed) Floriculture, ornamental and plant biotechnology: Advances and topical issues, 1st edn. Global Science Books, UK, pp 552–555

Yu K, Schafer U, Glavin TL, Goring DR, Rothstein SJ (1996) Molecular characterization of the S-locus in two self-incompatible *Brassica napus* lines. Plant Cell 8:2369–2380

Chapter 8
'Self' Pollen Rejection Through the Intersection of Two Cellular Pathways in the Brassicaceae: Self-Incompatibility and the Compatible Pollen Response

M.A. Samuel, D. Yee, K.E. Haasen, and D.R. Goring

Abstract The sporophytic self-incompatibility (SI) system, which operates in the Brassicaceae is primarily controlled by two multi-allelic loci, encoding the SP11/SCR pollen ligand, and the stigma-specific *S* Receptor Kinase (SRK). Haplotype-specific recognition of SP11/SCR by SRK triggers the activation of SRK's intracellular kinase domain. This is predicted to cause the phosphorylation-mediated recruitment of the ARC1 E3 ubiquitin ligase. ARC1 is predicted to inhibit its substrate by ubiquitination, and recent work suggests that Exo70A1 is a target of ARC1. Exo70A1 is predicted to regulate targeted secretion and is required in the stigma to promote compatible pollen hydration, germination and pollen tube growth. SRK is also known to interact with a number of other proteins, including the *M* locus protein kinase (MLPK), which may function with SRK to co-regulate ARC1. Here we review our present knowledge of the various cellular components that act in concert during the SI response. We also discuss the cellular mechanisms of how SI can cause pollen rejection through the inhibition of stigmatic factors that promote compatible pollen acceptance.

Abbreviations

ARC1	Arm repeat containing; an E3 ubiquitin ligase
ARM	Armadillo
BAK1	BRI1 Associated receptor Kinase 1
BnExo70A1	*Brassica napus* gene encoding a protein with sequence similarity to Exo70
BRI1	BRASSINOSTEROID-INSENSITIVE 1
BY2 cells	Bright Yellow 2 cells

M.A. Samuel, D. Yee, K.E. Haasen, and D.R. Goring
Department of Cell and Systems Biology, University of Toronto, Toronto, Ontario, Canada M5S 3B2, e-mail: d.goring@utoronto.ca

$[Ca^{2+}]_i$	Cytosolic free intracellular calcium concentration
ER	Endoplasmic reticulum
Exo70	A subunit of the exocyst complex
KAPP	Kinase-associated protein phosphatase
MLPK	*M* Locus Protein Kinase
mod	Modifier
mRNA	Messenger RNA
PCPs	Pollen coat proteins
RLCK	Receptor-like cytoplasmic kinase
RNAi	RNA interference
SCR/SP11	*S*-locus cysteine-rich protein (the pollen *S*-determinant in Brassica)
SI	Self-incompatibility
SLG	*S*-locus glycoprotein
S	locus Self-incompatibility locus
SLR1	*S*-locus related-1 (a homologue of SLG)
SRK	*S*-locus receptor kinase (the pistil *S*-determinant in Brassica)
THL1 and THL2	Thioredoxin *h* proteins
U-box	A motif present in a family of E3 ligases
UND	A novel N-terminal domain present in a number of predicted U-box/ARM proteins

8.1 Introduction

In the Brassicaceae, a conserved sporophytic self-incompatibility (SSI) system is present, and detailed genetic studies have resulted in the identification of highly polymorphic *S* genes that confer this trait. The SSI system has been best characterised in the genus *Brassica*, and is primarily controlled by a receptor–ligand system encoded in two tightly linked and multi-allelic genes: the *S* Receptor kinase (SRK), and the small cysteine-rich secreted protein, SP11/SCR (see Chap. 6 for a discussion of SRK and SP11/SCR polymorphism and Chap. 7 for an introduction to the discovery of the *S*-locus determinants and other components involved in *Brassica* SI). The co-evolved SRK and SP11/SCR alleles constitute different *S*-haplotypes, and 'self' pollen rejection occurs when the *S*-haplotype of the pollen parent matches the pistil *S*-haplotype (Boyes and Nasrallah 1993). The haplotype-specific interaction between the pollen-specific SP11/SCR and stigma-specific SRK elicits a rapid rejection response. Successful fertilisation can occur only if the *S*-haplotype of the pollen parent and the pistil do not match. Thus, by rejecting self pollen, the outcome of self-incompatibility is an increased propensity towards out-crossing.

Although the interactions between SRK and SP11/SCR has been well mapped out, the subsequent signalling pathway that leads to the rejection response is relatively less well known. In this chapter, we discuss the various intracellular signalling components identified downstream of SRK and elaborate on their possible cellular mechanisms. Recent research reinforces the concept that SI functions by blocking the compatible pollen response (Roberts et al. 1980; Dickinson 1995), and so it is appropriate to first start with a review of the literature on cellular factors and changes governing compatible pollen–pistil interactions. The early stages of interactions between the compatible pollen and the stigmatic surface at the top of the pistil involve pollen capture, adhesion and hydration as well as pollen tube penetration of the stigmatic barrier. These events have been largely studied in the *Brassica* and *Arabidopsis* genera and are reviewed below as these are the stages targeted by the *Brassica* SI response. Following these early stages, the compatible pollen tube enters the expanded stigmatic papillar cell wall, grows down to enter the intercellular space below the papilla and then grows to the ovary where fertilisation occurs (Elleman et al. 1992). The complex interactions that occur between the pollen tube and the pistil are reviewed in detail elsewhere (Edlund et al. 2004; Swanson et al., 2004).

8.2 The Early Stages of Compatible Pollen–Stigma Interactions in the Brassicaceae

When a random pollen grain lands on a stigma of another plant species, it cannot successfully complete fertilisation. This is due to an active recognition system in place to allow successful fertilisation following compatible pollinations. In plant species with dry stigmas, as found in the Brassicaceae, this recognition starts at the earliest stages of pollen capture and adhesion (Roberts et al. 1980; Dickinson 1995; Swanson et al. 2004).

8.2.1 Pollen Capture and Adhesion

Once a compatible pollen grain comes in contact with the stigma, the pollen is captured, and a strong connection forms between the pollen grain and the stigmatic papilla (Fig. 8.1a). This initial stage is mediated by the 'glue-like' surface properties of both the pollen grain and stigma (Clarke et al. 1979; Roberts et al. 1980; Zinkl et al. 1999). In the Brassicaceae, the surfaces of the stigmatic papillae are coated with a waxy cuticle and a thin proteinaceous pellicle. It is the pellicle that has the adhesive properties and has been proposed to be important for pollen–stigma recognition events, since removal of the pellicle disrupts these events (Mattison et al. 1974; Stead et al. 1980). The pollen grain has a highly sculptured exine outer layer with a pollen coat composed of lipids and proteins filling in the cavities of the exine (Mayfield et al. 2001; Murphy 2006). Pollen capture is mediated by the exine and does not appear to be dependent on the pollen coat, since this initial stage is not affected when the pollen coat is absent (Zinkl et al. 1999).

After pollen capture, the pollen coat mobilises and spreads towards the pollen–stigma interface to form a meniscus-shaped 'foot', and the lipidic and proteinaceous

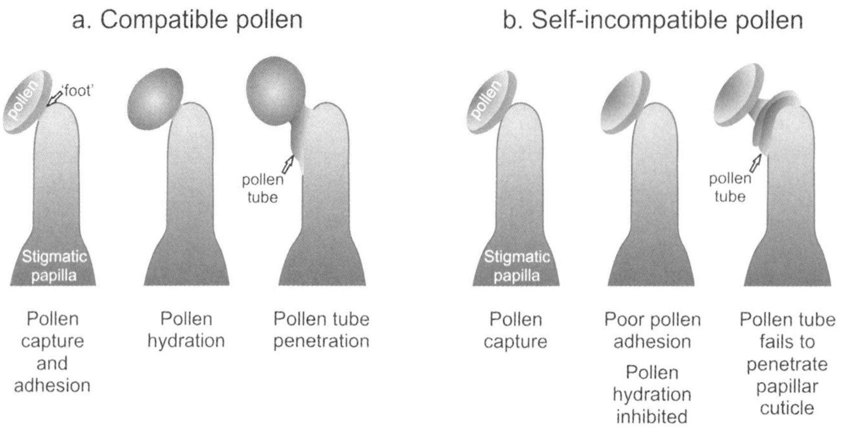

Fig. 8.1 Illustrations of early events following a compatible (**a**) or self-incompatible (**b**) pollination. Images are based on micrographs published in Ockendon (1972) and Dickinson (1995). See text for more details

contents of both surfaces mix (Elleman and Dickinson 1990; Preuss et al. 1993). It is at this stage where the pollen coat mediates pollen adhesion, and there is a cross-linking of the pollen grain to the stigmatic papilla (Elleman et al. 1992; Zinkl et al. 1999). Two stigma-specific proteins, the *S*-locus glycoprotein (SLG) and the *S*-locus Related-1 (SLR1) protein, have been proposed to be required for pollen adhesion in *Brassica* (Fig. 8.2a). Decreased pollen adhesion was observed in *Brassica* transgenic lines with reduced SLR1 expression through transgenic suppression, and when *Brassica* stigmas were pre-treated with antibodies to SLR1 and SLG (Luu et al. 1997b, 1999). Both SLG and SLR1 have been found to bind to Class A pollen coat proteins (PCPs) and are thought to mediate pollen adhesion through these interactions (Fig. 8.2a) (Doughty et al. 1993; Hiscock et al. 1995; Takayama et al. 2000b). Consistent with this model, the pre-treatment of *Brassica* stigmas with pollen coat proteins also decreased adhesion (Luu et al. 1999).

8.2.2 Pollen Hydration

Once the pollen grain has adhered to the stigmatic papilla, hydration follows (Fig. 8.1a). Since the Brassicaceae stigma is of the 'dry' type, pollen hydration occurs through the passage of water from the stigma to the pollen (Heslop-Harrison 1979; Roberts et al. 1984). Pollen grains are quiescent and desiccated upon maturation, and can only reactivate and germinate following rehydration on the stigma (Heslop-Harrison 1979; Zuberi and Dickinson 1985). In both *Brassica* and *Arabidopsis*, the cuticular layer of the dry stigma serves as a barrier to regulate water transfer during selective pollen hydration. As a result, this minimises the occurrence of indiscriminate pollination, allowing for species-specificity in this process (Sarker et al. 1988; Hulskamp et al. 1995; Zinkl et al. 1999).

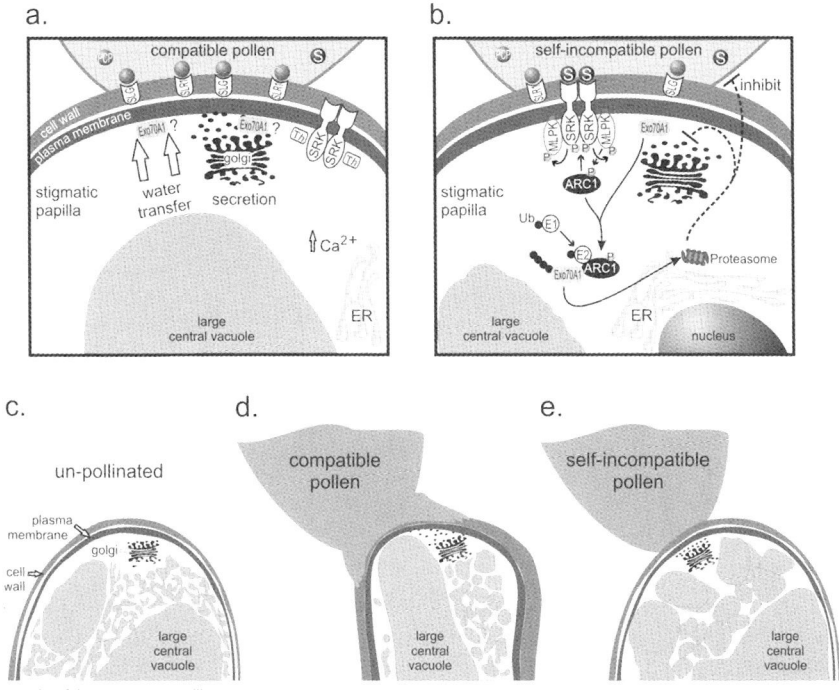

Fig. 8.2 Models of cellular signalling events following a compatible or self-incompatible pollination. (**a**) Compatible pollen-stigmatic papillar interactions: Pollen coat proteins (PCP) bind to SLG and SLR1 to promote pollen adhesion. This may lead to other signalling events, such as calcium signalling, in the stigmatic papilla. Exo70A1 may have a role in targeting the vacuolar network and/or secretory vesicles to the plasma membrane to promote water transfer for pollen hydration and enzyme secretion to loosen the papillar surface for pollen tube growth. SRK remains inactive due to the inhibitory effects of thioredoxin h (T*h*), and the absence of the haplotype-specific SP11/SCR (S) pollen ligand. (**b**) Self-incompatible pollen-stigmatic papillar interactions: The haplotype-specific SP11/SCR (S) pollen ligand binds to SRK and activates the SRK/MLPK complex. A phosphorylation cascade activates the ARC1 E3 ubiquitin ligase, leading to the ubiquitination (Ub) and inactivation of Exo70A1. Both water transfer and loosening of the papillar surface are blocked causing pollen rejection. (**c**) The vacular network in the un-pollinated stigmatic papilla. (**d**) Following a compatible pollination, the vacuolar network in the stigmatic papilla appears to become directed towards the pollen grain, possibly for water transfer. Secretory events lead to the expansion of the papillar cell wall prior to pollen tube penetration. (**e**) Following a self-incompatible pollination, the vacuolar network in the stigmatic papilla appears to be disrupted, possibly blocking pollen hydration. Vesicles are observed to be accumulating next to the unexpanded papillar cell wall, perhaps due to an inhibition of vesicle docking by the SRK signalling pathway shown in (**b**). Images in (**c–e**) are based on data and micrographs published in Iwano et al. (2007) and Elleman and Dickinson (1996)

The lipid-rich pollen coat and the cuticular layer of the stigma at the point of pollen contact are believed to form a unique hydraulic conduit for water flow from the stigma to the pollen grain (Elleman et al. 1992; Dickinson, 1995). The mechanism that allows hydrophobic lipids to form such a liquid conduit is not known, but numerous studies show that lipids from the pollen coat and stigma surface are both necessary and sufficient for proper pollen hydration. For example, the *Arabidopsis fiddlehead* mutant has altered cuticle properties from a β-ketoacyl CoA synthase mutation, which is thought to alter long-chain lipid biosynthesis, and results in aberrant organ fusion (Lolle et al. 1997; Yephremov et al. 1999; Pruitt et al. 2000). The *fiddlehead* mutants also show inappropriate pollen hydration on non-stigmatic surfaces indicating that proper lipid content is critical for the control of pollen hydration (Lolle and Cheung 1993). Other *Arabidopsis* mutants with defects in the long-chain lipid synthesis, such as the *cer* mutants, have altered pollen coat lipid contents and cannot hydrate properly (Preuss et al. 1993; Hulskamp et al. 1995).

8.2.3 Pollen Germination and Pollen Tube Penetration

Following hydration, the pollen grain germinates and a pollen tube emerges to grow into the hydrophobic interface corresponding to the foot (Fig. 8.1a). It is believed that the ability of the pollen grain to sense a water gradient through the lipid conduit determines the initial germination orientation and polarity. Thus, the site of the pollen–stigma contact becomes the site of pollen tube penetration into the stigmatic surface (Elleman et al. 1992; Dickinson 1995; Edlund et al. 2004). Secreted enzymes from either the pollen tube or the stigma are thought to selective breakdown the papillar surface, allowing pollen tube entry. Accordingly, both pollen and stigma serine esterases have been identified, and treatment of *Brassica* stigmas with a serine esterase inhibitor blocks pollen tube invasion (Hiscock et al. 2002). Various other enzymes, including cutinases, polygalacturonase and pectin esterases, have also been identified in *Brassica* pollen and/or stigma (Hiscock et al. 1994; Kim et al. 1996; Dearnaley and Daggard 2001). Similarly, cell wall modifying enzymes (e.g. pectin methylesterases) may also play a role in loosening the papillar cell wall in preparation for pollen tube penetration (Micheli 2001). The involvement of so many types of enzymes is not surprising considering that breaching the papillar surface would logically implicate the breakdown, or at the very least the modification, of the waxy cuticle, its surrounding proteinaceous pellicle and the underlying complex epidermal cell wall.

Calcium appears to serve as a second messenger during this stage during pollination in *Arabidopsis*, and may be related to secretory events associated with papillar cell wall loosening (Elleman and Dickinson 1996; Hiscock et al. 2002). In *Arabidopsis*, imaging of cytosolic-free intracellular calcium levels ($[Ca^{2+}]_i$) in the stigmatic papilla revealed three intervals of cytosolic calcium increases. The first occurred at the pollen attachment site following pollen hydration, followed by a second increase prior to pollen germination and a final third increase just prior to pollen tube penetration of the stigmatic surface (Iwano et al. 2004). Increases in $[Ca^{2+}]_i$ have also been observed following compatible pollinations in *Brassica* (Dearnaley et al. 1997; Elleman and Dickinson 1999).

8.3 The SI Response Causes Pollen Arrest at the Stigmatic Surface

In the Brassicaceae, self pollen rejection is controlled by the female side through an organised cascade of cellular events in the stigmatic papillae, leading to pollen arrest. With the dry type stigma, this response starts early by disrupting pollen adhesion, pollen hydration and stigma penetration by the pollen tube if necessary (Dickinson 1995). Following self pollination, pollen capture rapidly occurs and pollen adhesion is also initiated, but is diminished relative to compatible pollinations (Stead et al. 1980; Luu et al. 1997a). One could speculate that both the 'species'-compatible and self-incompatible pollen recognition events have been initiated, but a small 'delay' in relaying the SI signal to the stigmatic papilla allows pollen capture and some pollen adhesion to occur (Fig. 8.1b).

In terms of pollen hydration, axis expansion through hydration was significantly decreased in *Brassica* self pollen compared to cross-pollen (Roberts et al. 1980; Zuberi and Dickinson 1985). Considering that adhesion is reduced in self pollinations, a suitable lipid interface may not form between the pollen and the stigmatic papillae and as a result water transfer from the stigma is impaired. Hydration can occur in some *Brassica* self pollen, which may be attributable to environmental humidity, though some pollen may achieve sufficient adhesion to allow for water transfer from stigma (Zuberi and Dickinson 1985; Luu et al. 1999). Self pollen that successfully hydrate either do not germinate a pollen tube or germinate a pollen tube that curls at the stigmatic surface without penetrating the cuticle (Fig. 8.1b) (Ockendon 1972; Carter et al. 1975; Zuberi and Dickinson 1985). Stigma-derived compatibility factors, such as secreted enzymes (to breakdown the stigmatic papillar surface), could be targeted by SI to prevent pollen tube entry (Hiscock et al. 2002). As well, any stigma-to-pollen permissive signals needed for pollen tube growth could be potentially denied; for example, chemo-attractants such as GABA identified in *Arabidopsis* (Palanivelu et al. 2003). A key area in the field has focused on identifying the cellular events in the stigmatic papilla, which disrupt these pollen–stigma interactions during a SI response, and very exciting progress has been made in recent years in uncovering this pathway (Fig. 8.2).

8.3.1 The S Receptor Kinase Activates a Cellular Signalling Pathway in the Stigmatic Papilla to Trigger Self Pollen Rejection

The SSI system in the *Brassicaceae* has been best characterised in the genus *Brassica*, and is regulated by the *S* Receptor kinase (SRK) small secreted SP11/SCR protein (see Chap. 7 for a review of the discovery of the *S*-locus determinants and Chap. 6 for a discussion of SRK and SP11/SCR polymorphism). SRK is expressed solely in the stigma and is the female determinant of this system (Takasaki et al. 2000; Silva et al. 2001) while SP11/SCR is a pollen-specific coat protein and the male determinant of this system (Schopfer et al. 1999; Suzuki 1999; Takayama et al. 2000a, b). During early anther development, SP11/SCR is secreted from the anther tapetal cells into the locule, where it is integrated into the pollen coat deposited on the pollen surface (Iwano et al. 2003). As predicted for a plant receptor kinase, SRK is an integral plasma membrane protein with a functional serine/threonine kinase (Goring and Rothstein 1992; Stein and Nasrallah 1993; Giranton et al. 2000). SRK exists as a pre-formed dimer in unpollinated pistils and provides a high-affinity ligand binding site for SCR/SP11 (Giranton et al. 2000; Shimosato et al. 2007). Following a self-incompatible pollination, there is a S-haplotype-specific interaction between SP11/SCR and SRK (Kachroo et al. 2001; Takayama et al. 2001). An intracellular signalling pathway is then activated in the stigmatic papilla, which rapidly leads to self pollen rejection (Fig. 8.2b).

SP11/SCR and *SRK* genes have also been identified in other genera in the Brassicaceae, including *Arabidopsis lyrata* (Kusaba et al. 2001; Schierup et al. 2001), *Capsella grandiflora* (Paetsch et al. 2006) and *Raphanus sativus* (Sato et al. 2004). Interestingly, *Arabidopsis thaliana,* which is naturally self-fertilising, carries non-functional *SP11/SCR* and *SRK* genes (Kusaba et al. 2001). However, some ecotypes can become fully self-incompatible with the addition of the *Arabidopsis lyrata SP11/SCR* and *SRK* genes (Nasrallah et al. 2004). This indicates that while *Arabidopsis thaliana* has lost its SI due to inactivation of the *S* genes, the downstream signalling components required for conferring SI has remained intact in some instances. The nature of the downstream signalling events has become much clearer in recent years and involves a novel receptor kinase signalling pathway (Fig. 8.2).

8.3.2 The M Locus Protein Kinase acts Together with the S Receptor Kinase to Promote SI

The *M* locus protein kinase (MLPK) was discovered following positional cloning of a modifier (*mod*) mutation in the *Brassica rapa* yellow sarson variety (Murase et al. 2004). This recessive mutation led to a complete breakdown of the stigmatic SI response (Murase et al. 2004). MLPK encodes a predicted kinase belonging to the receptor-like cytoplasmic kinase (RLCK) family, and was found to have serine/threonine kinase activity. As well, transient expression of a functional MLPK in the stigmatic papillae of *Brassica rapa* var. yellow sarson was sufficient to restore the SI response (Murase et al. 2004).

Brassica MLPK, and its *Arabidopsis* orthologue, can exist as two isoforms produced through alternate transcriptional start sites, and both forms are targeted to the plasma membrane by either an N-terminal hydrophobic domain in one isoform or by myristoylation of the other isoform (Kakita et al. 2007b). Membrane localisation of MLPK is required for conferring SI, and this likely facilitates interactions with SRK, as demonstrated through bi-molecular fluorescence complementation in tobacco protoplasts (Kakita et al. 2007b). Finally, the active SRK kinase domain was shown to efficiently phosphorylate MLPK in vitro, further supporting that MLPK functions with SRK to promote SI (Kakita et al. 2007a).

Given MLPK's plasma membrane location and interactions with SRK, one can speculate that MLPK functions in a complex with SRK to promote downstream signalling events following SRK activation by the SP11/SCR ligand (Fig. 8.2b). The theme of receptor kinases functioning with 'cytoplasmic' kinases is quite common in animal receptor kinase systems (Pawson 2002). The SRK/MLPK complex may also have some parallels to the BRI1/BAK1 complex involved in brassinosteroid signalling (Gendron and Wang 2007).

8.3.3 The SRK Kinase Domain can Interact with a Range of Intracellular Proteins

Kinase domains of receptor kinases can often interact with a number of intracellular proteins to activate intracellular signalling pathways and regulate receptor signalling (Pawson 2002). Thus, it is not surprising that protein–protein interaction screens have identified a number of interacting partners for the SRK kinase domain. At least six interacting partners have been identified: two thioredoxin *h* proteins, THL1 and THL2 (Bower et al. 1996); the E3 ubiquitin ligase, ARC1 (Gu et al. 1998; Stone et al. 2003); the Kinase-Associated Protein Phosphatase (KAPP); a sorting nexin; and calmodulin (Vanoosthuyse et al. 2003). Of these six interacting partners, only ARC1 has been shown to act downstream of SRK as a positive regulator of SI (Stone et al. 1999). Thioredoxin *h* has been shown to be a negative regulator of SRK and SI (Cabrillac et al. 2001; Haffani et al. 2004). Both KAPP and sorting nexin are known to be negative regulators in other systems. KAPP has been implicated in the down-regulation of various receptor kinases following activation (Johnson and Ingram 2005), and sorting nexins are known to participate in the sorting of endosomal-localised receptors for either recycling or degradation (Carlton et al. 2005).

Another interesting feature is the nature of the interactions between these proteins and the SRK kinase domain in vitro. Both ARC1 and KAPP show phosphorylation-dependent interactions, suggesting that they may only bind to the SRK kinase domain following receptor activation (Gu et al. 1998; Vanoosthuyse et al. 2003). The thioredoxin *h* proteins (THL1/2), the sorting nexin and the calmodulin were able to bind both the active and inactive forms of the SRK kinase domain, indicating that their regulatory functions are likely unrelated to receptor activation (Bower et al. 1996; Vanoosthuyse et al. 2003). This feature implicates a complexity to the dynamics of SRK interactions during the SI response. In unpollinated stigmas, SRK is likely maintained in an inactive state with reduced basal activity, potentially through interactions with thioredoxin *h* and/or calmodulin. In response to self pollen, SRK is activated through conformational changes and released from these potential negative regulators. The phosphorylated SRK kinase domain recruits downstream signalling partners, such as ARC1, to activate the cellular response for pollen rejection. SRK returns to an inactive state when dephosphorylated by KAPP. Sorting nexin may also be involved at this stage if receptor endocytosis has occurred to regulate SRK-signalling through endosome-mediated degradation or recycling.

8.3.4 Thioredoxin h Inhibits SRK Activity in the Absence of Self Pollen

The thioredoxin *h* proteins, THL1 and THL2, were identified as SRK interactors through screening a *Brassica napus* pistil cDNA yeast two-hybrid library with SRK kinase domain (Bower et al. 1996). These proteins were able to bind both the active and inactive forms of the SRK kinase domains and required a cysteine at the end of the transmembrane domain for this interaction (Bower et al. 1996; Mazzurco et al. 2001). The role of thioredoxin *h* as an inhibitor of SRK was elegantly demonstrated by Cabrillac et al. (2001). They found that SRK was maintained in an inactive state in unpollinated pistils, yet became constitutively active when immuno-purified due to the loss of a soluble stigmatic inhibitor during the extraction procedure. Through a series of experiments, the inhibitor was identified as a thioredoxin-type protein, and the addition of recombinant thioredoxin *h* (THL1) was able to reconstitute this activity (Cabrillac et al. 2001). Subsequently, Haffani et al. (2004) demonstrated that transgenic *Brassica napus* Westar lines carrying an antisense-THL1 construct exhibited constitutive low levels of pollen rejection, possibly through the increased activity of an endogenous SRK. Interestingly, similar parallels have been observed in pathogen signalling with the tomato Cf-9 receptor system. A thioredoxin protein, CITRX, was found to interact with the C-terminal tail of the Cf-9 receptor, and function as a negative regulator of Cf-9 signalling (Rivas et al. 2004). Recently, it was shown that CITRX functions as an adaptor between Cf-9 and the ACIK cytoplasmic serine/threonine kinase, a positive regulator of disease resistance (Nekrasov et al. 2006). Whether a similar conserved arrangement exists between SRK-thioredoxin *h*-MLPK does remain to be seen. Given the function of thioredoxins in alleviating oxidative stress and their potential to be modified by redox imbalances (Vieira Dos Santos and Rey 2006), it is reasonable to postulate that a possible localised oxidative burst near the site of pollen attachment could modify thioredoxin *h* and relieve the inhibition of SRK (Fig. 8.2a, b).

8.3.5 ARC1 Functions Downstream of SRK to Promote SI

One of the most interesting SRK interactor to come from the yeast two-hybrid screen was the Armadillo repeat containing protein, ARC1 (Gu et al. 1998). ARC1 is a modular protein with a novel N-terminal domain (UND), followed by a U-box, and an ARM repeat domain (Samuel et al., 2006). UND is a conserved domain present in a number of predicted U-box/ARM proteins, though its function is unknown (Samuel et al. 2006). The U-box is a conserved motif present in the U-box family of E3 ubiquitin ligases, and is responsible for binding to the E2 ubiquitin conjugating enzyme as part of the ubiquitination reaction (Hatakeyama and Nakayama 2003). The ARM repeat domain mediates binding of ARC1 to the phosphorylated SRK kinase domain (Gu et al. 1998). ARM repeat domains are composed of tandemly repeated 42 amino acid ARM repeats, which form a super helix of alpha-helices, and provide binding sites for interacting partners (Samuel et al. 2006).

ARC1 is specifically expressed in the stigma, a pattern which mimics the expression of SRK, and indicates a function specific to this tissue (Gu et al. 1998). Antisense suppression of ARC1 in self-incompatible W1 plants resulted in a partial breakdown of SI, demonstrating a positive role for ARC1 during SI response (Stone et al. 1999). These ARC1 antisense W1 plants had functional *SP11/SCR* and *SRK* genes present, and thus, indicated that ARC1 was functioning downstream of SRK as predicted. The lack of complete breakdown in these ARC1 antisense W1 plants is likely attributed to the incomplete suppression of the ARC1 mRNA, though one can not rule out that other potential intracellular signalling proteins were partially compensating for the loss of ARC1 (Stone et al. 1999).

ARC1 can function as an E3 ubiquitin ligase, and increases in the levels of ubiquitinated proteins were observed 30 min following a self pollination (Stone et al. 2003). In keeping with the role of ARC1 as an E3 ligase during the SI response, pre-treatments of pistils with a proteasomal inhibitor also reduced the SI response, leading to a large increase in pollen adhesion and pollen tube growth down the pistil. Finally, transient expression of ARC1 and the active SRK kinase domain in tobacco BY2 cells resulted in targeting of ARC1 to ER-associated proteasomes in the perinuclear region (Stone et al. 2003). Altogether, these results point to a model where activated SRK directs ARC1 to re-localise to the proteasomes, where ARC1 targets various substrates for degradation (Fig. 8.2b).

Recently, we have found that MLPK can have a similar effect on ARC1 localisation in tobacco BY2 cells, and MLPK can also phosphorylate ARC1 very efficiently in vitro, at a much higher level than that seen with SRK (Samuel et al. 2008a). This is consistent with MLPK functioning in a complex with SRK to activate downstream events and raises the possibility that MLPK functions with SRK to activate ARC1 (Fig. 8.2b). ARC1 'activation' by phosphorylation could include directing ARC1 to a specific sub-cellular location (such as the proteasome) and/or causing ARC1 to have increased binding affinity for its substrate.

In keeping with the general model that SI functions by blocking compatible pollen–stigma interactions (Dickinson 1995), ARC1 substrates are hypothesised to be stigmatic factors, which would normally promote events required for compatible pollen. *Activated* ARC1 would inhibit these factors, possibly by proteasomal degradation, and cause pollen arrest on the stigma surface. The possibility of ARC1 functioning as an inhibitor of compatibility factors brings us to an important crossroad in understanding where SI meets the compatible pollen pathway.

8.3.6 BnExo70A1 is a Potential Substrate for ARC1 and is Required for Compatible Pollen–Stigma Interactions

To identify potential substrates for ARC1, the N-terminus of ARC1 was used to screen a *Brassica napus* pistil cDNA yeast two-hybrid library, and *Bn*Exo70A1 was identified as a potential interactor (Samuel et al. 2008b). The interaction between ARC1 and *Bn*Exo70A1 was confirmed through in vitro pull down assays and an in vitro ubiquitination assay of *Bn*Exo70A1 by ARC1. As well, transient expression of an active SRK kinase domain, ARC1 and *Bn*Exo70A1 in tobacco BY2 cells resulted in targeting of ARC1 and *Bn*Exo70A1 to ER-associated proteasomes in the perinuclear region. This indicated the possibility that *Bn*Exo70A1 could be targeted for inactivation by ARC1 during the SI response (Fig. 8.2b).

*Bn*Exo70A1 displays sequence similarity to the conserved eukaryotic protein, Exo70, a subunit of the exocyst complex. In yeast and mammalian cells, the exocyst complex is comprised of eight subunits (Sec3, 5, 6, 8, 10 and 15; Exo70 and 84), and functions in tethering and docking selected secretory vesicles to specific sites on the plasma membrane. Thus, the exocyst regulates specialised secretory events, and functions include polarised exocytosis during yeast budding and neurite outgrowth, insulin-stimulated trafficking of the Glut4 transporter, and selective tethering of vesicles to the apical or basolateral membranes in polarised epithelial cells (Munson and Novick 2006). Plant genomes contain all the predicted exocyst genes, and have specifically expanded the Exo70 gene family, but whether they function in a similar manner to other systems is not known. In *Arabidopsis*, exocyst subunit mutants have been identified with defects in root hair elongation, pollen germination and polar growth and development (Cole and Fowler 2006).

A potential role for *Bn*Exo70A1 in regulating polarised secretion is very consistent with previous studies documenting cellular changes in the stigmatic papilla following compatible pollinations, and raises a very exciting new development in our understanding of these processes. *Bn*Exo70A1 would be predicted to play a role in the stigmatic papilla promoting events such as pollen adhesion, pollen hydration or penetration of the pollen tube through the stigmatic surface, following a compatible pollination. Consistent with this, the RNAi suppression of *Bn*Exo70A1 in the stigma of compatible *Brassica napus* Westar plants resulted in a severe reduction in seed production (Samuel et al. 2008b). This was a consequence of decreased pollen adhesion and hydration of what normally should be compatible pollen. As well, any pollen tubes that formed were incapable of penetrating the papillar surface, essentially phenocopying the SI response. A role for Exo70A1 in promoting compatible pollen–stigma interactions was also found to be conserved in *Arabidopsis*, where null *At*Exo70A1 mutants were also found to display defects in pollen hydration and germination (Samuel et al. 2008b).

From our results, we propose that during compatible pollen–stigma interactions, Exo70A1 functions as a positive regulator by facilitating targeted secretion of stigmatic factors in response to compatible pollen (Fig. 8.2a). In contrast, during the SI response, SRK-MLPK activation leads to ARC1-mediated suppression of Exo70A1 function, resulting in pollen rejection (Fig. 8.2b). In concurrence with this hypothesis, previous microscopy studies have identified changes in the endomembrane system following self or cross pollinations, which may point to potential cellular roles for Exo70A1 (Fig. 8.2c–e).

8.3.7 Endomembrane Changes in the Stigmatic Papillae Following Compatible and Self-Incompatible Pollinations in the Brassicaceae

The *Brassica* papillar cytoplasm is a thin layer surrounding a large central vacuole as well as a network of smaller tubular or round vacuoles (Dearnaley et al. 1997; Iwano et al. 2007). The papillar cytoplasm also utilises small vesicles to secrete proteins into the cell wall (Roberts et al. 1984). Following contact with a compatible pollen grain, structural changes in the cell wall occur with the outer wall expanding beneath the pollen grain. This expansion has not been observed following self-incompatible pollinations (Elleman and Dickinson 1990; Dickinson et al. 2000). While changes in secretory vesicles could not be observed with the addition of compatible or self-incompatible *Brassica* pollen grains, the addition of isolated pollen coat to the stigma resulted in multiple secretory vesicles being targeted to the stigmatic cell wall from the cytoplasm (Elleman and Dickinson 1996). Following pollen coat treatment, there was a rapid expansion of the stigmatic outer cell wall, and the cytoplasm under the extended stigmatic cell wall region frequently had extensive ER and Golgi structures as well as plasma membrane vesiculation. Elleman and Dickinson (1996) proposed a localised secretion event in the zone of pollen–stigma interaction, where pre-loaded vesicles carrying hydration factors and cell wall loosening enzymes discharge at the plasma membrane, resulting in a localised loosening of the cell wall matrix to allow pollen tube penetration (Fig. 8.2a, d). This phase could be regulated by the exocyst by marking the pollen attachment site for polarised exocytosis. During the SI response, vesicles were observed to be accumulating by the unexpanded cell wall (Elleman and Dickinson 1996). If SI leads to a rapid inactivation of Exo70A1 by ARC1, this may result in the accumulation of secretory vesicles, which are unable to dock to the stigmatic plasma membrane at the pollen–stigma interface, and rapid pollen arrest occurs (Fig. 8.2b, e).

More recently, Iwano et al. (2007) found changes in the vacuolar network in *Brassica* stigmatic papillae following pollination. With compatible pollinations, the vacuolar network appears to be directed towards the pollen while self-incompatible pollinations appeared to disrupt the vacuolar network in the stigmatic papillae (Fig. 8.2d, e). Iwano et al. (2007) proposed that these changes were related to promoting water and ion transport to compatible pollen grains while restricting transport to incompatible pollen grains. Interestingly, the vacuolar changes were also linked to altered actin dynamics in the compatible and incompatible pollinations. Compatible pollinations led to an accumulation of actin bundles in the apical region of the stigmatic papilla adjacent to the pollen grain at approximately the start of pollen hydration. In contrast, self-incompatible pollinations resulted in decreased actin filaments in the apical region, suggesting actin depolymerisation was occurring (Iwano et al. 2007). Interestingly, Boyd et al. (2004) found that while yeast Sec3p and Exo70 arrive at the plasma membrane prior to exocyst assembly, the other six subunits are transported to this site with the secretory vesicles in an actin-dependent manner. The Iwano et al. (2007) study raises the intriguing question of whether the exocyst could play a role in regulating the vacuolar network during compatible pollinations (Fig. 8.2d, e). This would be a novel role for Exo70, but it might in part account for the observed plant-specific expansion of the Exo70 gene family (Cole and Fowler 2006).

8.4 Conclusions and Prospects

The emerging picture, in the pursuit of identifying the molecular mechanisms behind the SI response in the Brassicaceae, has brought us to a critical junction where this response intersects with and inhibits components of the compatibility pathway (with Exo70A1 identified as a key player). Although recent evidence has added to our general understanding of the signalling mechanism, a number of questions still remain unanswered. While the receptor–ligand interaction between SRK and SP11/SCR has been well studied, the biochemical role of MLPK in relation to SRK and ARC1 or how it participates in mediating downstream signalling is not clear. Since the loss of MLPK results in a complete failure of SI, SRK may impart its effects through MLPK and co-ordinately regulate ARC1. For example, activated SRK may bind ARC1 bringing it in close proximity to MLPK for phosphorylation and activation.

Similarly, a number of questions arise for ARC1 and Exo70A1 on their cellular roles and relationships: What is Exo70A1 regulating during a compatible pollination: secretory vesicles and/or the vacuolar network? The analysis of loss-of-function Exo70A1 plants suggests a role for Exo70A1 in the stigmatic papillae during multiple stages following compatible pollinations. This includes pollen hydration, pollen germination and pollen tube penetration of the stigmatic surface. If the vacuolar network is the target of Exo70A1, this may allow for the control of water release to compatible pollen grains. If Exo70A1 is targeting secretory vesicles for membrane fusion, what is the cargo in these vesicles: secretory enzymes for the modification of the papillar surface? Given Exo70A1's essential role in compatible pollen–stigma interactions, it does become a logical target for the SI response to elicit a rapid pollen rejection. How ARC1 regulates Exo70A1 in vivo during SI is still an outstanding question. Finally, a number of other SRK kinase-interactors have been identified such as calmodulin, KAPP and sorting nexin, but their in vivo biological functions in regulating SRK remain to be uncovered. Nevertheless, a number of exciting new players have been uncovered in recent years, and identifying how they participate in pollen–stigma interactions will continue to uncover some very exciting discoveries in the coming years.

Acknowledgements D.R.G. is supported by grants from the Natural Sciences and Engineering Research Council of Canada (NSERC) and a Canada Research Chair.

References

Bower MS, Matias DD, Fernandes-Carvalho E, Mazzurco M, Gu T, Rothstein SJ, Goring DR (1996) Two members of the thioredoxin-*h* family interact with the kinase domain of a *Brassica S*-locus receptor kinase. Plant Cell 8:1641–1650

Boyd C, Hughes T, Pypaert M, Novick P (2004) Vesicles carry most exocyst subunits to exocytic sites marked by the remaining two subunits, Sec3p and Exo70p. J Cell Biol 167:889–901

Boyes DC and Nasrallah JB (1993) Physical linkage of the SLG and SRK genes at the self-incompatibility locus of *Brassica oleracea*. Mol Gen Genet 236:369–373

Cabrillac D, Cock JM, Dumas C, Gaude T (2001) The *S*-locus receptor kinase is inhibited by thioredoxins and activated by pollen coat proteins. Nature 410:220–223

Carlton J, Bujny M, Rutherford A, Cullen P (2005) Sorting nexins-unifying trends and new perspectives. Traffic 6:75–82

Carter AL, Williams ST, McNeilly T (1975) Scanning electron-microscope studies of pollen behavior on immature and mature brussels sprout (*Brassica oleracea* var. *gemmifera*) stigmas. Euphytica 24:133–141

Clarke A, Gleeson P, Harrison S, Knox RB (1979) Pollen-stigma interactions: Identification and characterization of surface components with recognition potential. Proc Natl Acad Sci USA 76:3358–3362

Cole RA, Fowler JE (2006) Polarized growth: Maintaining focus on the tip. Curr Opin Plant Biol 9:579–588

Dearnaley JD, Levina NN, Lew RR, Heath IB, Goring DR (1997) Interrelationships between cytoplasmic Ca^{2+} peaks, pollen hydration and plasma membrane conductances during compatible and incompatible pollinations of *Brassica napus* papillae. Plant Cell Physiol 38:985–999

Dearnaley JDW, Daggard GA (2001) Expression of a polygalacturonase enzyme in germinating pollen of *Brassica napus*. Sex Plant Reprod 13:265–271

Dickinson H (1995) Dry stigmas, water and self-incompatibility in *Brassica*. Sex Plant Reprod 8:1–10

Dickinson HG, Elleman CJ, Doughty J (2000) Pollen coatings – chimaeric genetics and new functions. Sex Plant Reprod 12:302–309

Doughty J, Hedderson F, McCubbin A, Dickinson H (1993) Interaction between a coating-borne peptide of the *Brassica* pollen grain and stigmatic *S* (self-incompatibility)-locus-specific glycoproteins. Proc Natl Acad Sci USA 90:467–471

Edlund AF, Swanson R, Preuss D (2004) Pollen and stigma structure and function: The role of diversity in pollination. Plant Cell 16(Suppl):S84–S97

Elleman CJ, Dickinson HG (1990) The role of the exine coating in pollen–stigma interactions in *Brassica oleracea*. New Phytol 114:511–518

Elleman CJ, Dickinson HG (1996) Identification of pollen components regulating pollination-specific responses in the stigmatic papillae of *Brassica oleracea*. New Phytol 133:197–205

Elleman CJ, Dickinson HG (1999) Commonalties between pollen/stigma and host/pathogen interactions: Calcium accumulation during stigmatic penetration by *Brassica oleracea* pollen tubes. Sex Plant Reprod 12:194–202

Elleman CJ, Franklin-Tong V, Dickinson HG (1992) Pollination in species with dry stigmas: The nature of the early stigmatic response and the pathway taken by pollen tubes. New Phytol 121:413–424

Gendron JM, Wang ZY (2007) Multiple mechanisms modulate brassinosteroid signaling. Curr Opin Plant Biol 10:436–441

Giranton JL, Dumas C, Cock JM, Gaude T (2000) The integral membrane *S*-locus receptor kinase of *Brassica* has serine/threonine kinase activity in a membranous environment and spontaneously forms oligomers in planta. Proc Natl Acad Sci USA 97:3759–3764

Goring DR, Rothstein SJ (1992) The *S*-locus receptor kinase gene in a self-incompatible *Brassica napus* line encodes a functional serine/threonine kinase. Plant Cell 4:1273–1281

Gu T, Mazzurco M, Sulaman W, Matias DD, Goring DR (1998) Binding of an arm repeat protein to the kinase domain of the *S*-locus receptor kinase. Proc Natl Acad Sci USA 95:382–387

Haffani YZ, Gaude T, Cock JM, Goring DR (2004) Antisense suppression of thioredoxin h mRNA in *Brassica napus cv.* Westar pistils causes a low level constitutive pollen rejection response. Plant Mol Biol 55:619–630

Hatakeyama S, Nakayama KI (2003) U-box proteins as a new family of ubiquitin ligases. Biochem Biophys Res Commun 302:635–645

Heslop-Harrison J (1979) An interpretation of the hydrodynamics of pollen. Am J Bot 66:737–743

Hiscock SJ, Dewey FM, Coleman JOD, Dickinson HG (1994) Identification and localization of an active cutinase in the pollen of *Brassica napus L*. Planta 193:377–384

Hiscock SJ, Doughty J, Willis AC, Dickinson HG (1995) A 7-kDa pollen coating-borne peptide from *Brassica napus* interacts with *S*-locus glycoprotein and *S*-locus-related glycoprotein. Planta 196:367–374

Hiscock SJ, Bown D, Gurr SJ, Dickinson HG (2002) Serine esterases are required for pollen tube penetration of the stigma in *Brassica*. Sex Plant Reprod 15:65–74

Hulskamp M, Kopczak SD, Horejsi TF, Kihl BK, Pruitt RE (1995) Identification of genes required for pollen–stigma recognition in *Arabidopsis thaliana*. Plant J 8:703–714

Iwano M, Shiba H, Funato M, Shimosato H, Takayama S, Isogai A (2003) Immunohistochemical studies on translocation of pollen *S*-haplotype determinant in self-incompatibility of *Brassica rapa*. Plant Cell Physiol 44:428–436

Iwano M, Shiba H, Miwa T, Che FS, Takayama S, Nagai T, Miyawaki A, Isogai A (2004) Ca^{2+} dynamics in a pollen grain and papilla cell during pollination of *Arabidopsis*. Plant Physiol 136:3562–3571

Iwano M, Shiba H, Matoba K, Miwa T, Funato M, Entani T, Nakayama P, Shimosato H, Takaoka A, Isogai A, Takayama S (2007) Actin dynamics in papilla cells of *Brassica rapa* during self- and cross-pollination. Plant Physiol 144:72–81

Johnson KL, Ingram GC (2005) Sending the right signals: Regulating receptor kinase activity. Curr Opin Plant Biol 8:648–656

Kachroo A, Schopfer CR, Nasrallah ME, Nasrallah JB (2001) Allele-specific receptor-ligand interactions in *Brassica* self-incompatibility. Science 293:1824–1826

Kakita M, Shimosato H, Murase K, Isogai A, Takayama S (2007a) Direct interaction between *S*-locus receptor kinase and *M*-locus protein kinase involved in *Brassica* self-incompatibility signaling. Plant Biotechnol 24:185–190

Kakita M, Murase K, Iwano M, Matsumoto T, Watanabe M, Shiba H, Isogai A, Takayama S (2007b) Two distinct forms of *M* Locus Protein Kinase localize to the plasma membrane and interact directly with *S*-Locus Receptor Kinase to transduce self-incompatibility signaling in *Brassica rapa*. Plant Cell 19:3961–3973

Kim HU, Chung TY, Kang SK (1996) Characterization of anther-specific genes encoding a putative pectin esterase of Chinese cabbage. Mol Cell 6:334–340

Kusaba M, Dwyer K, Hendershot J, Vrebalov J, Nasrallah JB, Nasrallah ME (2001) Self-incompatibility in the genus *Arabidopsis*: Characterization of the *S*-locus in the outcrossing *A. lyrata* and its autogamous relative *A. thaliana*. Plant Cell 13:627–643

Lolle SJ, Cheung AY (1993) Promiscuous germination and growth of wildtype pollen from *Arabidopsis* and related species on the shoot of the *Arabidopsis* mutant, *fiddlehead*. Dev Biol 155:250–258

Lolle SJ, Berlyn GP, Engstrom EM, Krolikowski KA, Reiter WD, Pruitt RE (1997) Developmental regulation of cell interactions in the *Arabidopsis fiddlehead-1* mutant: A role for the epidermal cell wall and cuticle. Dev Biol 189:311–321

Luu DT, Heizmann P, Dumas C (1997a) Pollen-stigma adhesion in kale is not dependent on the self-incompatibility genotype. Plant Physiol 115:1221–1230

Luu DT, Heizmann P, Dumas C, Trick M, Cappadocia M (1997b) Involvement of *SLR1* genes in pollen adhesion to the stigmatic surface in *Brassicaceae*. Sex Plant Reprod 10:227–235

Luu DT, Marty-Mazars D, Trick M, Dumas C, Heizmann P (1999) Pollen-stigma adhesion in *Brassica spp* involves SLG and SLR1 glycoproteins. Plant Cell 11:251–262

Mattison O, Knox RB, Heslopha J, Heslopha Y (1974) Protein pellicle of stigmatic papillae as a probable recognition site in incompatible reactions. Nature 247:298–300

Mayfield JA, Fiebig A, Johnstone SE, Preuss D (2001) Gene families from the *Arabidopsis thaliana* pollen coat proteome. Science 292:2482–2485

Mazzurco M, Sulaman W, Elina H, Cock JM, Goring DR (2001) Further analysis of the interactions between the *Brassica S* receptor kinase and three interacting proteins (ARC1, THL1 and THL2) in the yeast two-hybrid system. Plant Mol Biol 45:365–376

Micheli F (2001) Pectin methylesterases: Cell wall enzymes with important roles in plant physiology. Trends Plant Sci 6:414–419

Munson M, Novick P (2006) The exocyst defrocked, a framework of rods revealed. Nat Struct Mol Biol 13:577–581

Murase K, Shiba H, Iwano M, Che FS, Watanabe M, Isogai A, Takayama S (2004) A membrane-anchored protein kinase involved in *Brassica* self-incompatibility signaling. Science 303:1516–1519

Murphy DJ (2006) The extracellular pollen coat of the *Brassicaceae*: Composition, biosynthesis, and functions in pollination. Protoplasma 228:31–38

Nasrallah ME, Liu P, Sherman-Broyles S, Boggs NA, Nasrallah JB (2004) Natural variation in expression of self-incompatibility in *Arabidopsis thaliana*: implications for the evolution of selfing. Proc Natl Acad Sci USA 101:16070–16074

Nekrasov V, Ludwig AA, Jones JD (2006) CITRX thioredoxin is a putative adaptor protein connecting Cf-9 and the ACIK1 protein kinase during the Cf-9/Avr9- induced defence response. FEBS Lett 580:4236–4241

Ockendon DJ (1972) Pollen tube growth and the site of the incompatibility reaction in *Brassica oleracea*. New Phytol 71:519–522

Paetsch M, Mayland-Quellhorst S, Neuffer B (2006) Evolution of the self-incompatibility system in the *Brassicaceae:* Identification of *S*-locus receptor kinase (SRK) in self-incompatible *Capsella grandiflora*. Heredity 97:283–290

Palanivelu R, Brass L, Edlund AF, Preuss D (2003) Pollen tube growth and guidance is regulated by POP2, an *Arabidopsis* gene that controls GABA levels. Cell 114:47–59

Pawson T (2002) Regulation and targets of receptor tyrosine kinases. Eur J Cancer 38(Suppl 5):S3–S10

Preuss D, Lemieux B, Yen G, Davis RW (1993) A conditional sterile mutation eliminates surface components from *Arabidopsis* pollen and disrupts cell signaling during fertilization. Gene Dev 7:974–985

Pruitt RE, Vielle-Calzada JP, Ploense SE, Grossniklaus U, Lolle SJ (2000) *FIDDLEHEAD*, a gene required to suppress epidermal cell interactions in *Arabidopsis*, encodes a putative lipid biosynthetic enzyme. Proc Natl Acad Sci USA 97:1311–1316

Rivas S, Rougon-Cardoso A, Smoker M, Schauser L, Yoshioka H, Jones JD (2004) CITRX thioredoxin interacts with the tomato Cf-9 resistance protein and negatively regulates defence. EMBO J 23:2156–2165

Roberts IN, Stead AD, Ockendon DJ, Dickinson HG (1980) Pollen stigma interactions in *Brassica oleracea*. Theor Appl Genet 58:241–246

Roberts IN, Harrod G, Dickinson HG (1984) Pollen–stigma interactions in *Brassica oleracea*. I. Ultrastructure and physiology of the stigmatic papillar cells. J Cell Sci 66:241–253

Samuel MA, Salt JN, Shiu SH, Goring DR (2006) Multifunctional arm repeat domains in plants. Int Rev Cytol 253:1–26

Samuel MA, Mudgil Y, Salt JN, Delmas F, Ramachandran S, Chilelli A, Goring DR (2008a) Interactions between the S-Domain receptor kinases and *At*PUB-ARM E3 ubiquitin ligases suggest a conserved signaling pathway in *Arabidopsis*. Plant Physiol Jun 13 [Epub ahead of print]

Samuel MA, Chong YT, Aldea-Brydges MG, Haasen KE, Stone SL, Goring DR (2008b) *Bn*Exo70A1 is essential for compatible pollinations and is regulated by *Brassica* self-incompatibility. *Submitted*

Sarker RH, Elleman CJ, Dickinson HG (1988) Control of pollen hydration in *Brassica* requires continued protein synthesis, and glycosylation in necessary for intraspecific incompatibility. Proc Natl Acad Sci USA 85:4340–4344

Sato Y, Okamoto S, Nishio T (2004) Diversification and alteration of recognition specificity of the pollen ligand SP11/SCR in self-incompatibility of *Brassica* and *Raphanus*. Plant Cell 16:3230–3241

Schierup MH, Mable BK, Awadalla P, Charlesworth D (2001) Identification and characterization of a polymorphic receptor kinase gene linked to the self-incompatibility locus of *Arabidopsis lyrata*. Genetics 158:387–399.

Schopfer CR, Nasrallah ME, Nasrallah JB (1999) The male determinant of self-incompatibility in *Brassica*. Science 286:1697–1700

Shimosato H, Yokota N, Shiba H, Iwano M, Entani T, Che FS, Watanabe M, Isogai A, Takayama S (2007) Characterization of the SP11/SCR high-affinity binding site involved in self/nonself recognition in *Brassica* self-incompatibility. Plant Cell 19:107–117

Silva NF, Stone SL, Christie LN, Sulaman W, Nazarian KA, Burnett LA, Arnoldo MA, Rothstein SJ, Goring DR (2001) Expression of the *S* receptor kinase in self-compatible *Brassica napus* cv. Westar leads to the allele-specific rejection of self-incompatible Brassica napus pollen. Mol Genet Genom 265:552–559

Stead AD, Roberts IN, Dickinson HG (1980) Pollen-stigma interaction in *Brassica oleracea*: the role of stigmatic proteins in pollen grain adhesion. J Cell Sci 42:417–423

Stein JC, Nasrallah JB (1993) A plant receptor-like gene, the *S*-locus receptor kinase of *Brassica oleracea* L., encodes a functional serine/threonine kinase. Plant Physiol 101:1103–1106

Stone SL, Arnoldo M, Goring DR (1999) A breakdown of *Brassica* self-incompatibility in ARC1 antisense transgenic plants. Science 286:1729–1731

Stone SL, Anderson EM, Mullen RT, Goring DR (2003) ARC1 is an E3 ubiquitin ligase and promotes the ubiquitination of proteins during the rejection of self-incompatible *Brassica* pollen. Plant Cell 15:885–898

Swanson R, Edlund AF, Preuss D (2004) Species specificity in pollen–pistil interactions. Annu Rev Genet 38:793–818

Takasaki T, Hatakeyama K, Suzuki G, Watanabe M, Isogai A, Hinata K (2000) The *S* receptor kinase determines self-incompatibility in *Brassica* stigma. Nature 403:913–916

Takayama S, Shiba H, Iwano M, Shimosato H, Che FS, Kai N, Watanabe M, Suzuki G, Hinata K, Isogai A (2000a) The pollen determinant of self-incompatibility in *Brassica campestris*. Proc Natl Acad Sci USA 97:1920–1925

Takayama S, Shiba H, Iwano M, Asano K, Hara M, Che FS, Watanabe M, Hinata K, Isogai A (2000b). Isolation and characterization of pollen coat proteins of *Brassica campestris* that interact with *S*-locus-related glycoprotein 1 involved in pollen-stigma adhesion. Proc Natl Acad Sci USA 97:3765–3770

Takayama S, Shimosato H, Shiba H, Funato M, Che FS, Watanabe M, Iwano M, Isogai A (2001) Direct ligand-receptor complex interaction controls *Brassica* self-incompatibility. Nature 413:534–538

Vanoosthuyse V, Tichtinsky G, Dumas C, Gaude T, Cock JM (2003) Interaction of calmodulin, a sorting nexin and kinase-associated protein phosphatase with the *Brassica oleracea S*-locus receptor kinase. Plant Physiol 133:919–929

Vieira Dos Santos C, Rey P (2006) Plant thioredoxins are key actors in the oxidative stress response. Trends Plant Sci 11:329–334

Yephremov A, Wisman E, Huijser P, Huijser C, Wellesen K, Saedler H (1999) Characterization of the *FIDDLEHEAD* gene of *Arabidopsis* reveals a link between adhesion response and cell differentiation in the epidermis. Plant Cell 11:2187–2201

Zuberi MI, Dickinson HG (1985) Pollen–stigma interaction in *Brassica*. III. Hydration of the pollen grains. J Cell Sci 76:321–336

Zinkl GM, Zwiebel BI, Grier DG, Preuss D (1999) Pollen–stigma adhesion in *Arabidopsis*: A species-specific interaction mediated by lipophilic molecules in the pollen exine. Development 126:5431–5440

Chapter 9
Molecular Biology of S-RNase-Based Self-Incompatibility

Y. Zhang and Y. Xue

Abstract Many flowering plants have developed self-incompatibility (SI) systems to avoid inbreeding and to promote out-crossing. Among the various SI systems, gametophytic SI (GSI) in the Solanaceae, Plantaginaceae and Rosaceae is believed to be the most common type, in which the specificity of SI response is controlled by a single polymorphic locus, termed the self-incompatibility *S*-locus. It has been shown that this locus is organised in a haplotype fashion carrying at least two genes determining the self and non-self pollen recognition specificity: *S-RNase* (*S-Ribonucleases*) expressed in pistil (pistil *S*) and *SLF* (*S*-Locus F-box)/*SFB*(*S*-haplotype-specific F-box) genes in pollen (pollen *S*). In this chapter, we present and discuss the current knowledge about molecular biology of S-RNase-based self-incompatibility.

Abbreviations

120K	120 kDa glycoprotein
AhSLF	*Antirrhinum hispanicum SLF*
C1 to C5	Conserved regions in the S-RNase sequence
CUL	Cullin
ECM	Extracellular Matrix
F-box	A protein motif; often components of SCF ubiquitin–ligase complexes
GSI	Gametophytic self-incompatibility
HT-B	H-Top Band; a small, novel asparagine-rich protein
HVa and HVb	Hypervariable regions in the S-RNase sequence

Y. Zhang and Y. Xue
Institute of Genetics and Developmental Biology, Chinese Academy of Sciences, and National Centre for Plant Gene Research, Beijing 100101, China, e-mail: ybxue@genetics.ac.cn

PhSBP1	*Petunia hybrida S-RNase binding protein1*
PiSLF	*Petunia inflata SLF*
PPMs	Pollen-part mutants
RNAi	RNA interference
rRNA	Ribosomal ribonucleic acid
SC	Self-compatible
SCF	Skp1-cullin-F-box complex
SFB/SLF	*S-Locus F-box/S-haplotype-specific F-box* (the pollen *S*-determinant in many GSI systems; *SLF* in *Antirrhinum* and *Petunia*; *SFB* in *Prunus*)
SI	Self-incompatibility
S-locus	Self-incompatibility locus
S-RNase	*S*-locus ribonuclease (the pistil *S*-determinant in many GSI systems)
SSK	SLF-*interacting SKP1-like*
TAC	Transformation-competent artificial chromosome

9.1 Introduction

Self-incompatibility (SI) is one of the most important systems adopted by many flowering plants to prevent self-fertilisation and thereby to generate and maintain genetic diversity within a species. Classic genetic studies established that SI in most species is controlled by a single polymorphic locus, the self-incompatibility *S*-locus. It is now aware that this locus contains at least two separate genes: one controlling the self and non-self pollen recognition specificity on the pistil side (pistil *S*), and the other on the pollen side (pollen *S*), thus the term *haplotype* is used to describe variants of the *S*-locus. Pollen inhibition occurs when the same *S*-haplotype is expressed by both pollen and pistil (de Nettancourt 2001).

The molecular nature of the *S*-locus has been extensively studied in several species with different genetic features, and accordingly, the SI systems have been classified into several types. Generally, we can define these types of SI as the Brassicaceae-type, Solanaceae-type and Papaveraceae-type SI, because the species from the three families were used first for molecular studies and subsequent findings revealed that they represent three different molecular types of the *S*-locus (McCubbin and Kao 2000; Sims 2005; Takayama and Isogai 2005). The Solanaceae-type SI appears to be the most phylogenetically widespread form of SI found in angiosperms and is shared by two more families, the Plantaginaceae (formerly placed in the Scrophulariaceae) (Xue et al. 1996; Olmstead et al. 2001) and Rosaceae (Sassa et al. 1996; Ishimizu et al. 1998). This type of SI is genetically determined as a single-locus gametophytic SI (GSI) system, of which the recognition specificity is determined by the haplotypes of the polymorphic *S*-locus: pollen tube growth is

9 Molecular Biology of S-RNase-Based Self-Incompatibility

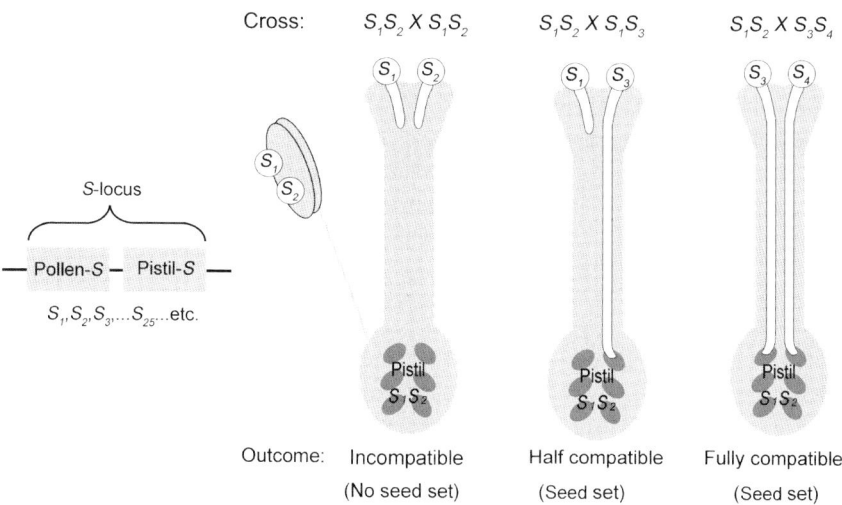

Fig. 9.1 Genetic control of gametophytic SI (GSI). Self-incompatibility is usually determined by a single polymorphic S-locus that is organised in a haplotype fashion carrying at least two genes: one controlling the self and non-self pollen recognition specificity on the pistil side (pistil S), and the other on the pollen side (pollen S). In GSI, the pollen SI phenotype is gametophytically controlled. Thus, half the pollen from an S_1S_2 plant is phenotypically S_1 and another half S_2. Pollen inhibition occurs on a 'like-matches-like' basis. When there is a match between the pollen S-haplotype and either of the two haplotypes present in the pistil, an incompatible reaction results and inhibition of the 'self' pollen occurs. This type of SI gives three possible classes of reaction: incompatible (all pollen is inhibited), half-compatible (50% inhibited, 50% grows normally) or compatible (all pollen grows normally) (modified from McClure and Franklin-Tong 2006, with kind permission from Springer)

inhibited when its S-haplotype is the same as either of the S-haplotypes of the diploid pistil; conversely, pollen with any S-haplotype not present in the pistil is compatible (Franklin-Tong et al. 2003; McClure and Franklin-Tong 2006) (Fig. 9.1). Much of the molecular information about this type of SI has been obtained from four genera of the Solanaceae (*Lycopersicon*, *Nicotiana*, *Petunia* and *Solanum*), three genera of the Rosaceae (*Malus*, *Prunus* and *Pyrus*) and one genus of the Plantaginaceae (*Antirrhinum*) (Kao and Tsukamoto 2004; McClure 2004, 2006). Since their female S-determinant (S-RNase) was the first gene identified in the S-locus, this type of SI is often referred to as the S-RNase-based SI. In this chapter, we present the current knowledge about the molecular biology of S-RNase-based SI. The readers are also referred to several excellent reviews about the earlier work in this area (McClure et al. 2000; Cruz-Garcia et al. 2003; Roalson and McCubbin 2003; Wang et al. 2003; Nasrallah 2005) and in this volume Chap. 10, which deals with mechanisms of S-RNase-based SI; Chaps. 4 and 6 deal with evolutionary aspects of S-RNase-based SI.

9.2 S-RNase Determines *S*-Specificity in Pistil

9.2.1 Isolation and Identification of S-RNase as the Pistil S

As the first step to understand the molecular mechanism underlying the SI response, it is imperative to identify the products of the *S*-locus. Putative specificity determinants must meet three criteria suggested by the biology and genetics of SI: linkage to the *S*-locus, polymorphism between different *S*-haplotypes and expression in the pollen or pistil. The searches for the pistil *S* and pollen *S* were essentially based on these three requirements.

Pistil *S*-allele products were first reported in *Nicotiana alata* as polymorphic basic glycoproteins that varied in molecular mass and isoelectric point and segregated with *S*-alleles (Bredemeijer and Blaas 1981). And a cDNA encoding a 32 kD stylar glycoprotein that segregated with the S_2-allele was isolated (Anderson et al. 1986). Similar stylar proteins and their cDNAs were also identified from members of Solanaceae, Rosaceae and Plantaginaceae (de Nettancourt 2001).

The biochemical nature of the pistil S-proteins remained unclear until their sequences were found to share similarity with RNase T_2 of *Aspergillus oryzae* (Kawata et al. 1988; McClure et al. 1989). Further studies confirmed the ribonuclease activity of the pistil S-proteins and thus they were referred to as S-RNases (McClure et al. 1990; Singh et al. 1991). S-RNases are primarily expressed in pistil and secreted into the extracellular matrix (ECM) of transmitting tract (Cornish et al. 1987; Anderson et al. 1989; Xue et al. 1996).

Although the *S-RNase* gene has satisfied the criteria expected for the pistil *S* determinant, direct proof that the S-RNases are necessary and sufficient for self-pollen rejection was obtained from transformation experiments using both loss- and gain-of-function approaches (Lee et al. 1994; Murfett et al. 1994). These studies demonstrated that introduction of a new *S-RNase* allele was sufficient to confer the transgenic plants the ability to reject pollen with the same haplotype as the *S-RNase* introduced. Conversely, suppression of expression of the *S-RNase* by antisense transformation abolished the ability of the transgenic plants to reject the pollen carrying the affected haplotype. Stylar-part mutations reported in Solanaceae (Royo et al. 1994) and Rosaceae (Sassa et al. 1997) reinforced the notion that the S-RNase is essential for expression of SI in pistil.

9.2.2 S-RNase Sequence Features and the Specificity Determinant

9.2.2.1 The Role of the Carbohydrate Moiety

Attempts have been made to identify where the *S*-specificity resides in the S-RNase sequences. Since S-RNases are secreted glycoproteins that vary in the number and position of N-linked glycan chains, the allelic specificity might lie in the carbohydrate moiety (Woodward et al. 1989, 1992; Oxley et al., 1996). However, the non-glycosylated S-RNase of *Petunia inflata*, with the only potential N-glycosylation site Asn replaced with Asp, functions as well as the wild-type glycosylated S-RNase in rejecting pollen carrying the same S-haplotype (Karunanandaa et al. 1994). Thus, the encoding of S-specificity does not reside in the glycan side chains.

9.2.2.2 The Role of Amino Acid Sequence

S-RNases are highly divergent, with amino acid sequence identity ranging from 38 to 98% (McCubbin and Kao 2000). Despite this high sequence diversity, five small conserved regions, C1 to C5, accounting for about 40 of the approximately 230 amino acids in a typical S-RNase, have been identified (Ioerger et al. 1991; Tsai et al. 1992). Similar structural features have been observed in the Rosaceae and the Plantaginaceae, with the exception that the position of C4 is occupied by conserved regions that share no homology with C4 or with each other (Xue et al. 1996; Ushijima et al. 1998). Most notably, the C2 and C3 regions, which share a high degree of sequence similarity with the corresponding regions of RNase T_2, contain conserved catalytic His residues, indicating that the RNase activity could be involved in self pollen inhibition (Fig. 9.2). Phylogenetic analysis of S-RNases suggested that this type of SI possessed by the species in these three distantly related dicot families shares a common origin, and might be exhibited by the ancestor of ~75% of all dicots (Xue et al. 1996; Igic and Kohn 2001). See also Chap. 5 for a review of this topic.

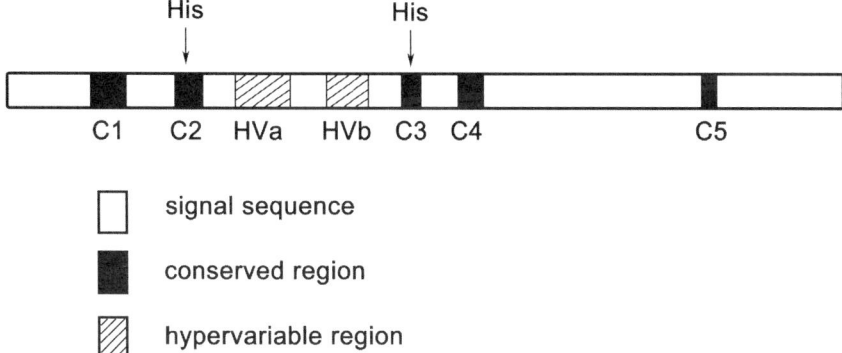

Fig. 9.2 Generalised features of S-RNases. A comparison of allelic S-RNases from several solanaceous species reveals a conserved structure among them. All contain a secretion signal sequence (*gray box*) and share five conserved domains (C1–C5; *black boxes*). Among them, C2 and C5 contain two histidine residues (His) that are required for ribonuclease activity as shown for related fungal RNases. Two hyper-variable regions (Hva and Hvb; *hatched boxes*) are also shown (modified form de Nettancourt 2001, with kind permission from Springer)

Sequence comparison of S-RNases from the Solanaceae also has revealed two hypervariable (HV) regions, HVa and HVb (Ioerger et al. 1991; Tsai et al. 1992), which were found subsequently to be the hypervariable regions of *Antirrhinum* S-RNases as well (Xue et al. 1996) and to correspond to two of the four regions of rosaceous S-RNases under positive selection (Ishimizu et al. 1998). The crystal structures of a solanaceous S-RNase and a rosaceous S-RNase show that HVa and HVb form a continuous solvent surface on one side of the protein (Ida et al. 2001; Matsuura et al. 2001), with a putative function to form a domain interacting with the pollen *S*-determinant. Therefore, HVa and HVb are considered the most likely candidates for the determinant of *S*-allele specificity. To assess this role, three sets of mosaic constructs among S-RNases had been utilised (Kao and McCubbin 1996; Zurek et al. 1997; Matton et al. 1997). However, these experiments led to different conclusions, and it remains possible that amino acids outside HVa and HVb are also involved in *S*-allele specificity (Verica et al. 1998). Nonetheless, it is clear that the HVa and HVb regions play a key role in the *S*-allele specificity determination.

9.2.3 The Role of S-RNase: A Cytotoxin Specifically Inhibits Self Pollen

9.2.3.1 S-RNase Cytotoxicity as the Underlying Cause of Pollen Tube Growth Inhibition

It has been shown that the two His residues important for ribonuclease activity are completely conserved in all S-RNases that are functional in pollen rejection (Ioerger et al. 1991), implicating that S-RNase enzymatic activity is involved in SI response. This has been supported by a series of experiments. A tracer experiment using ^{32}P-labelled pollen RNA showed that pollen rRNA are degraded after incompatible but not compatible pollination (McClure et al. 1990). In addition, catalytically inactive S-RNases due to the mutation of His residues lost the ability to reject self pollen (Huang et al. 1994; Royo et al., 1994). A cytotoxic model for S-RNase-based SI has been proposed accordingly (McClure et al. 1990). In this model, each *S*-haplotype encodes a unique S-RNase. In incompatible pollination, S-RNase gains access to the pollen tube cytoplasm where its ribonuclease activity causes a cytotoxic effect, whereas in compatible pollination, this cytotoxic effect is evaded.

Although rRNA degradation in incompatible pollen tubes is correlative with the expression of SI, it is difficult to discriminate whether it is the direct cause or secondary effect of incompatibility (McClure et al. 1990, 2000). And since no substrate specificity of S-RNase has been detected in vitro, it is still not clear how S-RNase activity could be restricted to self pollen tube (McClure et al. 1990). Moreover, incompatible pollen tubes could be 'rescued' if they were grafted onto compatible styles (i.e. they reverted to normal growth); thus the cytotoxic effect of S-RNase is not permanent. This seems inconsistent with the cytotoxic model, and had been interpreted that pollen tubes could synthesise rRNA continuously (Lush and Clarke 1997). Nevertheless, it has been widely accepted that S-RNase functions as a cytotoxin to inhibit self pollen, no matter how this inhibition is achieved through the intrinsic RNase activity or other effects.

9.2.3.2 S-RNases Enter the Pollen Tubes in a Non-Haplotype-Specific Manner

How do the S-RNases specifically exert their cytotoxicity specifically in self pollen tubes? It was originally proposed that the membrane or cell wall-bound receptors of pollen tubes function as specific gatekeepers to allow the S-RNase of the same *S*-haplotype, but not S-RNases of different *S*-haplotypes to enter, and thereby the growth of self pollen tubes is selectively inhibited. Although intuitively this appeared to be a most efficient or likely mechanism, subsequent evidence strongly suggested that this model was incorrect. S-RNases have been shown by immunolocalisation to be present in the cytoplasm of both self and non-self pollen tubes growing in the style, strongly indicating that uptake of S-RNases into pollen tubes is not *S*-haplotype specific (Luu et al. 2000).

9.3 F-Box Proteins Determine *S*-Specificity in Pollen

9.3.1 Clues from Pollen-Part Self Compatible Mutants

Early attempts to isolate the pollen *S* were focused on the analysis of pollen-part mutants (PPMs) (de Nettancourt 2001). For species from Solanaceae, the most frequent types of lesions were either translocations or small 'centric' fragments (short extra chromosomes) that carried a duplicated copy of an *S*-allele. Breakdown of the pollen SI response in those plants has mainly been ascribed to a 'competitive interaction' that enabled pollen with two different *S*-haplotypes (but not two identical *S*-haplotypes) to grow through an incompatible style. Apart from PPMs, competitive interaction was also thought to be the reason why tetraploids that were derived from self-incompatible diploids in the Solanaceae were often self-fertile (de Nettancourt 2001). Consistently, molecular studies have lent support for competitive interaction (Thompson et al. 1991; Golz et al. 1999, 2001). Golz et al. (1999, 2001) performed a large scale gamma ray mutagenesis in *N. alata*, with the expectation to isolate PPMs with the pollen *S* deletion; however, all PPMs acquired were found to be associated with duplicated *S*-haplotypes. Similar results were obtained from a transposon-based mutagenesis conducted in *Antirrhinum* (our unpublished data).

In Rosaceae, however, no SI breakdown mutants have been found to be associated with the presence of a centric fragment (de Nettancourt 2001), and the tetraploids can be either self-compatible or self incompatible (reviewed in Mable 2004). In fact, analyses of these mutants have provided foundation for the isolation and confirmation of the pollen *S*, and for the dissection of the molecular mechanisms underlying the SI response.

9.3.2 Isolation of the Pollen SI Determinant, SLF/SFB

During the past several years, extensive efforts have been made for the pollen S identification based on its predicted genetic features similar to the pistil determinant. This has included RNA differential display, subtractive hybridisation (Dowd et al. 2000; Li et al. 2000; McCubbin et al. 2000) and yeast two hybrid assay using HV domain of S-RNases as bait (Sims and Ordanic, 2001). However, no gene that segregated with *S-RNase* was identified through these routes. This was a real challenge since the pollen S protein appeared to be not as abundant as the S-RNase. Eventually, the pollen S gene candidates were successfully identified by a large-scale genomic DNA sequencing in the vicinity of *S-RNase* genes.

The first promising candidate was obtained from DNA sequence analysis of a $\sim$64 kb region containing the S_2-*RNase* of *Antirrhinum hispanicum* (Lai et al. 2002). This region contained a novel F-box protein gene, *AhSLF-S_2* (*A. hispanicum S*-locus F-box of S_2-haplotype), which was specifically expressed in pollen. More extensive searches of the adjacent regions of S_2-*RNase* revealed the existence of multiple paralogous genes that encoded similar F-box proteins. However, recombination analyses confirmed that *AhSLF-S_2* was the most adjacent haplotype-specific gene among those F-box-encoding genes linked to the *S*-locus, although their allelic products had much lower sequence diversity (ca. 97% amino acid sequence identity) than those of S-RNases (Zhou et al. 2003).

Similar experiments have been conducted in species of Rosaceae (Entani et al. 2003; Ushijima et al. 2003) and Solanaceae (Sijacic et al. 2004). Entani et al. (2003) analysed the $\sim$60 kb genomic region around the *S*-locus in self-incompatible *Prunus mume* (Japanese apricot) and found that the *S-RNase* gene was present in a region that contained at least four independent F-box genes. The F-box gene closest to the *S-RNase* gene was named *SLF*. It showed a high level of allelic diversity, with amino acid sequence identities of three alleles of *PmSLF* ranging from 78 to 81%, and was expressed exclusively in pollen.

At the same time, a $\sim$70 kb segment of the *S*-locus of the rosaceous species *P. dulcis* (almond) was sequenced completely (Ushijima et al. 2003). This region was found to contain two pollen-expressed F-box genes. One of them, named *SFB* (*S*-haplotype-specific F-box), was expressed specifically in pollen and showed a high level of *S*-haplotype-specific sequence polymorphism. Another showed little allelic sequence polymorphism and was expressed also in pistil and named *PdSLF* (*P. dulcis S*-locus F-box). It should be noted that *PdSLF* was so named to emphasise that it shows approximately the same low level of allelic sequence diversity as detected among *AhSLF* rather than to indicate that *PdSLF* is an ortholog of *AhSLF*. In fact, both *SFB* and *PdSLF* are less than 25% identical to *AhSLF* at the deduced amino acid sequence level (Kao and Tsukamoto 2004).

In *P. inflata*, a thorough search of the pollen *S*-determinant in a large *S*-locus region (328 kb BAC contig of S_2-haplotype) identified a polymorphic F-box gene, named *PiSLF*, 161 kb downstream of the *S-RNase* gene. Although the genomic region outside this contig contained two more polymorphic F-box genes that were genetically linked to the *S*-locus, the PiSLF exhibited the highest sequence diversity

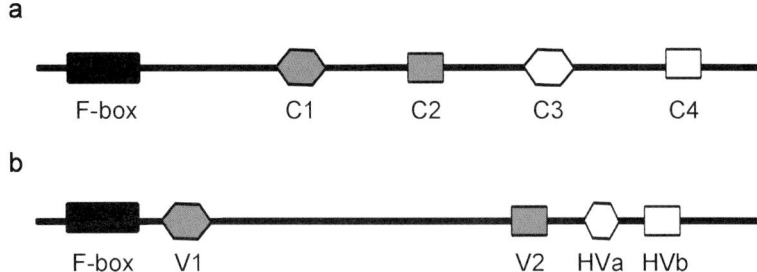

Fig. 9.3 Structural features of SLFs and SFBs. (**a**) Phylogenetic analyses of AhSLFs and their family members from *Arabidopsis* and several other plant species revealed that the *SLF* genes appeared to have had a monophyletic origin. Besides the F-box motif (*black box*), these SLFs contain four conserved domains (C1–C4; *gray boxes*) (Wang et al. 2004a). (**b**) Amino acid sequence alignment of *Prunus* SFBs reveals a conserved feature among them. All share an F-box domain (*black box*) and (hyper)variable domains (V1, V2, HV1 and HV2; *grey boxes*) (Ikeda et al. 2004)

(ca 90% in amino acid identity), but this was still much lower than that among *Prunus* SFBs (60–80% in amino acid identity) (Wang et al. 2004b).

9.3.3 Sequence Analysis of SLFs and SFBs

More alleles of *SLF* and *SFB* have been cloned by PCR, based on the conserved regions found among these genes (Ushijima et al. 2003; Yamane et al. 2003a, b; Zhou et al. 2003; Wang et al. 2004a; Tsukamoto et al. 2005; Cheng et al. 2006; Nunes et al. 2006; Vaughan et al. 2006). The phylogenetic analyses suggested that all of the SLF/SFBs from *Antirrhinum*, *Petunia* and *Prunus* belong to a monophyletic group sharing four conserved domains (Wang et al. 2004a) (Fig. 9.3a), implying that the common ancestor of many dicots possessed S-RNase/F-box-based GSI. This result provides further support on the monophyletic origin of the S-RNases (Xue et al. 1996; Igic and Kohn 2001). Nevertheless, all of the SFBs from the *Prunus* species formed a sub-lineage distinct from the functionally known SLF from *Antirrhinum* and *Petunia*, suggesting that they are likely diverged earlier during the S-RNase-based SI evolution (Ushijima et al. 2004; Wang et al. 2004a). See Chap. 5 for further discussion of the phylogeny of SLFs and SFBs. Based on amino acid sequence comparison, Ikeda et al. (2004) summarised the structural features for *Prunus* SFBs. All share an N terminal F-box domain, and four (hyper)variable regions, which were shown to be under positive selection (Fig. 9.3b).

9.3.4 Identification of SLF as the Pollen S

Direct evidence that *SLF* encodes the pollen *S*-determinant was finally obtained from transformation experiments in *P. inflata* (Sijacic et al. 2004) and *P. hybrida*

(Qiao et al. 2004a). To ascertain whether *PiSLF* encodes the pollen *S*-determinant, the phenomenon of competitive interaction, which occurs in heteroallelic pollen was utilised (see Sect. 3.1). Consistent with this phenomenon, the transformation of S_1S_1, S_1S_2 and S_2S_3 plants with the S_2-allele of *PiSLF* (*PiSLF2*) caused breakdown of the pollen function in SI. Furthermore, molecular genetic analyses of the progeny from self pollination of S_1S_2/*PiSLF2* and S_2S_3/*PiSLF2* revealed that only S_1 and S_3 pollen carrying the *PiSLF2* transgene (corresponding to heteroallelic pollen), but not S_2 pollen carrying *PiSLF2* (corresponding to homoallelic pollen), became self-compatible (Sijacic et al. 2004). These results conclusively demonstrated that *SLF* is the long-sought after pollen *S*-determinant in S-RNase-based SI. Similar experiments were conducted with the *AhSLF* genes (Qiao et al. 2004a). Both a pollen-expressed *AhSLF-S_2* gene constructs and a transformation-competent artificial chromosome (TAC) clone containing both S_2-*RNase* and *SLF-S_2* were transformed into SI *P. hybrida*, and SI breakdown on the pollen side was subsequently observed. Moreover, introduction of other F-box proteins in the vicinity of the *S*-locus of *A. hispanicum* had no effect on the pollen SI behaviour (our unpublished results). These findings showed that the *AhSLF* and *PiSLF* are orthologs, representing the sole pollen *S*-determinant. Moreover, consistently, duplication of the *SLF* was associated with breakdown of pollen SI in a natural population of self-incompatible *P. axillaris* (Tsukamoto et al. 2005), lending further support to the proposal that *SLF* encodes the pollen *S*.

9.3.5 SFBs from Rosaceae Likely Represent Another Class of F-Box Genes

Gain-of-function or loss-of-function transgenic experiments in *Prunus* tree species are hindered by the lack of an efficient transformation system and a long juvenile period. Functional dissection of *SFB* mainly came from analysis of naturally occurring pollen-part self compatible mutants. Intriguingly, in contrast with the *SLF*, whose deletion could be lethal (Sect. 9.3.1), deletion of *SFB* appeared to be associated with SI breakdown on the pollen side in *Prunus* (Ushijima et al. 2004; Sonneveld et al. 2005; Hauck et al. 2006a; Tsukamoto et al. 2006; Vilanova et al. 2006). In addition, the heteroallelic pollen of a tetraploid *Prunus* was found to be self-incompatible (Hauck et al. 2006b), contrary to the competitive interaction phenomena observed in the Solanaceae and Plantaginaceae. These differences between *SLF* and *SFB* could possibly reflect a mechanistic diversity, which will be discussed later (Sect. 9.6).

Although *SFB* fulfills the genetic requirements for the pollen *S*-determinant: linkage to *S-RNase*, haplotype-polymorphism and expression in pollen, recent findings call for more careful attention about the nature of *SFB*. The S_8-RNase from SI *P. tenella* was found to be identical to S_1-RNase from *P. avium*, whereas the corresponding SFB alleles were different. This finding was interpreted as the pollen *S* specificity is more tolerant of non-conservative replacements and this tolerance is important for SI evolution (Surbanovski et al. 2007). However, further functional dissection is needed to clarify whether these variants of *SFB* are the intermediate form of the pollen *S* during evolution or merely polymorphic genes linked to the *S*-locus rather than the pollen *S*. Furthermore, it is puzzling when the *S*-locus linked and pollen-expressed F-box genes from *Malus domestica* (apple), another species from Rosaceae, had been isolated by PCR and showed more sequence identity to the *SLFs* in the Solanaceae and Plantaginaceae rather than the *SFBs* in the Rosaceae (Cheng et al. 2006). What made the problem even more complicated was the finding that multiple pollen-specific F-box genes in apple (*M. domestica*) and Japanese pear (*Pyrus pyrifolia*), termed the *S*-locus F-box brothers (*SFBB*), display high homology to the *SLFs* isolated by Cheng et al. (2006), which also appear to meet the expected characteristics of the pollen *S* (Sassa et al. 2007). Although species from these three families (Solanaceae, Plantaginaceae and Rosaceae) use a similar molecule as the pistil *S*-determinant (S-RNase), the status of the pollen *S*-determinant in Rosaceae seems more complicated. It seems possible that there may be more than one gene product involved in S-RNase recognition (Sassa et al. 2007).

9.4 Other Genes That Modulate the SI Response

Although the *S*-locus encodes the determinants for *S*-haplotype specificity, there is evidence for the existence of other unlinked genes, termed modifier genes that are required for the SI response (reviewed in McClure et al. 2000, 2006; Cruz-Garcia et al. 2003; Kao and Tsukamoto 2004), which will be discussed here.

9.4.1 The Pistil Modifier Factors

9.4.1.1 HT-B

HT-B protein, a small asparagine-rich protein expressed late in style development, was originally identified by differential cDNA hybridisation to screen style genes that were expressed in self-incompatible species *N. alata* but not in *N. plumbaginifolia*, a closely related self-compatible species (McClure et al. 1999). Homologs of HT-B also have been identified in two other genera of the Solanaceae, *Lycopersicon* and *Solanum* (Kondo et al. 2002b; O'Brien et al. 2002). Down-regulation of *HT-B* expression by anti-sense transformation and RNAi deprived the transformants of the ability to reject self pollen, suggesting that it is essential for SI (McClure et al. 1999; O'Brien et al. 2002). In a comparative analysis of self-incompatible and self-compatible taxa of *Lycopersicon*, the expression of *HT-B* gene was not detected in all self-compatible taxa (Kondo et al. 2002a). Since no direct interaction between HT-B and S-RNases was detected, the exact role of HT-B for SI response was unclear until a recent immunolocalisation experiment (Goldraij et al. 2006).

This revealed that in self pollen tubes, HT-B was likely to facilitate S-RNase transport from an endomembrane compartment to the cytoplasm, where they could exert their cytotoxicity, leading to the arrest of pollen tube growth, while in compatible pollen tubes, HT-B level was significantly down regulated and the S-RNases were compartmentalised (Goldraij et al. 2006). See Chap. 10 for further details.

9.4.1.2 120K Glycoprotein

The 120 kDa glycoprotein (120K) is an abundant protein in the stylar ECM and taken up by the growing pollen tubes (Lind et al. 1994, 1996). The 120K protein binds to S-RNase in vitro and, like HT-B, suppression of its expression by RNAi prevented self pollen rejection (Cruz-Garcia et al. 2005; Hancock et al. 2005). In recent immunolocalisation experiments, antibodies to the 120 kDa glycoprotein were found to label the compartment membrane that surrounds the S-RNases inside the pollen tubes. However, since S-RNase uptake is normal in 120K protein defective plants, its precise role in SI is still not clear (Goldraij et al. 2006); see Chap. 10.

9.4.2 The Pollen Modifier Factors

9.4.2.1 SSK1

F-box proteins often serve as adaptors that bind specific substrate proteins to the SCF (Skp1-Cullin-F-box) E3 ubiquitin ligase complex (Petroski and Deshaies 2005). This raised the possibility of whether the SLF involved in the SI also participates in an SCF complex, mediating S-RNase ubiquitination. Identification of other components in such a putative complex is obviously necessary to address this question. *SSK1* (*SLF-interacting SKP1-like1*), a homolog of *SKP1*, was originally isolated in *A. hispanicum* through a yeast two-hybrid screening against a pollen cDNA library using *AhSLF-S_2* as bait (Huang et al. 2006). Pull-down assays suggested that AhSSK1 could be an adaptor that connects SLF to CUL1 protein. Therefore, it is thought that SLF and SSK1 are likely to be recruited to a canonical SCF complex, which could be responsible for S-RNase ubiquitination (Fig. 9.4a); see also Chap. 10.

9.4.2.2 SBP1

In an attempt to isolate the pollen *S*, Sims and Ordanic (2001) screened a yeast two-hybrid library from mature pollen of *P. hybrida* using *P. hybrida* S_1-RNase as bait, and identified a gene named *PhSBP1* (*S-RNase Binding Protein1*). Its homolog in *S. chacoense* was obtained based on a similar approach (O'Brien et al. 2004). However,

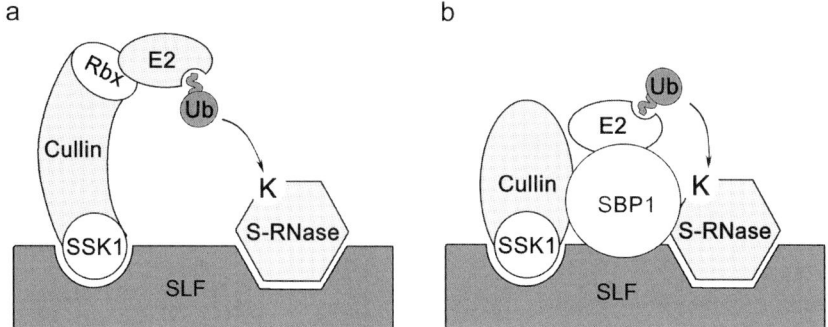

Fig. 9.4 Two proposed SCF^{SLF} recognition complexes in S-RNase-based self-incompatibility. Experiments described in the text demonstrate that the pollen S is an S-locus encoded F-box gene. F-box proteins are the recognition components of SCF E3 ubiquitin ligase complexes. AhSLF-S$_2$ from *A. hispanicum* has been shown to interact with S-RNases both in vitro and in vivo. AhSSK1, identified by yeast two-hybrid screening using AhSLF-S$_2$ as bait, is able to form a complex together with Cullin1 and S-RNase. These results suggest that SLF and SSK1 are components of a canonical SCF complex, which transfers ubiquitin from an E2 enzyme to S-RNase. PhSBP1, originally identified as a factor that binds to S-RNase in yeast two-hybrid assay, has been shown to be able to interact with PhSLF, Cullin1 and E2, with a possible role that replaces RBX1 as the RING protein of the complex. In either case, recognition of S-RNase by the SCF^{SLF} complex is predicted to lead to ubiquitination of the S-RNase protein

the *SBP1* gene displayed no haplotype polymorphism and was found to be expressed in almost all tissues. In addition, it was unlinked to the *S*-locus and therefore is unlikely to encode the pollen *S*-determinant. However, sequence analysis revealed that SBP1 contains a RING-finger domain, which is characteristic of E3 ubiquitin ligases (Kerscher et al. 2006), indicating a possible role for SBP1 in S-RNase ubiquitination and degradation (Sims and Ordanic 2001; O'Brien et al. 2004). Interestingly, *P. inflata* SBP1 (PiSBP1) has recently been shown to be able to interact with PiSLFs, Pi CUL1 and an ubiquitin-conjugating enzyme, and a novel E3 ligase complex has been suggested, with the possibility that PiSBP1 plays a combined role of SKP1 and RBX1 (Sims 2005; Hua and Kao 2006) (Fig. 9.4b); see also Chap. 10.

9.5 Molecular Mechanisms for S-RNase-based SI

9.5.1 Pollen S, the Positive or Negative Regulator of S-RNase? Clues from Genetic Evidence

It is now generally accepted that S-RNases exert their cytotoxicity inside the pollen tube. This raises the question of what the function of the pollen *S* product is. One possibility is that the pollen *S* product might promote or inhibit the activity of S-RNases. Genetic evidence has indicated possible distinct roles for *SLF* genes in

Solanaceae and Plantaginaceae and *SLF/SFB* genes in Rosaceae, and we briefly discuss this here.

9.5.1.1 SLF, the Inhibitor of S-RNase in Solanaceae and Plantaginaceae

It has been suggested that the pollen *S* product in Solanaceae and Plantaginaceae might be required for pollen tube growth in styles possessing functional S-RNases, possibly because its product acts as an RNase inhibitor (Sect. 9.3.1). A simple inhibitor model has been proposed in which the pollen component was assumed to be an RNase inhibitor inside the pollen tube, which was able to inhibit all S-RNases except the one of the corresponding *S*-haplotype (Dodds et al. 1996; Kao and McCubbin 1996; Luu et al. 2000). In this model, the pollen *S* encodes a protein that has two functions of both inhibiting non-self S-RNases and conferring specificity (Kao and Tsukamoto 2004) (Fig. 9.5a). The simple inhibitor model could readily explain the phenomenon of competitive interaction. When two different pollen *S*-alleles are expressed in the same pollen grain, their products could together inhibit the RNase activity of all S-RNases, thus resulting in the breakdown of SI (Kao and Tsukamoto 2004); also see Chap. 10.

The inhibitor effect of SLF could also provide an explanation for its low sequence divergence. Since SLF functions to inhibit all but self S-RNase, it would be fatal if the sequence varies to a higher degree. Moreover, relatively small amounts of sequence divergence can nevertheless result in allelic specificity, as is the case for the S_{11} and S_{13} alleles of *S. chacoense* (Saba-el-Leil et al. 1994). This possibility could be further tested by structural and functional studies on SLF.

9.5.1.2 SFBs in Rosaceae Appear to Protect Self S-RNases

In Rosaceae, however, another scenario is emerging. The deletion of *SFB* appears to have resulted in pollen-part self-compatibility in *Prunus* (Sect. 9.3.5). Loss-of-function mutations strongly indicate the presence of a general inactivation mechanism, with SFB acting to confer specificity by protecting self S-RNases from inactivation (Ushijima et al. 2004; Sonneveld et al. 2005; Hauck et al. 2006a). Such a mechanism was originally proposed by Luu et al. (2001) as a modified inhibitor model to explain the SI behaviour of a dual-specificity chimeric S-RNase they had generated (Matton et al. 1999) (Fig. 9.5b); see also Chap. 10.

9.5.2 The Fate of S-RNases: S-RNase Restriction is Likely to Involve Ubiquitination

Current knowledge concerning the molecular mechanism of S-RNase-based SI has mainly been obtained from the Solanaceae and Plantaginaceae. Genetic evidence

9 Molecular Biology of S-RNase-Based Self-Incompatibility

a. Simple inhibitor model

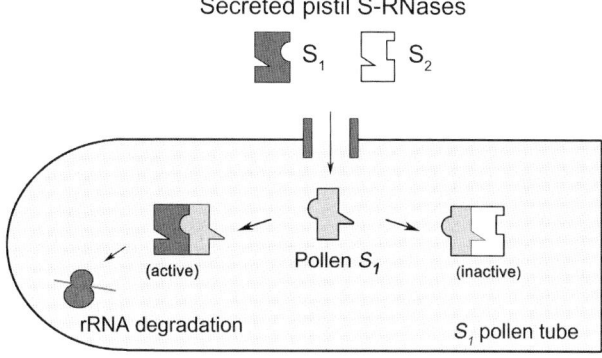

b. General inhibitor model

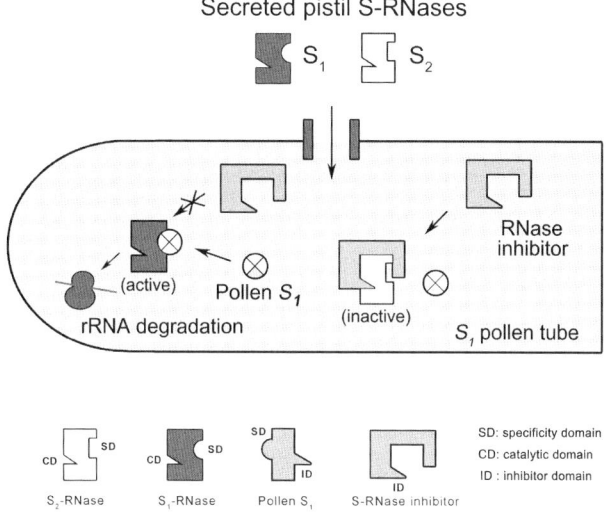

Fig. 9.5 Two models for the mechanism of S-RNase-based self-incompatibility. (**a**) Simple inhibitor model. Both S_1- and S_2-RNases are taken up by an S_1 pollen tube, but only S_1-RNase is active in degrading pollen RNA, because the pollen S_1-haplotype product specifically inhibits the RNase activity of S_2-RNase. Binding of S_1-RNase to the pollen S_1-haplotype product through their matching *S*-haplotype-specificity domains (SD) blocks binding of the inhibitor domain (ID) of the pollen S_1-haplotyoe product to the catalytic domain (CD) of S_1-RNase. Interaction between S_2-RNase and the pollen S_1-haplotype product is through the CD of the former and the ID of the latter, thus rendering S_2-RNase inactive. (**b**) General inhibitor model. This model differs from the simple inhibitor model in two major respects: the pollen *S*-haplotype products are homotetramers and only contain the SD and a general RNase inhibitor is assumed to be responsible for the inhibition of the RNase activity of S-RNases. The *S*-haplotype-specific inhibition of the RNase activity of S-RNases is achieved in a manner similar to that described for (**a**) (from Franklin-Tong and Franklin 2003, with kind permission from Elsevier)

has suggested that SLF in these two families functions to inhibit the S-RNase activity; therefore, the key questions now facing us are how the compatible pollen tube evades the S-RNase cytotoxicity and what the molecular function of pollen S product is during this process? It has been shown that SLF and SSK1, the components of an SCF complex, function to inhibit the S-RNase activity (Sect. 9.5.2.1). Since SCF is well known to act in conjunction with the E2 enzymes to ubiquitinate target proteins (Kerscher et al. 2006), it is assumed that the SCF^{SLF} could recognise and ubiquitinate S-RNase. This speculation has been tested and confirmed by two separate experiments in *A. hispanicum* and *P. inflata*, respectively, although no S-haplotype-specificity was observed in both cases (Qiao et al. 2004b; Hua and Kao 2006). Furthermore, compatible pollinations were specifically blocked after the treatments of the proteasomal inhibitors MG115 and MG132, which had little effect on incompatible pollination, indicating that ubiquitination/26S proteasomal activity is involved in compatible pollination processes. Based on these findings, the S-RNase degradation model was proposed, in which S-RNases are ubiquitinated by non-self SLF followed by proteasome-mediated degradation after compatible pollination, while the S-RNases remain active in self pollen tube due to an ineffective interaction of SLF and self S-RNase. This latter situation would be predicted to result in pollen tube growth arrest (Qiao et al. 2004b; Hua and Kao 2006) (Fig. 9.6).

There are still some questions remaining with respect to this model. For example, how is specificity attained? Why does the quantity of S-RNases show no significant decrease during compatible pollination though there is no detectable S-RNase activity within the pollen tube? Nevertheless, consistent with the genetic evidence and with an explanation for the competitive interaction, the inhibitor model has been widely accepted to interpret the molecular mechanism of SI response in the Solanaceae and Plantaginaceae; see also Chap. 10.

However, a recent study at a cytological level has provided a new vision. It has been suggested that the compartmentalisation of S-RNases in compatible pollen tube is the reason for the restriction of the S-RNase cytotoxicity, and the role of HT-B during this process has been discussed (Goldraij et al. 2006; McClure 2006). These two proposed mechanisms are not necessarily mutually exclusive. Immunolocalisation studies have suggested that the distribution of AhSLF appears associated with the membrane system, indicating that the SLF probably has a function in membrane trafficking (Wang and Xue 2005). Thus, it is possible that these two mechanisms could work together to inhibit non-self S-RNase activity during compatible pollination. A closer inspection of the molecular function of the pollen S protein will be helpful for a better understanding of the fate of S-RNases. See also Chap. 10 for further discussion of this perplexing problem.

9.5.3 Future Perspectives

During the last two decades, extensive efforts have been made to elucidate the molecular mechanism of S-RNase-based SI response, and the most breathtaking

a. Compatible pollination

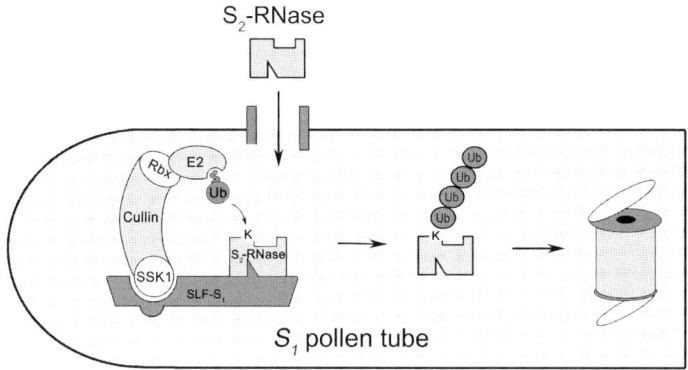

b. Incompatible pollination

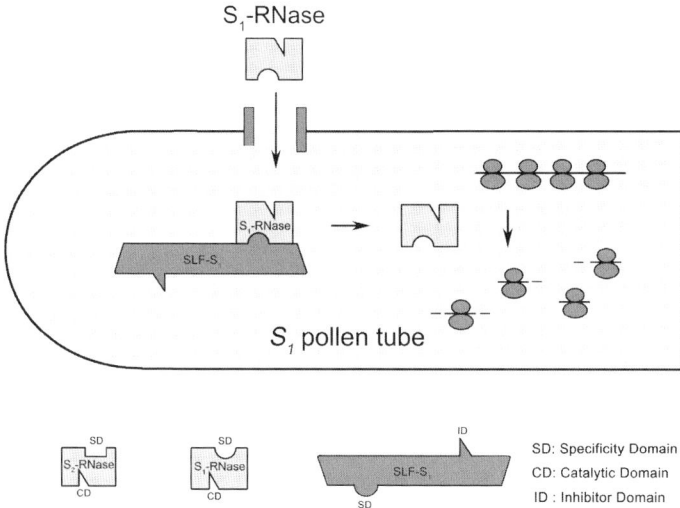

Fig. 9.6 S-RNase degradation model. After non-self S-RNase enters the pollen tube, its catalytic domain (CD) interacts with the inhibitor domain (ID), leading to formation of a functional SCF complex that targets the S-RNase for ubiquitination and degradation by the 26S proteasome, thus the pollen tube growth continues normally (**a**). By contrast, after self S-RNase enters the pollen tube, the specificity domains (SD) of S-RNase and SLF interact with each other, resulting in conformational change in SCF^{SLF} and thus the S-RNase is not ubiquitinated and degraded. S-RNase cytotoxicity results in RNA degradation and consequent pollen tube growth inhibition (**b**). However, it is unknown how the *S*-haplotype specificity is attained

advances were the molecular identification of the pistil and pollen S-determinants. Although several attractive working models have been proposed to explain how they interact to elicit the SI signalling, we are still some distance from understanding the nature of the molecular and cell biology of the interaction and how exactly this GSI system operates mechanistically. Several outstanding questions clearly need immediate attention. For example, we are still ignorant of the molecular function of SFBs in Rosaceae. Could these SFBs also interact and ubiquitinate S-RNases like the SLFs in the Solanaceae and Plantaginaceae do, or is there a distinct mechanism for non-self S-RNase inhibition in the Rosaceae? It is also unclear why so many F-box genes reside in the vicinity of the S-locus. No interaction between these F-box proteins and S-RNase or SSK1 were detected (Qiao et al. 2004b; Huang et al. 2006) and they did not appear to have a direct role for S-RNase inhibition. What roles, if any, do they have? Further experiments, with the integration of the genetic, biochemical and cytological techniques, will help us to get a better understanding of this unique cell recognition and rejection system in flowering plants.

Acknowledgements We thank the members of the Xue lab for their comments on the manuscript and apologise to those colleagues whose works were not cited due to the space limitation. The work was supported by Chinese Academy of Sciences and National Natural Science Foundation of China (30221002).

References

Anderson MA, Cornish EC, Mau SL, Williams EG, Hogart R, Atkinson A, Bonig I, Grego B, Simpson R, Roche PJ, Haley JD, Penshow JD, Niall HD, Tregear GW, Coghlan JP, Crawford RJ, Clarke AE (1986) Cloning of cDNA for a stylar glycoprotein associated with expression of self-incompatibility in *Nicotiana alata*. Nature 321:38–44

Anderson MA, McFadden GI, Bernatzky R, Atkinson A, Orpin T, Dedman H, Tregear G, Fernley R, Clarke AE (1989) Sequence variability of three alleles of the self-incompatibility gene of *Nicotiana alata*. Plant Cell 1:483–491

Bredemeijer GMM, Blaas J (1981) S-Specific proteins in styles of self-incompatible *Nicotiana alata*. Theor Appl Genet 59:185–190

Brewbaker JL, Natarajan AT (1960) Centric fragments and pollen-part mutation of self-incompatibility alleles in *Petunia*. Genetics 45:699–704

Cheng J, Han Z, Xu X, Li T (2006) Isolation and identification of the pollen-expressed polymorphic F-box genes linked to the S-locus in apple (*Malus × domestica*). Sex Plant Reprod 19:175–183

Cornish EC, Pettitt JM, Bonig I, Clarke AE (1987) Developmentally controlled expression of a gene associated with self-incompatibility in *Nicotiana alata*. Nature 326:99–102.

Cruz-Garcia F, Hancock CN, McClure BA (2003) S-RNase complexes and pollen rejection. J Exp Bot 54:123–130

Cruz-Garcia F, Hancock CN, Kim D, McClure BA (2005) Stylar glycoproteins bind to S-RNase in vitro. Plant J 42:295–304

de Nettancourt D (2001) Incompatibility and incongruity in wild and cultivated plants, 2 edn. Springer, Berlin Heidelberg New York

Dodds PN, Clarke AE, Newbigin E (1996) A molecular perspective on pollination in flowering plants. Cell 85:141–144

Dowd PE, McCubbin AG, Wang X, Verica JA, Tsukamoto T, Andos T, Kao T-H (2000) Use of *Petunia inflata* as a model for the study of solanaceous type self-incompatibility. Ann Bot 85(Suppl A):87–93

Entani T, Iwano M, Shiba H, Che FS, Isogai A, Takayama S (2003) Comparative analysis of the self-incompatibility *S*-locus region of *Prunus mume*: Identification of a pollen-expressed *F-box* gene with allelic diversity. Gene Cell 8:203–213

Franklin-Tong VE, Franklin FCH (2003) Gametophytic self-incompatibility inhibits pollen tube growth using different mechanisms. Trends Plant Sci 8:598–605

Goldraij A, Kondo K, Lee CB, Hancock CN, Sivaguru M, Vazquez-Santana S, Kim S, Phillips TE, Cruz-Garcia F, McClure BA (2006) Compartmentalization of S-RNase and HT-B degradation in self-incompatible *Nicotiana*. Nature 439:805–810

Golz JF, Su V, Clarke AE, Newbigin E (1999) A molecular description of mutations affecting the pollen component of the *Nicotiana alata S*-locus. Genetics 152:1123–1135

Golz JF, Oh HY, Su V, Kusaba M, Newbigin E (2001) Genetic analysis of Nicotiana pollen-part mutants is consistent with the presence of an S-ribonuclease inhibitor at the *S*-locus. Proc Natl Acad Sci USA 98:15372–15376

Hancock CN, Kent L, McClure BA (2005) The stylar 120 kDa glycoprotein is required for *S*-specific pollen rejection in *Nicotiana*. Plant J 43:716–723

Hauck NR, Ikeda K, Tao R, Iezzoni AF (2006a) The mutated S_1-haplotype in sour cherry has an altered *S*-haplotype-specific F-box protein gene. J Hered 97:514–520

Hauck NR, Yamane H, Tao R, Iezzoni AF (2006b) Accumulation of nonfunctional *S*-haplotypes results in the breakdown of gametophytic self-incompatibility in tetraploid *Prunus*. Genetics 172:1191–1198

Hua Z, Kao T-H (2006) Identification and characterization of components of a putative *petunia S*-locus F-box-containing E3 ligase complex involved in S-RNase-based self-incompatibility. Plant Cell 18:2531–2553

Huang J, Zhao L, Yang Q, Xue Y (2006) AhSSK1, a novel SKP1-like protein that interacts with the *S*-locus F-box protein SLF. Plant J 46:780–793

Huang S, Lee HS, Karunanandaa B, Kao T-H (1994) Ribonuclease activity of *Petunia inflata* S-proteins is essential for rejection of self-pollen. Plant Cell 6:1021–1028

Ida K, Norioka S, Yamamoto M, Kumasaka T, Yamashita E, Newbigin E, Clarke AE, Sakiyama F, Sato M (2001) The 1.55 Å resolution structure of *Nicotiana alata* S_{F11}-RNase associated with gametophytic self-incompatibility. J Mol Biol 314:103–112

Igic B, Kohn JR (2001) Evolutionary relationships among self-incompatibility RNases. Proc Natl Acad Sci USA 98:13167–13171

Ikeda K, Igic B, Ushijima K, Yamane H, Hauck NR, Nakano R, Sassa H, Iezzoni AF, Kohn JR, Tao R (2004) Primary structural features of the *S*-haplotype-specific F-box protein, SFB, in *Prunus*. Sex Plant Reprod 16:235–243

Ioerger TR, Gohlke JR, Xu B, Kao T-H (1991) Primary structural features of the self-incompatibility protein in Solanaceae. Sex Plant Reprod 4:81–87

Ishimizu T, Shinkawa T, Sakiyama F, Norioka S (1998) Primary structural features of rosaceous S-RNases associated with gametophytic self-incompatibility. Plant Mol Biol 37:931–941

Kao T-H, McCubbin AG (1996) How flowering plants discriminate between self and non-self pollen to prevent inbreeding. Proc Natl Acad Sci USA 93:12059–12065

Kao T-H, Tsukamoto T (2004) The molecular and genetic bases of S-RNase-based self-incompatibility. Plant Cell 16(Suppl):S72–S83

Karunanandaa B, Huang S, Kao T-H (1994) Carbohydrate moiety of the *Petunia inflata* S_3 protein is not required for self-incompatibility interactions between pollen and pistil. Plant Cell 6:1933–1940

Kawata Y, Sakiyama F, Tamaoki H (1988) Amino-acid sequence of ribonuclease T_2 from *Aspergillus oryzae*. Eur J Biochem 176:683–697

Kerscher O, Felberbaum R, Hochstrasser M (2006) Modification of proteins by ubiquitin and ubiquitin-like proteins. Annu Rev Cell Dev Biol 22:159–180

Kondo K, Yamamoto M, Itahashi R, Sato T, Egashira H, Hattori T, Kowyama Y (2002a) Insights into the evolution of self-compatibility in *Lycopersicon* from a study of stylar factors. Plant J 30:143–153

Kondo K, Yamamoto M, Matton DP, Sato T, Hirai M, Norioka S, Hattori T, Kowyama Y (2002b) Cultivated tomato has defects in both *S-RNase* and *HT* genes required for stylar function of self-incompatibility. Plant J 29:627–636

Lai Z, Ma W, Han B, Liang L, Zhang Y, Hong G, Xue Y (2002) An F-box gene linked to the self-incompatibility *S*-locus of *Antirrhinum* is expressed specifically in pollen and tapetum. Plant Mol Biol 50:29–42

Lee HS, Huang S, Kao T-H (1994) S proteins control rejection of incompatible pollen in *Petunia inflata*. Nature 367:560–563

Lewis D (1943) Physiology of incompatibility on plants. III. Autopolyploids. J Genet 45:171–185

Li JH, Nass N, Kusaba M, Dodds PN, Treloar N, Clarke AE, Newbigin E (2000) A genetic map of the *Nicotiana alata S*-locus that includes three pollen-expressed genes. Theor Appl Genet 100:956–964

Lind JL, Bacic A, Clarke AE, Anderson MA (1994) A style-specific hydroxyproline-rich glycoprotein with properties of both extensins and arabinogalactan proteins. Plant J 6:491–502

Lind JL, Bönig I, Clarke AE, Anderson MA (1996) A style-specific 120 kDa glycoprotein enters pollen tubes of *Nicotiana alata* in vivo. Sex Plant Reprod 9:75–86

Lush WM, Clarke AE (1997) Observations of pollen tube growth in *Nicotiana alata* and their implications for the mechanism of self-incompatibility. Sex Plant Reprod 10:27–35

Luu DT, Qin X, Morse D, Cappadocia M (2000) S-RNase uptake by compatible pollen tubes in gametophytic self-incompatibility. Nature 407:649–651

Luu DT, Qin X, Laublin G, Yang Q, Morse D, Cappadocia M (2001) Rejection of *S*-heteroallelic pollen by a dual-specific S-RNase in *Solanum chacoense* predicts a multimeric SI pollen component. Genetics 159:329–335

Mable BK (2004) Polyploidy and self-incompatibility: Is there an association? New Phytol 162:803–811

Matsuura T, Sakai H, Unno M, Ida K, Sato M, Sakiyama F, Norioka S (2001) Crystal structure at 1.5-Å resolution of *Pyrus pyrifolia* pistil ribonuclease responsible for gametophytic self-incompatibility. J Biol Chem 276:45261–45269

Matton DP, Maes O, Laublin G, Xike Q, Bertrand C, Morse D, Cappadocia M (1997) Hypervariable domains of self-incompatibility RNases mediate allele-specific pollen recognition. Plant Cell 9:1757–1766

Matton DP, Luu DT, Xike Q, Laublin G, O'Brien M, Maes O, Morse D, Cappadocia M (1999) Production of an S-RNase with dual specificity suggests a novel hypothesis for the generation of new *S* alleles. Plant Cell 11:2087–2097

McClure BA (2004) *S-RNase* and *SLF* determine *S*-haplotype-specific pollen recognition and rejection. Plant Cell 16:2840–2847

McClure BA (2006) New views of S-RNase-based self-incompatibility. Curr Opin Plant Biol 9:639–946

McClure BA, Franklin-Tong VE (2006) Gametophytic self-incompatibility: Understanding the cellular mechanisms involved in "self" pollen tube inhibition. Planta 224:233–245

McClure BA, Haring V, Ebert PR, Anderson MA, Simpson RJ, Sakiyama F, Clarke AE (1989) Style self-incompatibility gene products of *Nicotiana alata* are ribonucleases. Nature 342:955–957

McClure BA, Gray JE, Anderson MA, Clarke AE (1990) Self-incompatibility in *Nicotiana alata* involves degradation of pollen rRNA. Nature 347:757–760

McClure BA, Mou B, Canevascini S, Bernatzky R (1999) A small asparagine-rich protein required for *S*-allele-specific pollen rejection in *Nicotiana*. Proc Natl Acad Sci USA 96:13548–13553

McClure BA, Cruz-Garcia F, Beecher BS, Sulaman W (2000) Factors affecting inter- and intra-specific pollen rejection in *Nicotiana*. Ann Bot 85:113–123

McCubbin AG, Kao T-H (2000) Molecular recognition and response in pollen and pistil interactions. Annu Rev Cell Dev Biol 16:333–364

McCubbin AG, Wang X, Kao T-H (2000) Identification of self-incompatibility *S*-locus linked pollen cDNA markers in *Petunia inflata*. Genome 43:619–627

Murfett J, Atherton TL, Mou B, Gasser CS, McClure BA (1994) *S-RNase* expressed in transgenic *Nicotiana* causes *S*-allele-specific pollen rejection. Nature 367:563–566

Nasrallah JB (2005) Recognition and rejection of self in plant self-incompatibility: Comparisons to animal histocompatibility. Trends Immunol 26:412–418

O'Brien M, Kapfer C, Major G, Laurin M, Bertrand C, Kondo K, Kowyama Y, Matton DP (2002) Molecular analysis of the stylar-expressed *Solanum chacoense* small asparagine-rich protein family related to the HT modifier of gametophytic self-incompatibility in *Nicotiana*. Plant J 32:985–996

O'Brien M, Major G, Chantha S-C, Matton DP (2004) Isolation of S-RNase binding proteins from *Solanum chacoense*:identification of an SBP1 (RING finger protein) ortholog. Sex Plant Reprod 17:81–87

Olmstead RG, de Pamphilis CW, Wolfe AD, Young ND, Elisons WJ, Reeves PA (2001) Disintegration of the Scrophulariaceae. Am J Bot 88:348–361

Oxley D, Munro SL, Craik DJ, Bacic A (1996) Structure of N-glycans on the S_3- and S_6-allele stylar self-incompatibility ribonucleases of *Nicotiana alata*. Glycobiology 6:611–618

Petroski MD, Deshaies RJ (2005) Function and regulation of cullin-RING ubiquitin ligases. Nat Rev Mol Cell Biol 6:9–20

Qiao H, Wang F, Zhao L, Zhou J, Lai Z, Zhang Y, Robbins TP, Xue Y (2004a) The F-box protein AhSLF-S_2 controls the pollen function of S-RNase-based self-incompatibility. Plant Cell 16:2307–2322

Qiao H, Wang H, Zhao L, Zhou J, Huang J, Zhang Y, Xue Y (2004b) The F-box protein AhSLF-S_2 physically interacts with S-RNases that may be inhibited by the ubiquitin/26S proteasome pathway of protein degradation during compatible pollination in *Antirrhinum*. Plant Cell 16:582–595

Roalson EH, McCubbin AG (2003) S-RNases and sexual incompatibility: Structure, functions, and evolutionary perspectives. Mol Phylogenet Evol 29:490–506

Royo J, Kunz C, Kowyama Y, Anderson MA, Clarke AE, Newbigin E (1994) Loss of a histidine residue at the active site of *S*-locus ribonuclease is associated with self-compatibility in *Lycopersicon peruvianum*. Proc Natl Acad Sci USA 91:6511–6514

Saba-el-Leil MK, Rivard S, Morse D, Cappadocia M (1994) The S_{11} and S_{13} self incompatibility alleles in *Solanum chacoense* Bitt. are remarkably similar. Plant Mol Biol 24:571–583

Sassa H, Nishio T, Kowyama Y, Hirano H, Koba T, Ikehashi H (1996) Self-incompatibility (*S*) alleles of the rosaceae encode members of a distinct class of the T_2/S ribonuclease superfamily. Mol Gen Genet 250:547–557

Sassa H, Hirano H, Nishio T, Koba T (1997) Style-specific self-compatible mutation caused by deletion of the *S-RNase* gene in Japanese pear (*Pyrus serotina*). Plant J 12:223–227

Sassa H, Kakui H, Miyamoto M, Suzuki Y, Hanada T, Ushijima K, Kusaba M, Hirano H, Koba T (2007) *S*-locus F-box brothers: multiple and pollen-specific F-box genes with *S*-haplotype-specific polymorphisms in apple and Japanese pear. Genetics 175:1869–1881

Sijacic P, Wang X, Skirpan AL, Wang Y, Dowd PE, McCubbin AG, Huang S, Kao T-H (2004) Identification of the pollen determinant of S-RNase-mediated self-incompatibility. Nature 429:302–305

Sims TL (2005) Pollen recognition and rejection in different self-incompatibility systems. Recent Res Dev Plant Mol Biol 2:31–62

Sims TL, Ordanic M (2001) Identification of a S-ribonuclease-binding protein in *Petunia hybrida*. Plant Mol Biol 47:771–783

Singh A, Ai Y, Kao T-H (1991) Characterization of ribonuclease activity of three *S*-Allele-associated proteins of *Petunia inflata*. Plant Physiol 96:61–68

Sonneveld T, Tobutt KR, Vaughan SP, Robbins TP (2005) Loss of pollen-*S* function in two self-compatible selections of *Prunus avium* is associated with deletion/mutation of an *S*-haplotype-specific F-box gene. Plant Cell 17:37–51

Surbanovski N, Tobutt KR, Konstantinovic M, Maksimovic V, Sargent DJ, Stevanovic V, Bošković R (2007) Self-incompatibility of *Prunus tenella* and evidence that reproductively isolated species of *Prunus* have different *SFB* alleles coupled with an identical *S-RNase* allele. Plant J 50:723–734

Takayama S, Isogai A (2005) Self-incompatibility in plants. Annu Rev Plant Biol 56:467–489

Thompson RD, Uhrig H, Hermsen JGT, Salamini F, Kaufmann H (1991) Investigation of a self-compatible mutation in *Solanum tuberosum* clones inhibiting *S*-allele activity in pollen differentially. Mol Gen Genet 226:283–288

Tsai DS, Lee HS, Post LC, Kreiling KM, Kao T-H (1992) Sequence of an S-protein of *Lycopersicon peruvianum* and comparison with other solanaceous S-proteins. Sex Plant Reprod 5:256–263

Tsukamoto T, Ando T, Watanabe H, Marchesi E, Kao T-H (2005) Duplication of the *S*-locus F-box gene is associated with breakdown of pollen function in an *S*-haplotype identified in a natural population of self-incompatible *Petunia axillaris*. Plant Mol Biol 57:141–153

Tsukamoto T, Hauck NR, Tao R, Jiang N, Iezzoni AF (2006) Molecular characterization of three non-functional *S*-haplotypes in sour cherry (*Prunus cerasus*). Plant Mol Biol 62:371–383

Ushijima K, Sassa H, Tao R, Yamane H, Dandekar AM, Gradziel TM, Hirano H (1998) Cloning and characterization of cDNAs encoding S-RNases from almond (*Prunus dulcis*): Primary structural features and sequence diversity of the S-RNases in Rosaceae. Mol Gen Genet 260:261–268

Ushijima K, Sassa H, Dandekar AM, Gradziel TM, Tao R, Hirano H (2003) Structural and transcriptional analysis of the self-incompatibility locus of almond: Identification of a pollen-expressed F-box gene with haplotype-specific polymorphism. Plant Cell 15:771–781

Ushijima K, Yamane H, Watari A, Kakehi E, Ikeda K, Hauck NR, Iezzoni AF, Tao R (2004) The *S*-haplotype-specific F-box protein gene, *SFB*, is defective in self-compatible haplotypes of *Prunus avium* and *P. mume*. Plant J 39:573–586

Verica JA, McCubbin AG, Kao T-H (1998) Are the hypervariable regions of S RNases sufficient for allele-specific recognition of pollen? Plant Cell 10:314–317

Vilanova S, Badenes ML, Burgos L, Martinez-Calvo J, Llacer G, Romero C (2006) Self-compatibility of two apricot selections is associated with two pollen-part mutations of different nature. Plant Physiol 142:629–641

Wang H, Xue Y (2005) Subcellular localization of the *S*-locus F-box protein AhSLF-S_2 in pollen and pollen tubes of self-incompatible *Antirrhinum*. J Integr Plant Biol 47:76–83

Wang L, Dong L, Zhang Y, Zhang Y, Wu W, Deng X, Xue Y (2004a) Genome-wide analysis of *S*-Locus F-box-like genes in *Arabidopsis thaliana*. Plant Mol Biol 56:929–945

Wang Y, Wang X, Skirpan AL, Kao T-H (2003) S-RNase-mediated self-incompatibility. J Exp Bot 54:115–122

Wang Y, Tsukamoto T, Yi KW, Wang X, Huang S, McCubbin AG, Kao T-H (2004b) Chromosome walking in the *Petunia inflata* self-incompatibility *S*-locus and gene identification in an 881-kb contig containing S_2-*RNase*. Plant Mol Biol 54:727–742

Woodward JR, Bacic A, Jahnen W, Clarke AE (1989) N-linked glycan chains on S-allele-associated glycoproteins from *Nicotiana alata*. Plant Cell 1:511–514

Woodward JR, Craik D, Dell A, Khoo KH, Munro SLA, Clarke AE, Bacic A (1992) Structural analysis of the N-linked glycan chains from a stylar glycoprotein associated with expression of self-incompatibility in *Nicotiana alata*. Glycobiology 2:241–250

Xue Y, Carpenter R, Dickinson HG, Coen ES (1996) Origin of allelic diversity in *Antirrhinum* S-locus RNases. Plant Cell 8:805–814

Yamane H, Ikeda K, Ushijima K, Sassa H, Tao R (2003a) A pollen-expressed gene for a novel protein with an F-box motif that is very tightly linked to a gene for S-RNase in two species of cherry, *Prunus cerasus* and *P. avium*. Plant Cell Physiol 44:764–769

Yamane H, Ushijima K, Sassa H, Tao R (2003b) The use of the *S*-haplotype-specific F-box protein gene, *SFB*, as a molecular marker for *S*-haplotypes and self-compatibility in Japanese apricot (*Prunus mume*). Theor Appl Genet 107:1357–1361

Zhou J, Wang F, Ma W, Zhang Y, Han B, Xue Y (2003) Structural and transcriptional analysis of *S*-locus F-box (*SLF*) genes in *Antirrhinum*. Sex Plant Reprod 16:165–177

Zurek DM, Mou B, Beecher B, McClure BA (1997) Exchanging sequence domains between S-RNases from *Nicotiana alata* disrupts pollen recognition. Plant J 11:797–808

Chapter 10
Comparing Models for S-RNase-Based Self-Incompatibility

B. McClure

Abstract S-RNase-based self-incompatibility (SI) is known to occur in three families of flowering plants: Solanaceae, Rosaceae and Plantaginaceae. It is the most widely distributed SI system described so far. It is a single-locus gametophytic system. Gene pairs at the S-locus determine S-specificity on the pistil and the pollen sides of SI. This chapter describes these genes (*S-RNase* on the pistil side and *S-locus F-box* (*SLF/SFB*) on the pollen side), the non-S-locus genes that contribute to their functions and models for S-specific pollen rejection.

Abbreviations

120 K	120 kDa glycoprotein
AGP	Arabinogalactan protein
CUL	Cullin
ECM	ExtraCellular matrix
ERAD	Endoplasmic reticulum associated protein degradation
Natrxh	*N. alata thioredoxin h*
PELPIII	Pistil extensin-like like protein III
SBP	S-RNase binding protein
SC	Self-compatibility or self-compatible
SCF	Skp1-Cullin-F-box complex
SI	Self-incompatibility or self-incompatible
SLF/SFB	*S*-locus F-box
S-RNase	S-ribonuclease

B. McClure
Department of Biochemistry, Interdisciplinary Plant Group, Christopher S. Bond Life Sciences Center, 240a Christopher S. Bond Life Sciences Center, 1201 East Rollins Street, Columbia, Missouri 65211, USA, e-mail: mcclureb@missouri.edu

SSK1   SLF-interacting Skp1-like1
TTS   Transmitting tract specific glycoprotein
Y2H   Yeast two-hybrid

10.1 The Biology of S-RNase-Based SI

In S-RNase-based SI, S-specific recognition and rejection occur after pollen germination, as pollen tubes grow through the style (de Nettancourt 1977). This form of SI is found in relatives of petunia, tomato, potato, snapdragon, apple and cherry. Pollen rejection occurs gradually and may even be reversible. Compatible and incompatible pollen tubes are indistinguishable for the first few hours after pollination. Changes in incompatible pollen tube morphology and growth rate become visible as much as 6–8 h after pollination (Lind et al. 1994; Geitmann et al. 1995; Lush and Clarke 1997). Changes in morphology include heavy deposits of callose, uneven walls and swollen tips (de Nettancourt 1977). It is not clear whether incompatible pollen tubes die or whether their growth merely slows as time passes (Lush and Clarke 1997).

10.1.1 Genetic Breakdown

The behaviour of self-compatible (SC) tetraploids has been especially important for developing models of SI (de Nettancourt 2001). Breakdown occurs in pollen, but pistil function is normal. For example, SC *Petunia axillaris* tetraploids (e.g. $S_1S_1S_2S_2$) reject pollen from their diploid progenitor (i.e. S_1S_2), a response that indicates normal pistil function. SI breakdown in pollen was indicated since pollen from the tetraploid grows in both the SI diploid S_1S_2 and SC tetraploid $S_1S_1S_2S_2$ pistils (Stout and Chandler 1942). Significantly, SI breakdown occurs only in diploid heteroallelic pollen (e.g. diploid S_1S_2 pollen), an effect known as the competitive interaction (de Nettancourt 2001). The competitive interaction is consistent with an inhibitory function for pollen S. For example, if the pollen S_1 product inhibits all S-RNases except S_1-RNase and if pollen S_2 inhibits all but S_2-RNase, then cross-protection in diploid S_1S_2 pollen inhibits all S-RNases. Alternatively, the competitive effect could also be explained if pollen S functions as a multimer, but heteroallelic multimers are incapable of recognising self-S-RNase. Both alternatives are viable, but in the cross-protection scenario recognition occurs normally, while under the latter scenario, SI breakdown is caused by a breakdown in recognition.

SI breakdown in pollen also can be induced by mutagenesis. Older studies showed that breakdown on the pollen side is often associated with centric fragments that also effectively render pollen heteroallelic (de Nettancourt 2001). More recently, Golz et al. (1999, 2000) induced pollen-part mutations in *N. alata* and analysed them using molecular markers. The results were consistent with interpretations of older experiments: All the mutations could be explained as duplications or translocations of the S-locus that generate heteroallelic pollen (Golz et al. 1999; 2001).

10.2 S-RNase and *S*-locus F-box Proteins

In the accompanying chapter (see Chap. 9, this volume), Zhang and Xue provide an overview of the identification of S-RNase and the *S*-locus F-box proteins SLF and SFB; see also Chap. 5, with respect to S-RNase genealogies, and Chap. 3, with respect to models for evolutionary modifications of S-RNases. The focus here will be on issues of special interest for current models of SI.

10.2.1 S-RNase Structure and Specificity

S-RNase protein structure is consistent with its having a dual role as both a cytotoxin and a recognition protein. Sequence comparisons among S-RNases revealed a remarkable degree of polymorphism, as was expected from the early observation that allelic *S-RNase* genes rarely cross-hybridise (Anderson et al. 1989). The typical S-RNase is 200 amino acids long. About 40 residues occur in five conserved regions (C1 to C5; Ioerger et al. 1991). Regions C2 and C3 contain histidine residues thought to participate in catalysis. The crystal structure of S_{F11}-RNase from *N. alata* (Ida et al. 2001) shows that C1, C2 and C5 comprise three strands of anti-parallel β-sheet in the core of the protein. The sheet forms the floor of a shallow cleft with C3 forming a helix situated to one side. The histidine-side chains in C2 and C3 protrude into the active site cleft. Thus C1, C2, C3 and C5 show an obvious connection to the ribonuclease function of S-RNase. C4, on the other hand, is more enigmatic: it is less conserved. S-RNases from the Rosaceae display a C4 variant that is sufficiently different from those in the Solanaceae that a separate designation is used (RC4; Ishimizu et al. 1998). In the S_{F11}-RNase structure, C4 (KDKFDLL) contributes hydrophobic residues interacting with the backside of the β-sheet. The lysine residues in C4 are of special interest because they have a potential role in ubiquitination, a key process in a favoured model for SI (Sect. 4.4). However, Qin et al. (2005) showed that these lysines are not essential for either *S*-specific pollen rejection or compatibility.

Residues outside the conserved regions vary between allelic S-RNases and confer a unique *S*-specificity to each S-RNase. The degree of variability at each site is not constant throughout the sequence, and attempts have been made to identify regions that determine *S*-specificity (Ioerger et al. 1991; Ishimizu et al. 1998; Igic et al. 2007). Ioerger (1991) analysed solanaceous S-RNase sequences using a normed variability index and designated two regions between C2 and C3 as being hypervariable regions: HVA and HVB. This nomenclature has been widely adopted.

S-specificity also has been addressed experimentally. Zurek et al. (1997) used in vitro mutagenesis to create a series of nine chimeric S-RNases exchanging single domains as well as combinations of two, three or four contiguous domains from *N. alata* S_{A2}- and S_{C10}-RNases. All the chimeras were active ribonucleases, but none were recognised as self-S-RNases by either S_{A2}- or S_{C10}-pollen (Zurek et al. 1997). When Kao and McCubbin (1996) exchanged the hypervariable domains in *P. inflata*

S_3-RNase, they found that the chimera functioned as a non-self-S-RNase. Matton et al. (1997) exchanged residues between *S. chacoense* S_{11}- and S_{13}-RNases, which differ at only ten amino acid residues. These authors found that converting just four residues in the hypervariable region of S_{11}-RNase to the corresponding residues in S_{13}-RNase changed the protein sufficiently that it was recognised as S_{13}- instead of S_{11}-RNase (Matton et al. 1997; see also Chap. 9, this volume). Thus, the hypervariable regions are important and were sufficient to distinguish these two closely related S-RNases, but the domain swaps in *N. alata* and *P. inflata* showed that S-specific recognition information is spread throughout the S-RNase sequence.

Analysis of the sequence and gene structure of plant T2-type RNases reveals the broad evolutionary history of S-RNase (Igic and Kohn 2001). All S-RNases fall into a single clade, indicating that they share a common ancestor. This finding is remarkable given that *S-RNase* sequence diversity is extremely high, which makes recovering their evolutionary history challenging. Igic and Kohn (2001) note that the last common ancestor of the three families known to possess S-RNase-based SI was also the ancestor of about 75% of all extant angiosperms. Since S-RNases are monophyletic, the authors infer that this ancestor was SI, raising the possibility that the system may be more widely distributed than currently appreciated; (see also Chap. 5, this volume). This interpretation is attractive, but the analysis only reveals the history of the *S-RNase* gene (or the gene it is derived from), i.e. just the pistil determinant, and not necessarily the entire S-RNase-based SI system. Moreover, analysis of the *S*-locus F-box genes raises some intriguing questions (see Chap. 5, this volume). Analysis of other genes implicated in SI (e.g. *HT-B, 120K*) should further illuminate the question of whether S-RNase-based SI is monophyletic, when taken as a whole.

10.2.1.1 S-RNase Uptake

S-RNase uptake into pollen tubes has been studied using immunolocalisation and biochemical analyses. Gray et al. (1991) used ^{3}H-S_2-RNase from *N. alata* to follow its uptake and fate in pollen tubes growing in vitro. Autoradiography and EM-level immunolocalisation showed that S-RNase enters pollen of any *S*-haplotype. Moreover, ^{3}H-S-RNase recovered from pollen tubes co-migrated with untreated S-RNase; there was no evidence of smearing or degradation in pollen tubes. Luu et al. (2000) and Goldraij et al. (2006) used S-RNase-specific antibodies to show that S-RNase enters both compatible and incompatible pollen tubes in *S. chacoense* and *N. alata,* respectively. Both studies showed S-RNase staining inside pollen tubes that was similar to staining in the transmitting tract, suggesting that large amounts of S-RNase enter pollen tubes. However, the authors differ about whether S-RNase is associated with the pollen endomembrane system. Goldraij et al. (2006) focused on regions of the pollen tube with large well-developed vacuoles and used vacuolar pyrophosphatase and aleurain as vacuole markers; whereas, Luu et al. (2000) focused on regions near the tip and did not use markers for the endomembrane system or methods that preserve membranes. Together, these three studies clearly show that large amounts of S-RNase enter pollen tubes and provide no evidence for its degradation. Thus, in compatible pollen tubes, S-RNase's cytotoxic activity must be inhibited by regulating RNase activity or by controlling its access to its RNA substrate in the cytosol.

10.2.2 S-locus F-Box Genes

The evidence that *SLF* and *SFB* are pollen *S* genes is strong (Lai et al. 2002; Entani et al. 2003; Sijacic et al. 2004; Qiao et al. 2004b; Ushijima et al. 2004; Wheeler and Newbigin 2007). The genes are physically very close to *S-RNase*, they are expressed appropriately, mutations cause SC and the transformation results obtained by Sijacic et al. (2004) appear compelling; see also Chaps. 5 and 9, this volume. Nevertheless, some questions remain. The sequence polymorphism of *SLF* genes is surprisingly low. Protein-level alignment of $AhSLF_1$, $AhSLF_2$, $AhSLF_4$ and $AhSLF_5$ shows that only 12 of 376 residues are variable. Many of the predicted substitutions are conservative, and there are no positions that differ in every sequence (Zhou et al. 2003). Likewise, alignment of three *P. inflata* SLF proteins showed only two unique positions (Sijacic et al. 2004). Even when sequences from *P. axillaris* are added to the alignment (Tsukamoto et al. 2005), it is the sequence similarity that is most noteworthy. The striking sequence divergence characteristic of S-RNases is not observed in solanaceous SLF proteins. It is not clear why this is. *SFB* genes in Rosaceae are much more variable, however. In a comparison of four *P. dulcis* SFB proteins, Ushijima et al. (2004) found 68–76% sequence identity, which is in the range observed for *Prunus* S-RNases. Like S-RNases, the *Prunus* SFB genes show *S*-haplotype-specific hybridisation patterns.

The multiplicity of *S*-locus F-box genes is also striking. It appears to be a widespread feature S-RNase-based SI. At least two genes encoding F-box proteins are linked to S-RNase in *Petunia* (Wang et al. 2003). Each of the four *S*-haplotypes examined in *Antirrhinum* contain two or three F-box genes (Zhou et al. 2003). Wheeler and Newbigin (2007) reported ten F-box protein genes linked to the *S*-locus in *Nicotiana*, five of which could not be excluded as pollen *S* candidates. The situation in Rosaceae is more complex. Entani et al. (2003) reported four F-box genes near *S-RNase* in *P. mume*. In *P. dulcis*, both *SFB* genes and one or more *SLF-like* genes are found near *S-RNase* (Ushijima et al. 2004). In *Malus* and *Pyrus*, there are multiple *SFB* genes that show signs of positive selection (Sassa et al. 2007). Wheeler and Newbigin (2007) note that, unlike *S-RNase* genes, *SLF* and *SFB* genes do not form a single clade. Also, F-box genes that have no role in SI cluster with those that are thought to function in SI. Furthermore, the markedly lower sequence polymorphism of *SLF* genes is not suggestive of the type of long history of balancing selection experienced by *S-RNase* genes (see Chap. 5, this volume).

Taken together, these observations paint a confusing picture. The traditional view would suggest that the pollen and the pistil *S*-specificity determinants evolve as stable gene pairs. Yet, *S-RNase* does not seem to share a common evolutionary history with the adjacent F-box genes (see Chap. 5, this volume). This stable-gene-pair expectation is grounded in a model in which the *S*-specificity determinants interact in a simple lock-and-key manner. Either this model is incorrect, or we do not really understand the *S*-specificity determinants. Sassa et al. (2007) and Wheeler and Newbigin (2007) discuss the possibility that *S*-specificity may be determined on the pollen side by interactions with multiple proteins. This alternative is consistent with the genetics of SI if the proposed *S*-specificity determinants do not recombine, but it is awkward biochemically. Since the biochemical mechanism of S-RNase-based SI is not yet known, there still may be some surprising discoveries concerning the details of *S*-specificity.

10.3 Non-*S*-Specific Genes

Non-*S*-specific genes are required for S-RNase-based SI (see Chap. 9, this volume). By definition, the *S*-locus encodes genes that determine *S*-specificity. This genetic simplicity implies that a limited number of factors determine specificity itself. It does not imply a simple, single-step pollen rejection mechanism. Non-*S*-specific genes, or modifier genes, fulfill other functions that act upstream or downstream of the specificity determining step. There are many examples in the literature providing evidence for these modifier genes (Anderson and de Winton 1931; East 1932; Mather 1943; Martin 1961, 1968; Ai et al. 1991; Murfett et al. 1996; Tsukamoto et al. 2003a). Two genes, *HT-B* and *120K*, have definitely been implicated in S-RNase-based SI (McClure et al. 1999; Hancock et al. 2005); others have been identified genetically or suggested as modifier gene candidates (McClure et al. 2000; Juarez-Diaz et al. 2006).

Categorising the functions of modifier genes is helpful for conceptualising their contributions to SI. *Group 1* modifiers directly affect the expression of the *S*-specificity genes. Tsukamoto et al. (1999) reported such a modifier in an Uruguayan population of *Petunia axillaris*. The effect of Group 1 modifiers is easy to understand: if the *S*-specificity determinants are not properly expressed, SI breaks down.

Group 2 modifiers have a specific effect on SI, but no wider role in pollination. For example, SC *P. hybrida* cv Strawberry Daddy expresses S_x- and S_o-RNases (Ai et al. 1991). The latter proved to be defective, but SI plants expressing S_x-RNase could be recovered after crossing to SI *P. inflata*. Thus, SC Strawberry Daddy has both a defective *S*-haplotype and a recessive defect in a Group 2 modifier. A *Nicotiana* modifier called *4936-factor* displays a similar phenotype; S-RNase is expressed normally, but pollen with any *S*-haplotype is compatible (McClure et al. 2000; Goldraij et al. 2006). Many of the modifiers in the older literature probably fall into this group, but few have been studied in detail. Two Group 2 modifiers, *120K* and *HT-B* (McClure et al. 1999; Hancock et al. 2005), have been characterised and will be discussed later.

10.3.1 HT Genes

Inter-specific pollination studies provided compelling evidence of a Group 2 factor requirement in *Nicotiana* SI and some types of interspecific incompatibility (Murfett et al. 1996). Differential cDNA cloning identified a candidate gene that was given a trivial name (i.e. *HT*, clone H-Top band) because its sequence did not suggest a biochemical function. An antisense experiment tested whether HT is required for SI in *Nicotiana*. Progeny of five independent *HT*-antisense transformed plants showed no detectable HT-protein and also lost *S*-specific pollen rejection (McClure et al. 1999). Since S-RNase levels are not significantly affected, *HT* is classed as a Group 2 modifier.

An HT gene family is now recognised in Solanaceae. The 78-residue mature *Nicotiana* HT protein includes a 20-residue ND domain near the C-terminus comprised solely of asparagine and aspartic acid (McClure et al. 1999). Two cysteine motifs, CXXCXC and CXXXCC, flank the ND-domain, the latter sequence forming the C-terminus. In *Solanum*, two or more *HT* genes are present, *HT-A* and *HT-B* (Kondo et al. 2002a; O'Brien et al. 2002). *HT-B* genes contain a sequence similar to PSXPLLX about seven residues from the N-terminus; the original *Nicotiana* gene contains the sequence PSISLLE and is now regarded as an *HT-B* gene. Kondo et al. (2002b) discovered a close correlation between *HT-B* expression and SI in the tomato clade (i.e. formerly *Lycopersicon*, now incorporated into *Solanum*) and suggested that *HT-B* mutations may have caused the shift from SI to SC in the tomato clade. O'Brien et al. (2002) tested for a function in *S. chacoense* SI using RNAi constructs. An *HT-B* RNAi construct was quite specific, mainly affecting *HT-B* transcript levels, with smaller changes in *HT-A*. Two plants were recovered that showed little or no *HT-B* expression, and as in *Nicotiana*, had lost SI. The antisense and RNAi results in *Nicotiana* and *Solanum* suggest that even very low levels of HT-B protein are sufficient for SI because an effect is seen only when expression is suppressed to the limit of detection.

Petunia and *Nicotiana* also express *HT*-family genes that lack an ND-domain. *P. inflata* expresses two *HT*-like transcripts, *HTL-A* and *HTL-B*, that are produced from a single gene by alternative splicing (Sassa and Hirano 2006). The predicted polypeptides are about 90 amino acids long and contain a secretion signal strikingly similar to other HT-family members. Both contain an HT-B-like PXXPLLX sequence and cysteine motifs similar to HT-A and HT-B. Plants with undetectable HTL-A and HTL-B transcript levels were produced by RNAi but showed no change in SI, leading to the conclusion that they are not essential for SI (Sassa and Hirano 2006). Similar genes are expressed in several *Nicotiana* species (K. Kondo, unpublished).

10.3.1.1 HT-B Degradation

The results just described provide good evidence that HT-B has a role in SI. Experiments in *Nicotiana* show that HT-B is degraded after pollination, and that degradation occurs to a much greater extent after compatible pollinations compared to incompatible pollinations (Goldraij et al. 2006). Immunolocalisation experiments were performed to follow HT-B uptake and localisation in pollen tubes, and nearly 500 pollen tubes were scored for HT-B staining intensity. Ninety percent of compatible pollen tubes had little or no detectable HT-B, whereas 83% of incompatible pollen tubes had high levels of detectable HT-B (Goldraij et al. 2006). By following HT-B protein in style extracts after compatible and incompatible pollinations in *N. alata*, Goldraij et al. (2006) found that HT-B levels dropped by 50% after incompatible pollinations and by 75–97% after compatible pollinations. These results suggest that compatible pollen tubes either degrade HT-B or convert it into a form that is not detected by the HT-B antibody. This is significant because the antisense and RNAi experiments show that SI breaks down when HT-B is absent. Although it is not known how the stability of HT-B is differentially controlled in compatible vs. incompatible pollen tubes, the control mechanism clearly must respond to recognition between the pollen and pistil *S*-specificity determinants in some fashion, or there would be no difference between compatible and incompatible pollination.

10.3.2 S-RNase Binding Proteins in the Transmitting Tract Extracellular Matrix

S-RNase binds to a group of arabinogalactan proteins (AGPs) that are abundant in the transmitting tract extracellular matrix (ECM). Binding proteins were identified using immobilised S-RNase binding assays (Cruz-Garcia et al. 2003). An 11 kDa copper-binding protein similar to chemocyanin and S-RNase were among the bound proteins, but AGPs were the most prominent proteins recovered. The AGPs include at least three abundant components of the transmitting tract ECM (Cruz-Garcia et al. 2003): pistil extensin-like protein III (PELPIII), transmitting tract specific glycoprotein (TTS) and the 120 kDa glycoprotein (*120K*). All three AGPs contain a proline-rich region and a cysteine-rich C-terminal domain (CTD), and all three interact with pollen tubes. Since PELPIII, TTS and *120K* interact with pollen tubes in both SC and SI species, they likely have a role in supporting compatible pollen tube growth. Binding of S-RNase to the AGPs therefore suggests some crosstalk between SI and other functions of the pistil (McClure et al. 2000; Cruz-Garcia et al. 2003; 2005).

10.3.2.1 *120K*

120K was first characterised as an abundant component of the *N. alata* transmitting tract ECM (Lind et al. 1994). It is especially interesting because it is taken up from the ECM into the pollen tube's cytoplasm (Lind et al. 1996; Goldraij et al. 2006). It is an extremely basic protein with both arabinogalactan and extensin-like tetra-arabinosyl carbohydrate. Analysis of *120K*-like proteins in a variety of SC and SI

Nicotiana species has shown that *120K*-like proteins are highly variable, with larger forms associated with SI (Hancock et al. 2005). A *120K* RNAi construct was used to test for a role in *S*-specific pollen rejection (Hancock et al. 2005). Progeny from four independent transformants showed reduced *120K* expression while retaining expression of NaPELPIII and NaTTS. Plants showing *120K* levels suppressed below the limit of detection failed to show a normal *S*-specific pollen rejection. This finding supports a specific role for *120K* in SI, since these plants expressed near normal levels of NaPELPIII and NaTTS.

120K is taken up from the transmitting tract ECM by pollen tubes and (at least a portion of it) has a similar fate as S-RNase. Using an antibody against deglycosylated *120K* in EM-level immunolocalisation experiments, Lind et al. (1996) observed diffuse labelling in compatible pollen tubes, near the tip region. In older regions containing lipid droplets, labelling was observed in vesicular structures. Goldraij et al. (2006) used confocal microscopy for immunolocalisation experiments with an anti-peptide antibody specific for the C-terminal domain of *120K*. S-RNase and the callosic pollen tube wall were imaged simultaneously. These experiments showed that the C-terminal domain *120K* antibody labels a vacuolar membrane that also contains S-RNase. Since S-RNase binds to *120K* (Cruz-Garcia et al. 2005), it is possible that movement of the two proteins is coordinated. However, S-RNase transport to the vacuole was normal in *120K* RNAi plants. The essential role of *120K* in SI is therefore at a later step, not during transport to the pollen tube vacuole, but probably afterwards. However, as is true when other pistil-side SI factors are missing, S-RNase remains in the vacuole when *120K* is absent (Goldraij et al. 2006). Thus, it is possible that vacuolar localisation of both *120K* and S-RNase is required for incompatibility.

10.3.2.2 Thioredoxin h

N. alata expresses another S-RNase binding protein—a novel thioredoxin *h* that is secreted into the transmitting tract ECM. Thioredoxins act as reducing agents and it has been shown that thioredoxin *h* is capable of reducing S-RNase. Juarez-Diaz et al. (2006) identified *Natrxh* (*N. alata thioredoxin h*) using a cDNA screen similar to the one used to identify *HT-B*. *Natrxh* belongs to a subgroup of thioredoxin *h* proteins, with an N-terminal extension of unknown function. The Natrxh protein accumulates in the transmitting tract ECM even though it does not contain a canonical secretion signal. Natrxh-S-RNase binding is resistant to detergent and high salt, indicating a tight association. Natrxh reduction of S-RNase exposes one or more thiols (Juarez-Diaz et al. 2006). Thus, it is possible that Natrxh alters the physical form or mobility of S-RNase in the ECM, but its role in SI has not been directly tested using antisense or RNAi.

10.3.3 Non-S-Specific Factors in Pollen

Efforts to identify non-*S*-specific factors required on the pollen side have used SLF proteins as baits for interacting proteins. Searching for components of a Skp-cullin-F-box (SCF) E3 ubiquitin ligase complex was a natural point of departure. A canonical SCF complex consists of an F-box protein and Skp1, cullin1, and Rbx proteins. The F-box protein binds specific proteins destined for ubiquitination and degradation (Vierstra 2003). The conserved F-box domain provides the interface for binding to Skp1, which in turn, binds a cullin1 that acts as a scaffold.

SLF binding partners have been investigated in *P. inflata* and *Antirrhinum* (Qiao et al. 2004b; Hua and Kao 2006; Huang et al. 2006; see also Chap. 9, this volume). In Y2H assays, $PiSLF_2$ does not bind to the three major Skp1 proteins expressed in *P. inflata* pollen nor to the representatives of the seven major groups of *Arabidopsis* Skp1 (Hua and Kao 2006). However, a previously identified S-RNase binding protein gene, *SBP1*, was recovered repeatedly in a Y2H cDNA screen, and the $PiSLF_2$–PiSBP1 interaction was verified with pull-down assays (Hua and Kao 2006). SBP1 is a RING domain protein identified in *S. chacoense* and *P. hybrida* by virtue of its binding to S-RNase (Sims and Ordanic 2001; O'Brien et al. 2004). It also has been identified as a weak CONSTANS-Like1 interacting protein in tomato (Ben-Naim et al. 2006). SBP1 is likely to have a general function, given it is expressed in most tissues (Sims and Ordanic 2001; O'Brien et al. 2004). PiSBP1 was further shown to bind specifically to one of the two cullin-1 proteins (Hua and Kao 2006). Thus, the results with $PiSLF_2$ are not consistent with formation of a canonical SCF complex. Hua and Kao (2006) proposed that two potential S-RNase E3 ligase complexes are active in *P. inflata* pollen: an *S*-specific complex (S-RNase-SLF-SBP1-CUL1-G) and a non-*S*-specific complex (S-RNase-SBP1-CUL1-G). SBP1 is proposed to fulfill the roles of both Skp1 and Rbx1 in the former complex and, presumably, of the F-box protein, Skp1, and Rbx1 in the latter complex. It is somewhat surprising that PiSLF does not appear to interact with any of the Skp1 proteins tested, as this is thought to be the function of the F-box.

Antirrhinum SLF proteins, in contrast, do appear to interact with Skp1-like proteins that, in turn, interact with a CUL1-like partner. AhSSK1 (*Antirrhinum hispanicum* <u>S</u>LF-interacting <u>S</u>kp1-like1) was identified in pollen using $AhSLF_2$ as a Y2H bait. AhSSK1 has a seven residue C-terminal tail that distinguishes it from other plant Skp1-like proteins. It interacts with $AhSLF_2$ and $AhSLF_5$ but not $AhSLF_1$ or $AhSLF_4$ (Huang et al. 2006). This interaction pattern is striking since the F-box domains of the latter two proteins differ at only three or four residues from $AhSLF_2$ and $AhSLF_5$ (Zhou et al. 2003). AhSSK1 also interacts with an 85 kDa CUL1-like protein. Thus, the *Antirrhinum* results are consistent with formation of an SSK1-CUL1-SLF complex. Qiao et al. (2004b) presented results showing that S-RNase binds to AhSLF proteins and suggested that an S-RNase-SCF complex is directly involved in SI (see Chap. 9, this volume).

10.4 Comparing Models for S-RNase-Based SI

10.4.1 How do Compatible Pollen Tubes Resist S-RNase Cytotoxicity?

The discovery that the *S*-specificity determinant in the pistil is a ribonuclease dramatically shifted thinking about the SI mechanism. Before this discovery, the challenge was to rationalise the rejection of incompatible pollen tubes. With the finding that the pistil *S*-determinant was an RNase, a new cytotoxic hypothesis was proposed that explained this aspect of SI, and the challenge then became understanding how compatible pollen tubes evade destruction by the S-RNases.

10.4.1.1 Model Types

Two types of models were proposed to account for resistance to S-RNase cytotoxicity in compatible pollen tubes: the receptor model and the inhibitor model (Kao and Tsukamoto 2004). The receptor model proposed that pollen *S* allows self-S-RNase into pollen tubes, but not non-self-S-RNase. As research progressed, it became clear that S-RNase enters both compatible and incompatible pollen tubes. Inhibitor models proposed that both self- and non-self-S-RNases gain entry into both compatible and incompatible pollen tubes and that compatible pollen tubes evade rejection by inhibiting non-self-S-RNase. This makes sense genetically; selection will maintain a protein that facilitates growth by inhibiting a toxin. Biochemically, pollen *S* could inhibit S-RNase by binding to it to inactivate its cytotoxic activity, by sequestering it from its substrate, or by degrading it. All three mechanisms have been proposed (see Chap. 9, this volume).

10.4.2 Is There a Separate Mechanism in the Rosaceae?

Citing three lines of evidence, Hauck et al. (2006) suggested that pollen *S* functions differently in *Prunus* than in the Solanaceae and Plantaginaceae. Unlike the behaviour just described (Sect. 10.3.1.2), tetraploid sour cherry remains SI unless at least two defective *S*-haplotypes are present (Hauck et al. 2006). Thus, the competitive effect does not occur in this species. Also, while the only known pollen-side mutants in the Solanaceae are duplications (or potential duplications) that include pollen *S* genes, truncated SFB proteins and radiation-induced deletions that include the whole *SFB* gene have been found in *Prunus* (Tsukamoto et al. 2003b; Ushijima et al. 2004; Sonneveld et al. 2005). Finally, sequence variability is higher for *SFB* genes than the corresponding *SLF* genes. It may also be significant that the pattern of positively selected sites in rosaceous *S-RNase* genes is different from those in other families (Vieira et al. 2007). There are clearly important differences between SI in *Prunus* and the systems in the Solanaceae and Plantaginaceae (see Chap. 5 and 9, this volume). Whether these differences imply a different mechanism cannot be fully resolved until the mechanism is better defined in all systems. The broad outline of the systems is, however, the same (i.e. specificity defined by S-RNase on one side and an F-box protein on the other). Therefore, where possible, it makes sense to attribute differences between the systems to differences in the specific properties of the components (S-RNase and F-box proteins) rather than to fundamentally different mechanisms.

10.4.3 Inhibiting S-RNase Enzyme Activity

Enzyme inhibitor models draw analogy to well-characterised RNase inhibitor proteins such as barstar (Luu et al. 2001). Although pollen S could, conceivably, bind and inactivate non-self-S-RNase similarly, this effect would require each allelic pollen S protein to 'specifically' bind numerous non-self-S-RNases, while failing to bind a single self-S-RNase. Luu et al. (2001) proposed a modified inhibitor model in which a general inhibitor acts on all S-RNases. In incompatible pollen, pollen S prevents the general inhibitor's action by binding specifically to self-S-RNase, thereby permitting pollen rejection. The general inhibitor is depicted as a peptide similar to barstar, and pollen S is depicted as a separate polypeptide that binds S-RNase. The form of this model (i.e. non-S-specific inhibition system whose activity is prevented by pollen S) appears in the other current models for SI, albeit with entirely different details. Luu et al. (2001) proposed their model to account for the behaviour of diploid heteroallelic pollen (i.e. $S_{11}S_{13}$ pollen from a $S_{11}S_{11}S_{13}S_{13}$ tetraploid) on pistils expressing a 'dual-specific' S_{apb}-RNase. The dual-specific S_{apb}-RNase causes rejection of both S_{11}- and S_{13}-pollen and was created by mutating three residues in S_{11}-RNase to the corresponding residues in S_{13}-RNase (Matton et al. 1999). As described above, the diploid heteroallelic pollen was compatible on an $S_{11}S_{13}$ pistil but, intriguingly, incompatible on pistils expressing S_{apb}-RNase. They hypothesised that incompatibility occurs when pollen S forms tetramers that bind to self-S-RNase, thereby preventing its inhibition and that recognition breaks down in diploid heteroallelic pollen. Luu et al. (2001) suggest that heterotetramers are unable to bind either of the two cognate S-RNases, leaving them available for inhibition, and thus causing SC. Although other explanations are possible, Luu's results support the conclusion that the pollen S protein functions in a multimeric form.

10.4.4 S-RNase Degradation

As evidence accumulated that F-box proteins were the determinants of S-specificity, models emerged that explained compatibility as a result of degradation of non-self-S-RNase (Entani et al. 2003; Qiao et al. 2004a,b; Sijacic et al. 2004; Ushijima et al. 2004). In these models, the inhibitory action of pollen S is a consequence of removing non-self-S-RNase from the pollen tube. Self-S-RNase is not degraded and is therefore able to cause rejection by degrading pollen tube RNA.

Ushijima et al. (2004) presented an early example of an S-RNase degradation model. In this model, SFB is part of an SCF E3 ligase complex (SCF^{SFB}). S-RNase and SFB each display a non-S-specific and an S-specific binding site. The S-RNase-SFB interaction through the non-S-specific cite corresponds to a non-self-interaction (e.g. S_1-RNase-SFB$_2$), while the S-specific interaction is a self-interaction through parts of the molecules that determine S-specificity. If SCF^{SFB} binds S-RNase at the non-S-specific site, then S-RNase becomes ubiquitinated and subsequently degraded. Binding at the S-specific site does not lead to S-RNase degradation. Instead, its cytotoxicity is expressed, causing pollen rejection. Huang et al. (2006) describe an SCF-like complex in *Antirrhinum* that could function as required under this kind of model.

S-RNase degradation models account for many of the central features of S-RNase-based SI. They account for the competitive interaction effect in tetraploids through cross-protection. For example, in S_1S_2 diploid pollen, SFB$_1$ eliminates all S-RNases except S_1-RNase, and SFB$_2$ eliminates all S-RNases except S_2-RNase. If the F-box proteins function as multimers, as has been described in some instances (Suzuki et al. 2000), then the competitive effect also could be explained by multimers that cannot recognise self-S-RNase (e.g. the S-specific site is buried in an SFB$_1$–SFB$_2$ multimer). Hauck et al. (2006) reported that the tetraploid *Prunus cerasus* accessions are SI unless at least two non-functional S-haplotypes are present. This observation may, as some have suggested (Hauck et al. 2006; Hua and Kao 2006), reflect a fundamentally different SI mechanism, but it could also be explained by subtly different properties of the F-box proteins present from different species. For example, SFB from *P. cerasus* may be less likely to form hetero-multimers than F-box proteins from other species. Given the variability of S-locus proteins, different properties should be expected.

Hua and Kao (2006) proposed a modified S-RNase degradation model that took account of the interactions they observed between SLF, SBP1, CUL1 and S-RNase (Fig. 10.1a). The form of this model is the same as the model proposed by Luu et al. (2001) in that non-S-specific inhibition (light lines, Fig. 10.1a) is coupled to an S-specific interaction that prevents inhibition of self-S-RNase or enhances degradation of non-self-S-RNase (dark lines, Fig. 10.1a). Hua and Kao's model differs from Luu's, in that it postulates ubiquitination and S-RNase degradation. Additionally, in Hua and Kao's model, self-S-RNase escapes inhibition because it does not bind efficiently to SLF. In pull-down assays, Hua and Kao (2006) observed more binding between SLF and non-self-S-RNase than between self proteins. They propose that S-RNase is ubiquitinated by a non-canonical SLF-SBP1-CUL1 E3 ubiquitin ligase complex and also by a separate SBP1-CUL1 complex. The SBP1-CUL1 E3 complex acts as a general inhibitor by tagging all S-RNases for degradation. In vitro results showing non-S-specific ubiquitination and degradation by the 26S proteasome are consistent with this proposal. Since SBP1 and CUL1 are widely expressed (Sims and Ordanic 2001; O'Brien et al. 2004), this mechanism is entirely independent of SI and reflects an aspect of physiology common to all cell types. Hua and Kao (2006) propose that an SLF-SBP1-CUL1 complex actively ubiquiti-

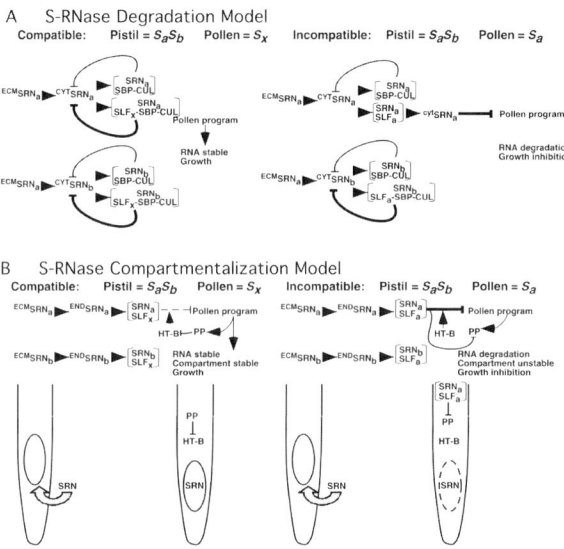

Fig. 10.1 Current models for S-RNase-based SI. (*Left*) A fully compatible pollination $S_aS_b \times S_xS_x$. (*Right*) Incompatible pollination $S_aS_b \times S_aS_a$. S-RNase location is shown by a superscript: ECM, extracellular matrix; CYT, cytosol; END, endomembrane. Complexes are shown in brackets, but they may be transient. (**a**) S-RNase degradation model based on Hau and Kao (2006). S-RNase is inhibited by a non-*S*-specific pathway (lighter lines) initiated by binding to an SBP1-CUL1 complex. The dominant S-RNase degradation pathway is provided by a second complex that includes SLF, SBP1 and CUL1, which preferentially binds non-self-S-RNase and initiates its degradation (heavy lines). Self-S-RNase does not stably enter this complex, leaving it free in the cytosol where it prevents expression of the pollen gene expression program. (**b**) S-RNase compartmentalisation model based on Goldraij et al. (2006). S-RNase and HT-B enter the endomembrane (END) system by endocytosis. Most S-RNase moves to a vacuole by retrograde transport, but a portion moves to the ER where it can exit to the cytosol. HT-B is required for S-RNase to exert its cytotoxic activity, but it is degraded as part of the pollen gene expression program by an unknown protein, PP in a compatible interaction (*left*). The self-S-RNase-SLF interaction inhibits the pollen protein responsible for HT-B degradation. Thus, the self-interaction suppresses HT-B degradation, and S-RNase cytotoxic activity prevents expression of the pollen gene expression program (*right*). The model is shown diagrammatically below. For clarity, only one S-RNase is shown, but both S-RNases in an S-heterozygote are taken up by pollen tubes

nates non-self-S-RNase because of its preferential binding to SLF, compared with self-S-RNase.

Some observations are not easily accommodated by S-RNase degradation models. Preferential degradation of non-self-S-RNase is a clear prediction of these models, but it is not observed in vitro or *in planta* (Goldraij et al. 2006; Hua and Kao 2006). The most recent S-RNase degradation model proposes that only a very small amount of non-self-S-RNase is degraded, an amount that cannot be detected against the massive amount of S-RNase present in the pistil (Hua and Kao 2006). This cannot be excluded, but neither can it be experimentally verified. Also, in so far as they postulate S-RNase-SFB interactions as the sole basis for resistance to S-RNase in compatible pollen tubes, S-RNase degradation models predict that pollen *S* null mutants cannot be tolerated. Indeed, Golz et al. (1999, 2001) found no evidence for pollen *S* deletions. Sonneveld et al. (2005), however, reported a pollen *S* deletion mutant in *Prunus avium*. A final problem is posed by specific recognition of non-self-S-RNases. In Ushijima's model, this function occurs through a non-*S*-specific recognition site; in Hua and Kao's model, it occurs by a preferential binding of non-self-S-RNase, compared to self-S-RNase (Ushijima et al. 2004; Hua and Kao 2006). As discussed in Sect. 10.2.1, conserved regions C1, C2 and C5 form the three β-strands at the core of the protein (Ida et al. 2001). The most conserved residues in regions C3 and C4 also project to the protein interior. All residues outside the conserved regions are subject to variation; there is no readily identifiable exposed surface that could act as a conserved non-*S*-specific binding site. Moreover, domain swaps between allelic S-RNases that generate novel chimeric S-RNases function as non-self-S-RNases (Zurek et al. 1997). Therefore, if there is a 'non-self' recognition surface on pollen *S* (Qiao et al. 2004b; Hua and Kao 2006), it must be able to recognise infinite, novel S-RNases.

10.4.5 S-RNase Compartmentalisation

The compartmentalisation model for S-RNase-based SI consists of three steps: (1) nonspecific uptake of S-RNase and other components of the transmitting tract ECM into the pollen tube endomembrane system; (2) an *S*-specific recognition step occurring in the pollen tube cytoplasm; and (3) a self-reinforcing cytotoxic step (Goldraij et al. 2006; McClure 2006). In this, model compatibility results from sequestration of S-RNase from the cytosol and incompatibility results from release of S-RNase. Degradation of HT-B also plays a role in this model.

Traffic of S-RNase from the ECM through the pollen tube endomembrane system to the pollen tube cytosol may be similar to other cytotoxins such as ricin (McClure 2006). Like S-RNase, ricin must penetrate and enter the cytosol to exert its cytotoxic effect. Ricin is a binary ribosome inactivating protein synthesised in *Ricinis* cotyledons. Plant ribosomes are sensitive to the toxin, but it is sequestered from the cytosol in protein-storage vacuoles. Ricin's lectin B-chain binds to surface glycoproteins on mammalian cells. Endocytosis and retrograde transport sequester 95% of the ricin in mammalian cells in vacuoles (van Deurs et al. 1988). Some ricin follows a different pathway and is transported to the ER (Wesche et al. 1999). There, the reducing environment favours dissociation, and the toxic A-chain exits the ER through the SEC61 channel. The A-chain is subject to degradation by the endoplasmic reticulum associated protein degradation (ERAD) pathway, and cytotoxicity is attributed to the small proportion of ricin A-chain that evades degradation (Wesche et al. 1999; Roberts and Smith 2004).

In the S-RNase compartmentalisation model (Fig. 10.1b), S-RNase is taken up non-specifically into the luminal compartments of the endomembrane system where, sequestered from the RNA in the cytosol, it cannot exert its cytotoxic activity. In compatible pollen tubes, S-RNase remains stably sequestered, and this is regarded as a sufficient explanation for compatibility. Sequestered S-RNase cannot effectively inhibit the pollen growth program; pollen RNA is stable, and growth continues. HT-B is also taken up, but is degraded by a hypothetical pollen protein, PP. S-RNase and HT-B are also taken up by incompatible pollen tubes (Fig. 10.1b, right). However, the self-S-RNase-SLF interaction results in stabilisation of HT-B and ultimately breakdown of the vacuole with concomitant release of S-RNase. Perhaps, the self-S-RNase-SLF interaction destabilises PP, but this is speculatory, as PP itself is hypothetical. Incompatibility is seen as self-reinforcing, because as more S-RNase is released the pollen tube's ability to maintain the integrity of the endomembrane system would be diminished. In any pollination (compatible or incompatible), some pollen tubes display normal, thin walls, while others display thicker, distorted walls. As pollen tube growth proceeds, compatible pollen tubes maintain a normal appearance, while an increasing number of incompatible pollen tubes adopt the distorted appearance (Sect. 10.1). Distorted pollen tubes usually do not show a well-organised compartment labelled by anti-*120K*, and S-RNase is present throughout the cytosol; 36 h after pollination, 96% of compatible *N. alata* pollen tubes show normal morphology and an intact S-RNase compartment, while 74% of incompatible pollen tubes show distorted morphology and a disrupted compartment (Goldraij et al. 2006). Thus, release of S-RNase is correlated with the classic changes in pollen tube morphology associated with SI. It is also correlated with pollen RNA degradation. Incompatible pollen tubes show little evidence of RNA degradation 12 h post-pollination but a progressive increase in RNA degradation is detectable after 18, 24, 36 and 48 h (McClure et al. 1990).

Any difference between compatible and incompatible pollen tubes must, by definition, result from interaction of pistil and pollen *S*-specificity determinants. The compartmentalisation model focuses on three *S*-specific effects: compartment integrity, HT-B degradation and RNA degradation. If S-RNase first enters the luminal compartments of the endomembrane system, then it must exit to the cytosol to interact with SLF, which is found in the cytosol associated with the ER (Wang and Xue 2005). S-RNase may follow the same pathway as ricin and exit the ER through the SEC61 channel, and may also be subject to ERAD. S-RNase in the cytosol could interact with SLF and control the three *S*-specific effects just mentioned.

Stable S-RNase compartmentalisation is observed in compatible pollen tubes as well as in otherwise incompatible pollen tubes in *HT-B* antisense, *120K* RNAi, or *4936-factor* defective pistils (Goldraij et al. 2006). Thus, vacuolar breakdown does not depend solely on interaction between pollen *S* and S-RNase but requires additional factors from the pistil-side. HT-B is of special interest, since it is required for SI and its stability is regulated by self-recognition. Given that S-RNase remains stably compartmentalised in its absence, HT-B probably acts after S-RNase uptake but before recognition. For example, HT-B may facilitate S-RNase's exit from the ER (Fig. 10.1b). In the model put forward by Goldraij et al. (2006), an unknown pollen protein, PP, causes degradation of HT-B. PP is constitutively active in a compatible pollination; HT-B is degraded, and whatever function it contributes to incompatibility is denied. In incompatible pollen tubes, self-S-RNase-SLF recognition inactivates PP, resulting in the stabilisation of HT-B (Fig. 10.1b). This results in further S-RNase release, degradation of pollen RNA and a diminished ability to grow and repair damage. Self-reinforcing loops are implied: degradation of HT-B reinforces compatibility, suppressing HT-B degradation reinforces pollen rejection. Thus, leakage of S-RNase from the endomembrane system may be both a cause and a consequence of SI.

This model accounts for most of the facts known about S-RNase-based SI. It is compatible with the competitive interaction if SLF forms multimers that are incapable of self-recognition. Other F-box proteins such as β-TrCP (Suzuki et al. 2000) are known to function as multimers. The compartmentalisation and S-RNase degradation models are not mutually exclusive. Pollen tubes could easily have a multi-layered resistance mechanism for S-RNase where compartmentalisation makes pollen tubes resistant to the bulk of the S-RNase they encounter and an *S*-specific degradation mechanism operates on S-RNase that escapes the endomembrane system. More thorough testing for *S*-specific degradation will determine whether a mixed model combining aspects of Fig. 10.1a and b is better than either model alone. This is challenging, but feasible. Moreover, additional components of the S-RNase-based SI system are sure to be discovered and characterised, and so there may well be aspects of the system that have been completely missed. This prospect makes the system all the more exciting.

Acknowledgements Funding has been provided by US National Science Foundation grant 0614962. The author thanks Melody Kroll for editorial assistance.

References

Ai Y, Kron E, Kao T-H (1991) *S*-alleles are retained and expressed in a self-compatible cultivar of *Petunia hybrida*. Mol Gen Genet 230:353–358

Anderson E, de Winton D (1931) The genetic analysis of an unusual relationship between self-sterility and self-fertility in *Nicotiana*. Ann Mo Bot Gard 18:97–116

Anderson M, McFadden G, Bernatzky R, Atkinson A, Orpin T, Dedman H, Tregear G, Fernley R, Clark AE (1989) Sequence variability of three alleles of the self-incompatibility gene of *Nicotiana alata*. Plant Cell 1:483–491

Ben-Naim O, Eshed R, Parnis A, Teper-Bamnolker P, Shalit A, Coupland G, Samach A, Lifschitz E (2006) The CAAT binding factor can mediate interactions between CONSTANS-like proteins and DNS. Plant J 46:462–476

Cruz-Garcia F, Hancock CN, McClure B (2003) S-RNase complexes and pollen rejection. J Exp Bot 53:123–130

Cruz-Garcia F, Hancock CN, Kim D, McClure B (2005) Stylar glycoproteins bind to S-RNase *in vitro*. Plant J 42:295–304

de Nettancourt D (1977) Incompatibility in angiosperms. Springer, Berlin Heidelberg New York

de Nettancourt D (2001) Incompatibility and incongruity in wild and cultivated plants. Springer, Berlin Heidelberg New York

East EM (1932) Studies of self-sterility. IX. The behavior of crosses between self-sterile and self-fertile plants. Genetics 17:175–202

Entani T, Iwano M, Shiba H, Che FS, Isogai A, Takayama S (2003) Comparative analysis of the self-incompatibility *S*-locus region of *Prunus mume*: Identification of a pollen-expressed F-box gene with allelic diversity. Genes Cells 8:203–213

Geitmann A, Huda'k J, Vennigerholz F, Walles B (1995) Immunoglod localization of pectin and callose in pollen grains and pollen tubes of *Brugmansia suaveolens* – Implications for the self-incompatibility reaction. J Plant Physiol 147:225–235

Goldraij A, Kondo K, Lee CB, Hancock CN, Sivaguru M, Vazquez-Santana S, Kim S, Phillips TE, Cruz-Garcia F, McClure B (2006) Compartmentalization of S-RNase and HT-B degradation in self-incompatible *Nicotiana*. Nature 439:805–810

Golz JF, Su V, Clarke AE, Newbigin E (1999) A molecular description of mutations affecting the pollen component of the *Nicotiana alata* S-locus. Genetics 152:1123–1135

Golz JF, Oh H-Y, Su V, Kusaba M, Newbigin E (2001) Genetic analysis of *Nicotiana* pollen-part mutants is consistent with the presence of an S-ribonuclease inhibitor at the S-locus. Proc Nat Acad Sci USA 98:15372–15376

Gray JE, McClure BA, Bönig I, Anderson MA, Clarke AE (1991) Action of the style product of the self-incompatibility gene of *Nicotiana alata* (S-RNase) on *in vitro*-grown pollen tubes. Plant Cell 3:271–283

Hancock CN, Kent L, McClure BA (2005) The 120 kDa glycoprotein is required for S-specific pollen rejection in *Nicotiana*. Plant J 43:716–723

Hauck NR, Yamane H, Tao R, Iezzoni AF (2006) Accumulation of nonfunctional S-haplotypes results in the breakdown of gametophytic self-incompatibility in tetraploid *Prunus*. Genetics 172:1191–1198

Hua Z, Kao T-H (2006) Identification and characterization of components of a putative *Petunia* S-locus F-box-containing E3 ligase complex involved in S-RNase-based self-incompatibility. Plant Cell 18:2531–2553

Huang J, Zhao L, Yang Q, Xue Y (2006) AhSSK1, a novel SKP1-like protein that interacts with the S-locus F-box protein SLF. Plant J 46:780–793

Ida K, Norioka S, Yamamoto M, Kumasaka T, Yamashita E, Newbigin E, Clarke AE, Sakiyama F, Sato M (2001) The 1.55 A resolution structure of *Nicotiana alata* SF11-RNase associated with gametophytic self-incompatibility. J Mol Biol 314:103–112

Igic B, Kohn JR (2001) Evolutionary relationships among self-incompatibility RNases. Proc Nat Acad Sci USA 98:13167–13171

Igic B, Smith WA, Robertson KA, Schaal BA, Kohn JR (2007) Studies of self-incompatibility in wild tomatoes: I. S-allele diversity in *Solanum chilense* Dun. (Solanaceae). Heredity 2001:1–9

Ioerger TR, Gohlke JR, Xu B, Kao T-H (1991) Primary structural features of the self-incompatibility protein in Solanaceae. Sex Plant Reprod 4:81–87

Ishimizu T, Endo T, Yamaguchi-Kabata Y, Nakamura KT, Sakiyama F, Norioka S (1998) Identification of regions in which positive selection may operate in S-RNase of Rosaceae: Implication for S-allele-specific recognition sites in S-RNase. FEBS Lett 440:337–342

Juarez-Diaz JA, McClure B, Vazquez-Santana S, Guevara-Garcia A, Leon-Mejia P, Marquez-Guzman J, Cruz-Garcia F (2006) A novel thioredoxin *h* is secreted in *N. alata* and reduces S-RNase *in vitro*. J Biol Chem 281:3418–4324

Kao T-H, McCubbin AG (1996) How flowering plants discriminate between self and non-self pollen to prevent inbreeding. Proc Natl Acad Sci USA 93:12059–12065

Kao T-H, Tsukamoto T (2004) The molecular and genetic bases of S-RNase-based self-incompatibility. Plant Cell 16:S72–S83

Kondo K, Yamamoto M, Itahashi R, Sato T, Egashira H, Hattori T, Kowyama Y (2002a) Insights into the evolution of self-compatibility in *Lycopersicon* from a study of stylar factors. Plant J 30:143–153

Kondo K, Yamamoto M, Matton DP, Sato T, Masashi H, Norioka S, Hattori T, Kowyama Y (2002b) Cultivated tomato has defects in both S-RNase and HT genes required for stylar function of self-incompatibility. Plant J 29:627–636

Lai Z, Ma W, Han B, Liang L, Zhang Y, Hong G, Xue Y (2002) An F-box gene linked to the self-incompatibility S-locus of *Antirrhinum* is expressed specifically in pollen and tapetum. Plant Mol Biol 50:29–42

Lind JL, Bacic A, Clarke AE, Anderson MA (1994) A style-specific hydroxyproline-rich glycoprotein with properties of both extensins and arabinogalactan proteins. Plant J 6:491–502

Lind JL, Bönig I, Clarke AE, Anderson MA (1996) A style-specific 120 kDa glycoprotein enters pollen tubes of *Nicotiana alata in vivo*. Sex Plant Reprod 9:75–86

Lush WM, Clarke AE (1997) Observations of pollen tube growth in *Nicotiana alata* and their implications for the mechanism of self-incompatibility. Sex Plant Reprod 10:27–35

Luu D-T, Xike Q, Morse D, Cappadocia M (2000) S-RNase uptake by compatible pollen tubes in gametophytic self-incompatibility. Nature 407:649–651

Luu D-T, Qin X, Laublin G, Yang Q, Morse D, Cappadocia M (2001) Rejection of *S*-heteroallelic pollen by a dual-specific S-RNase in *Solanum chacoense* predicts a multimeric SI pollen component. Genetics 159:329–335

Martin FW (1961) The inheritance of self-incompatibility in hybrids of *Lycopersicon esculentum* Mill. x *L. chilense* Dun. Genetics 46:1443–1454

Martin FW (1968) The behavior of *Lycopersicon* incompatibility alleles in an alien genetic milieu. Genetics 60:101–109

Mather K (1943) Specific differences in *Petunia* I. Incompatibility. J Genet 45:215–235

Matton DP, Maes O, Laublin G, Xike Q, Bertrand C, Morse D, Cappadocia M (1997) Hypervariable domains of self-incompatibility RNases mediate allele-specific pollen recognition. Plant Cell 9:1757–1766

Matton DP, Luu D-t, Xike Q, Laublin G, O'Brien M, Maes O, Morse D, Cappadocia M (1999) Production of an S RNase with dual specificity suggests a novel hypothesis for the generation of new *S* alleles. Plant Cell 11:2087–2097

McClure B (2006) New views of S-RNase-based self-incompatibility. Curr Opin Plant Biol 6:639–646

McClure BA, Gray JE, Anderson MA, Clarke AE (1990) Self-incompatibility in *Nicotiana alata* involves degradation of pollen rRNA. Nature 347:757–760

McClure BA, Mou B, Canevascini S, Bernatzky R (1999) A small asparagine-rich protein required for S-allele-specific pollen rejection in *Nicotiana*. Proc Nat Acad Sci USA 96:13548–13553

McClure BA, Cruz-Garcia F, Beecher BS, Sulaman W (2000) Factors affecting inter- and intraspecific pollen rejection in *Nicotiana*. Ann Bot 85:113–123

Murfett JM, Strabala TJ, Zurek DM, Mou B, Beecher B, McClure BA (1996) *S* RNase and interspecific pollen rejection in the genus *Nicotiana*: Multiple pollen rejection pathways contribute to unilateral incompatibility between self-incompatible and self-compatible species. Plant Cell 8:943–958

O'Brien M, Kapfer C, Major G, Laurin M, Bertrand C, Kondo K, Kowyama Y, Matton DP (2002) Molecular analysis of the stylar-expressed *Solanum chacoense* asparagine-rich protein family related to the HT modifier of gametophytic self-incompatibility in *Nicotiana*. Plant J 32:1–12

O'Brien M, Major G, Chanta S-C, Matton DP (2004) Isolation of S-RNase binding proteins from *Solanum chacoense*: Identification of an SBP1 (RING finger protein) orthologue. Sex Plant Reprod 17:81–87

Qiao H, Wang F, Zhao L, Zhou J, Lai Z, Zhang Y, Robbins TP, Xue Y (2004a) The F-box protein AhSLF-S_2 controls the pollen function of S-RNase-based self-incompatibility. Plant Cell 16:2307–2322

Qiao H, Wang H, Zhao L, Zhou J, Huang J, Zhang Y, Xue Y (2004b) The F-box protein AhSLF-S_2 physically interacts with S-RNases that may be inhibited by the ubiquitin/26S proteasome pathway of protein degradation during compatible pollination in *Antirrhinum*. Plant Cell 16:582–595

Qin X, Soulard J, Laublin G, Morse D, Cappadocia M (2005) Molecular analysis of the conserved C4 region in S_{11}-RNase of *Solanum chacoense*. Planta 221:531–537

Roberts LM, Smith DC (2004) Ricin: The endoplasmic reticulum connection. Toxicon 44:469–472

Sassa H, Hirano H (2006) Identification of a new class of pistil-specific proteins of *Petunia inflata* that is structurally similar to, but functionally distinct from, the self-incompatibility factor *HT*. Mol Gen Genom 275:97–104

Sassa H, Kakui M, Miyamoto Y, Suzki T, Hanada K, Ushijima K, Kasaba M, Hirano H, Koba T (2007) *S-locus-F-box brothers*: Multiple and pollen-specific F-box genes with *S*-haplotype-specific polymorphisms in apple and Japanese pear. Genetics 175:1869–1881

Sijacic P, Wang X, Skirpan AL, Wang Y, Dowd PE, McCubbin AG, Huang S, Kao T-H (2004) Identification of the pollen determinant of S-RNase-mediated self-incompatibility. Nature 429:302–305

Sims TL, Ordanic M (2001) Identification of a S-ribonuclease-binding protein in *Petunia hybrida*. Plant Mol Biol 47:771–783

Sonneveld T, Tobutt KR, Vaughan SP, Robbins TP (2005) Loss of pollen-*S* function in two self-compatible selections of *Prunus avium* is associated with deletion/mutation of an *S*-haplotype-specific F-box gene. Plant Cell 17:37–51

Stout AB, Chandler C (1942) Hereditary transmission of induced tetraploidy and compatibility in fertilization. Science 96:257–258

Suzuki H, Chiba T, Suzuki T, Fujita T, Ikenoue T, Omata M, Furuichi K, Shikama H, Tanaka K. (2000) Homodimer of two F-box proteins βTrCP1 or βTrCP2 binds to IκBα for signal dependent ubiquitination. J Biol Chem 275:2877–2884

Tsukamoto T, Ando T, Kokubun H, Watanabe H, Sato T, Masada M, Marchesi E, Kao T-h (2003a) Breakdown of self-incompatibility in a natural population of *Petunia axillaris* caused by a modifier locus that suppresses the expression of an S-RNase gene. Sex Plant Reprod 15:255–263

Tsukamoto T, Ando T, Takahashi K, Omori T, Watanabe H, Kokubun H, Marchesi E, Kao T-h (2003b) Breakdown of self-compatibility in a natural population of *Petunia axillaris* caused by loss of pollen function. Plant Physiol 131:1903–1912

Tsukamoto T, Ando T, Watanabe A, Marchesi E, Kao T-H (2005) Duplication of the *S*-locus F-box gene is associated with breakdown of pollen function in an *S*-haplotype identified in a natural population of self-incompatible *Petunia axillaris*. Plant Mol Biol 57:141–163

Ushijima K, Yamane H, Watari A, Kakehi E, Ikeda K, Hauck NR, Iezzoni AF, Tao R (2004) The *S*-haplotype-specific F-box protein gene, *SFB*, is defective in self-compatible haplotypes of *Prunus avium* and *P. mume*. Plant J 39:573–586

van Deurs B, Sandvig K, Peterson OW, Olsnes S, Simons K, Griffiths G (1988) Estimation of the amount of internalized ricin that reaches the *trans*-golgi network. J Cell Biol 106:253–267

Vieira J, Morales-Hojas R, Santos RAM, Vieira CP (2007) Different positively selected sites at the gametophytic self-incompatibility pistil *S-RNase* gene in the Solanaceae and Rosaceae (*Prunus*, *Pyrus*, and *Malus*). J Mol Evol 65:175–185

Vierstra RD (2003) The ubiquitin/26S proteasome pathway: The complex last chapter in the life of many plant proteins. Trends Plant Sci 8:135–142

Wang H-Y, Xue Y (2005) Subcellular localization of the *S*-locus F-box protein AhSLF-S2 in pollen and pollen tubes of self-incompatible *Antirrhinum*. J Integr Plant Biol 47:76–83

Wang Y, Wang X, McCubbin AG, Kao T-H (2003) Genetic mapping and molecular characterization of the self-incompatibility *S*-locus in *Petunia inflata*. Plant Mol Biol 53:565–580

Wesche J, Rapak A, Olsnes S (1999) Dependence of ricin toxicity on translocation of the toxin A-chain from the endoplasmic reticulum to the cytosol. Proc Nat Acad Sci USA 274:34443–34449

Wheeler D, Newbigin E (2007) Expression of ten S class *SLF-like* genes in *Nicotiana alata* pollen and its implications for understanding the pollen factor of the *S*-locus. Genetics 177:2171–2180

Zhou J, Wang F, Ma W, Zhang Y, Han B, Xue Y. (2003) Structural and transcriptional analysis of *S*-locus F-box genes in *Antirrhinum*. Sex. Plant Reprod. 16:165–177

Zurek DM, Mou B, Beecher B, McClure B. (1997) Exchanging sequence domains between S-RNases from *Nicotiana alata* disrupts pollen recognition. Plant J. 11:797–808

Chapter 11
Self-Incompatibility in *Papaver Rhoeas*: Progress in Understanding Mechanisms Involved in Regulating Self-Incompatibility in *Papaver*

V.E. Franklin-Tong

Abstract Over the last 20 years or so, our knowledge of what is involved in the rejection of 'self' pollen in *Papaver rhoeas* has expanded tremendously. From initial studies of the population genetics of the *S*-locus polymorphism, and identification of the pistil S-determinant, the focus has moved to elucidating the signals and mechanisms involved in mediating the inhibition of incompatible pollen tube growth. A key finding was the involvement of a Ca^{2+}-dependent signalling network. This led to the discovery of several SI-induced events, including depolymerisation of the actin cytoskeleton and phosphorylation of a soluble inorganic pyrophosphatase, Pr-p26.1, which are involved in the rapid inhibition of pollen tube growth. Programmed cell death is also triggered; this provides a neat way to destroy self pollen. Recent studies have begun to unravel components involved in this important event, involving activation of several caspase-like activities. Here we review some of the key findings in recent years.

Abbreviations

ABPs	Actin-binding proteins
$[Ca^{2+}]_i$	Intracellular cytosolic free calcium concentration
DEVD	Ac-DEVD-CHO A commercial cell-permeable tetrapeptide comprising aspartic acid, glutamic acid, valine, aspartic acid
DEVDase	A caspase-3 activity
GFP	Green fluorescent protein
HR	Hypersensitive Response
Jasp	Jasplakinolide

V.E. Franklin-Tong
School of Biosciences, University of Birmingham, Edgbaston, Birmingham B15 2TT, U.K.,
e-mail: v.e.franklin-tong@bham.ac.uk

LatB	Latrunculin B
LEVDase	A caspase-4 activity
MAPKKK	Mitogen activated protein kinase
MAPKKK	Mitogen activated protein kinase kinase kinase
NO	Nitric oxide
p56	A 56 kDa phosphoprotein, so named because it migrates to 56 kDa on DS-PAGE
p56-MAPK	p56 identified as a MAPK
PARP	Poly (ADP-ribose) polymerase
PCD	Programmed Cell Death
Pi	Inorganic phosphate
PPi	Inorganic pyrophosphate
Pr-p26.1a/b	*Papaver rhoeas* phosphoprotein p26.1a or p26.1b, two soluble inorganic pyrophosphatases
PrpS	*Papaver rhoeas* pollen S (the pollen S-determinant)
PSIL	Papaver Self-Incompatibility-Like
RLK	Receptor-like kinase
ROS	Reactive oxygen species
SBP	S protein Binding Protein
SI	Self-incompatibility
SPH	S protein homologue
sPPase	Soluble inorganic pyrophosphatase
S protein	Self-incompatibility protein (the pistil S-determinant)
S-RNase	S-ribonuclease
U0124	The negative analogue of U0126
U0126	A drug used specifically to inhibit MAPK cascades
VEIDase	A caspase-6 activity
YVAD	Ac-YVAD-CHO A commercial cell-permeable tetrapeptide comprising tyrosine, valine, alanine, aspartic acid
YVADase	A caspase-1 activity

11.1 Introduction

11.1.1 Genetics and Cell Biology of Self-Incompatibility in Papaver

As mentioned earlier (see Chaps. 4 and 7), self-incompatibility (SI) is probably one of the most important systems used by flowering plants to prevent self-fertilisation. Use of classical genetic studies, making controlled crosses has established that SI is generally controlled by a single polymorphic locus, the self-incompatibility *S*-locus, though there are exceptions (see Chap. 13). The *S*-locus has a minimum of two genes: pistil *S* and pollen *S*. As the locus is highly polymorphic, carrying many *S* alleles, the term *haplotype* is used to describe the *genotype*. A self-incompatible response is triggered when pollen expressing an *S*-haplotype interacts with a pistil expressing the same *S*-haplotype (see Chap. 9; Fig. 9.1).

As can be seen from other accompanying chapters in this book, the molecular basis of the *S*-locus has been extensively studied in several species. Within the gametophytic SI systems, two major different classes of SI have been identified: the *Solanaceae-type* (or S-RNase type) and the *Papaveraceae-type* SI, due to the species in which they were first identified see (Takayama and Isogai 2005) Chaps. 9 and 10, this volume, for further details. *Papaver rhoeas L.* (the field poppy; see Fig. 11.1a) has been established as having a single-locus gametophytic SI (GSI) system (Lawrence 1975). The haplotypes of the polymorphic *S*-locus control the recognition specificity, and pollen is rejected when it lands on a pistil carrying one of the *S*-haplotypes that matches its *S*-haplotypes (see Chap. 9; Fig. 9.1). Thus, analysis at a genetical level does not distinguish between the different types of SI systems.

Despite this, examination of the cell biology of SI responses reveals quite striking differences between those in the S-RNase-based SI and in *Papaver* [and also in the grasses (see Chap. 13)]. In all the species examined to date exhibiting S-RNase-based SI, the *S*-specific rejection usually occurs several hours after pollen germination, as pollen tubes grow through the style (de Nettancourt 1977), with heavy deposition of callose; see Chap. 10 for further details of this type of SI. In contrast, the SI response in *Papaver* superficially looks rather similar to that seen in *Brassica* and *Ipomoea* (see Chaps. 8 and 12, this volume). Inhibition is very rapid, and although germination sometimes occurs, generally callose is deposited at the tip shortly after polarity is established (see Fig. 11.1b).

11.1.2 How Studies on Self-Incompatibility in Papaver Started

Research on the genetical basis of SI in *P. rhoeas L.* (the field poppy) was initiated in the 1970s by Mike Lawrence. He analysed sets of reciprocal pollinations between members of families of *P. rhoeas* and enforced self-pollinations, using aniline blue staining. The simplest interpretation of the data obtained was that SI is determined by a single gene and that genetical control of the pollen phenotype is gametophytic (Lawrence 1975). The use of aniline blue staining had the important advantage of confirming the numerical data by allowing direct observations of half-compatible pollinations, which contained a mixture of compatible pollen tubes and incompatible pollen grains (Lawrence 1975). Thus, it was established that *Papaver* had a gametophytic SI system, which was, at least at the level of genetical control, identical to that found in the Solanaceae and other species that have the S-RNase SI system (see Chaps. 4 and 9, this volume).

Mike Lawrence was responsible for initiating all of the research on *Papaver* SI, which has been carried out at Birmingham University since then. His major interest was the population genetics of the *S*-locus polymorphism. He collected and generated an extensive collection of *P. rhoeas S*-haplotypes from wild populations and from the cultivated Shirley variety. He generated full-sib families and used these

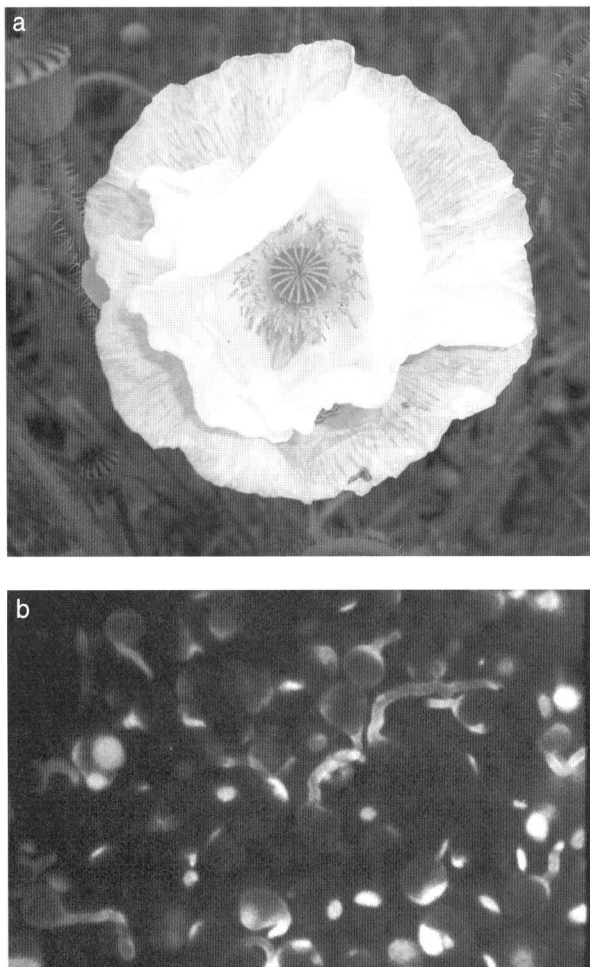

Fig. 11.1 (**a**) *Papaver rhoeas* L., the field poppy. Note that the stigma comprises a disc carrying a set of stigmatic rays and that there is no style. (**b**) Aniline blue staining showing the rapid inhibition of incompatible pollen soon after polarity is established or during germination on the stigma surface

to carry out an extensive programme using classical half-diallel pollinations and assessing SI using aniline blue staining, over several decades. He established the number of *S*-haplotypes present in *P. rhoeas* and estimated their frequencies in four populations (see, e.g. Campbell and Lawrence 1981; Lane and Lawrence 1993; Lawrence et al., 1993). These studies revealed that, contrary to expectation (Wright, 1939), there are a large number of *S*-haplotypes and that the frequencies of *S*-haplotypes are highly unequal (Lawrence and O'Donnell, 1981). Moreover, quite widely separated populations share a large proportion of the *S*-haplotypes (Lane and Lawrence 1993). For discussion of the population genetics and phylogeny of other self-incompatible species see Chap. 5, this volume.

In the early 1980s, with the advent of molecular biology becoming feasible in plants, he initiated collaborations to start investigations into the molecular basis of SI in *Papaver*. At that time, researchers were identifying pistil proteins that segregated with *S*-haplotypes in attempts to identify the pistil *S*-locus component (see, e.g. Bredemeijer and Blaas 1981; Hinata et al. 1982; Mau et al. 1982; see also Chaps. 7 and 12, this volume). We took a similar approach, but also fortuitously, developed an *in vitro* bioassay for SI in *Papaver* (Franklin-Tong et al. 1988). This has proved to be a mainstay of these studies, as it has allowed us to identify and analyse events triggered in incompatible pollen in considerable detail. This has led to a much better understanding of mechanisms involved in *Papaver* SI than in other SI systems, though progress in other areas has lagged behind. Here we attempt to outline some of the major areas where we have made progress in understanding the basis of SI in this species.

11.1.3 The Papaver *in Vitro SI System*

Because this is such a central feature to many of our investigations, it is worth briefly outlining the main features of the *in vitro* bioassay for SI in *Papaver*. The principle is quite simple: pistil S proteins are added to pollen growing *in vitro*, and depending on the combination of *S*-haplotypes used, the pollen will either grow (in a compatible situation) or be inhibited (i.e. where an incompatible is achieved). Half compatible situations can also be achieved by the appropriate combination of *S*-haplotypes. The initial demonstration (Franklin-Tong et al. 1988) used stigmatic extracts added to pollen. Later experiments used addition of recombinant S proteins expressed in *E. coli* (see, e.g. Foote et al. 1994). Because inhibition of incompatible pollen in *Papaver* is rapid (soon after germination), *in vitro* SI can easily be scored, either by eye, or usually by counting or measuring inhibited pollen tubes. Moreover, the set-up can be scaled down to visualise events in individual pollen tubes undergoing the SI response, using live-cell imaging (see Bosch and Franklin-Tong, 2007; Franklin-Tong et al., 1993, 1997) or in fixed pollen tubes (Geitmann et al. 2000, 2004). It can also be scaled up to extract proteins for analysis to detect or measure *S*-specific changes, either using a proteomic approach (see Rudd et al. 1996) or by analysing changes of specific proteins (see Rudd et al. 2003; Snowman et al., 2002). Drugs can be added to pollen growing *in vitro* prior to SI being induced, to investigate the potential role of specific components in the SI response (see, e.g. Li et al. 2007; Thomas and Franklin-Tong, 2004). The *in vitro* bioassay is also amenable to manipulation to test gene function, for example by addition of antisense oligonucleotides, or expression of GFP-tagged proteins using particle bombardment as a transient expression system (de Graaf et al. 2006). Thus, it has proved to be a robust and adaptable system for analysis of events triggered by SI.

The use of this SI *bioassay* has led to a much better understanding of mechanisms involved in *Papaver* SI than in other SI systems, though progress in other areas has lagged behind. Here we attempt to outline progress in understanding the basis of SI in this species.

11.2 *S* Proteins Determine *S*-Specificity in the Pistil

11.2.1 *Identification of Pistil* S-*locus Components*

The cloning and identification of the *Papaver S*-locus components has not been straightforward. In contrast to the other SI systems characterised at a molecular level to date, the pistil components are rather low abundance. This made their identification and isolation difficult, particularly with the technology available at the time. N-terminal amino acid sequence data obtained from stigmatic proteins segregating with the S_1 haplotype allowed the cloning of the *Papaver* pistil S_1 cDNA. As expected, expression was tissue-specific and developmentally controlled (Foote et al. 1994). The cDNA encodes a predicted secreted, hydrophilic protein of ~15 kDa. As poppy is not easily transformable, the *in vitro* bioassay was used to demonstrate that the recombinant protein exhibited the expected *S*-haplotype-specific pollen inhibitory activity, thereby demonstrating that the product of the proposed pistil *S* gene cloned had the function expected of it (Foote et al. 1994).

The *Papaver* pistil *S* gene is single copy and has no obvious close homologues (Foote et al. 1994), though subsequently, detailed analysis of *Arabidopsis* genome revealed the presence of a large gene family with low homology to the *Papaver* pistil *S* genes. They were named *S protein homologues* (*SPHs*), with similar predicted secondary structures to the pistil *S* genes (Ride et al. 1999). Subsequently, a family of Arabidopsis genes (*Papaver Self-Incompatibility-Like; PSIL genes*) expressed in embryo-sac, anthers or pollen have been found to have homology to the *Papaver* pistil *S* genes and which probably represent part of the SPH family (Jones-Rhoades et al. 2007). Because the *Papaver* pistil *S* gene encoded a small, secreted unique protein, it was proposed to act as a signalling ligand. This was subsequently confirmed (see Sect. 11.4.1). Interestingly, it appears that the *SPH*s may be signalling ligands with a variety of different functions, as they are expressed in a range of tissues in Arabidopsis (Wheeler et al. unpublished data).

11.2.2 Pistil S-Protein Sequence Information and Residues Required for Function

Subsequently, several other *S*-haplotypes were cloned (Kurup et al., 1998;Walker et al. 1996), which enabled comparisons to be made between the sequences. The pistil *S*-alleles, as expected, exhibit a high degree of sequence polymorphism, ranging from 51.3 to 63.7% at the amino acid level. However, they are predicted to have very similar secondary structures (Kurup et al. 1998). Two studies to date have attempted to define the amino acid residues within that participate in *S*-haplotype-specific inhibition of incompatible pollen (Jordan et al. 1999; Kakeda et al. 1998), but as amino acid sequence variation was not present in hypervariable blocks, this hampered studies. Nevertheless, construction of mutant derivatives of S proteins, using site-directed mutagenesis and testing their biological activity against pollen enabled the identification of several amino acid residues that are required for S-specific inhibition of incompatible pollen. Both variable and conserved amino acids in a region of the S protein corresponding to a hydrophilic predicted surface loop were identified as key elements that play a role in the recognition and inhibition of incompatible pollen (Kakeda et al. 1998). Further studies will require structural information, as the variable amino acids are highly dispersed. However, the low solubility of the recombinant pistil S proteins has proved problematic.

11.3 Identification of the Pollen *S*-Determinant

The identification of the *Papaver* pollen *S*-determinant (like similar studies in other species; see also Chaps. 7 and 9) has been difficult. To date, data definitively demonstrating the nature of this component has not been published, though a good candidate has now been identified. Because of what we know about the pistil component (which is a small, unique secreted protein) and the pollen response (see Sect. 11.4), the pollen *S* gene product is predicted to be a receptor that interacts with the stigmatic *S* protein in an *S* specific manner. Studies identified a pollen membrane-located protein that binds stigmatic S proteins, although apparently not in an *S*-haplotype-specific manner. This was named S protein binding protein (SBP) (Hearn et al. 1996). It was suggested that this may act as an accessory receptor, perhaps mediating presentation of S proteins to the authentic pollen *S* receptor. Analysis of several mutant derivatives of the S_1 protein revealed that mutations to key amino acid residues caused significant reductions in their SBP-binding activity and a concomitant reduction in their ability to inhibit incompatible pollen (Jordan et al. 1999). As interaction between the pistil S proteins and SBP appears to be crucial for inhibition of incompatible pollen, it suggests a key role for SBP in SI.

We have recently identified a pollen gene, $PrpS_1$, which is tightly linked to the pistil *S* gene. It displays the polymorphism typical of an *S*-locus component and appears to be a good candidate for the pollen *S*-locus determinant. *PrpS* encodes a small, predicted transmembrane protein, and so is likely to be transmembrane receptor, as predicted. It has (as expected) no homology to proteins in existing databases. Moreover, data indicate that it has the *S*-specific pollen inhibitory biological function expected (Wheeler et al. unpublished data). Recent investigations using a heterologous expression system have provided data demonstrating that *PrpS* functions as a channel (Stevens et al. unpublished), which fits very nicely with data relating to the pollen response (see Sect. 11.4.1).

11.4 Mechanisms Involved in SI in the *Papaver* System

11.4.1 Calcium Signalling Mediates Papaver SI

Ca^{2+}-dependent signalling networks are universally used to generate responses to a huge variety of stimuli in both animal and plant cells. Calcium imaging has provided a useful tool to establish a role for Ca^{2+} as a key second messenger by visualising alterations in intracellular cytosolic-free calcium concentration ($[Ca^{2+}]_i$). This approach has shown that transient increases in $[Ca^{2+}]_i$ are generated in response to many abiotic and biotic stimuli (Hetherington and Brownlee 2004; Rudd and Franklin-Tong 1999, for reviews). Moreover, the 'calcium signature' generated (the temporal and spatial alterations in $[Ca^{2+}]_i$) is generally quite specific to the interaction. As the *Papaver* pistil S proteins had no known homologues suggesting function, speculative studies investigated whether they might be possible candidates for signalling ligands, as they were novel, relatively small secreted proteins. These experiments laid the foundation for future work, as they implicated that the S proteins functioned as signalling ligands.

Ca^{2+} imaging established that $[Ca^{2+}]_i$ acted as a second messenger in *Papaver* pollen, triggered specifically by interaction with incompatible S proteins and preceding pollen tube tip growth arrest (Franklin-Tong et al. 1993, 1995). Subsequent studies confirmed this and revealed important spatial and quantitative information about these alterations (Franklin-Tong et al. 1997). Intriguingly, the increases in $[Ca^{2+}]_i$ appeared as a Ca^{2+} wave in the 'shank' of the pollen tube in the vicinity of the nuclear region, and not at the tip. This was surprising, as tip growth was thought to be the major target for the signals. It was later shown that at least some of these increases in $[Ca^{2+}]_i$ were from extracellular sources (Franklin-Tong et al. 2002), implicating Ca^{2+} influx being triggered by an incompatible SI response. Again, channels in the shank of the pollen tube were implicated, rather than channels at the tip, which were implicated in regulating tip growth (Malho et al. 1995).

These early studies were pivotal in determining the direction of future studies and formed the basis of a long-standing hypothesis, proposing that the pollen S-determinant is a transmembrane receptor that interacts with secreted S proteins that act as signalling ligands, in an *S*-specific manner. Thus, only in an incompatible situation would receptor–ligand binding be expected to occur, triggering a

Ca^{2+}-dependent signalling cascade in incompatible pollen. The identification of the pollen S-determinant, *PrpS*, as a transmembrane protein and preliminary evidence suggesting that the pollen S-determinant functions as a channel (see Sect. 11.3) in this context is extremely exciting as it fits expectations very well. As Ca^{2+} imaging implicated a receptor–ligand interaction triggering downstream signalling events leading to incompatible pollen tube inhibition, this possibility was pursued, and investigations into the possible targets were initiated. Ca^{2+} signalling often results in altered protein kinase activity, resulting in post-translational modification such as phosphorylation. One of the first studies identified elicitor-triggered phosphorylation implicated in defence signalling (Dietrich et al. 1990). It was therefore an obvious next step to investigate whether alterations in protein phosphorylation were stimulated by SI in incompatible *Papaver* pollen.

11.4.2 A Role for Soluble Inorganic Pyrophosphatases (sPPases) in Papaver SI

11.4.2.1 Identification of sPPases as Targets for Phosphorylation

Use of a phospho-proteomic-type approach revealed rapid S-specific phosphorylation of several proteins in incompatible pollen. One of these was a 26 kDa cytosolic protein named Pr-p26.1, which was phosphorylated in a Ca^{2+}-dependent manner (Rudd et al. 1996). Subsequent analysis and cloning identified two pollen genes: *Pr-p26.1a/b* as soluble inorganic pyrophosphatases (sPPases) (de Graaf et al. 2006). sPPases are important enzymes involved in hydrolysis of inorganic pyrophosphate (PPi). During biopolymer synthesis PPi is generated, and sPPase hydrolysis of PPi generates inorganic phosphate (Pi), providing a thermodynamic pull, driving biosynthesis (Kornberg 1962).

The *Papaver* pollen sPPases have classic sPPase activities, which are Mg^{2+}-dependent. Moreover, sPPase activity was inhibited by both Ca^{2+} and by phosphorylation (de Graaf et al. 2006). Although it is well known that sPPases can be inhibited by Ca^{2+}, it had not been previously established that phosphorylation could modify sPPase activity. Thus, identification of sPPases as an early target of SI-induced Ca^{2+}-dependent signaling, and kinase-dependent phosphorylation revealed novel mechanisms for regulating sPPase activity.

11.4.2.2 Evidence for sPPase Involvement in Regulating Pollen Tube Growth

The identification of sPPases as targets for SI signals in incompatible pollen also suggested a possible mechanism to inhibit incompatible pollen tube growth. As mentioned earlier, sPPases are key enzymes that regulate biosynthesis; in principle, high sPPase activity results in faster biosynthesis, and low sPPase activity will result in a decrease in biosynthesis. SI, which results in both high $[Ca^{2+}]_i$ and phosphorylation of *Pr-p26.1a/b*, would be expected to inhibit sPPase activity, which would therefore be predicted to inhibit pollen tube growth, which requires extensive biosynthesis of membrane and cell wall components.

This hypothesis was tested using antisense oligonucleotides based on *Pr-p26.1a/b* sequences to down-regulate Pr-p26.1. Examining their effect on pollen tube growth revealed that the *Pr-p26.1a/b* antisense oligonucleotides resulted in significantly inhibited pollen tubes, suggesting that, as expected, Pr-p26.1 sPPases are necessary for pollen tube growth (de Graaf et al. 2006). To investigate whether sPPases were involved in SI, PPi levels ([PPi]) in pollen were measured after SI induction. Since sPPases hydrolyse PPi, cellular [PPi] is predicted to increase if sPPase activity is inhibited. These experiments established that SI-induced incompatible pollen tubes had increased [PPi], providing good evidence for a functional link between alterations in sPPase activity and SI induction in vivo in pollen tubes (de Graaf et al. 2006).

Together, these studies identified sPPases as novel, early targets for the Ca^{2+}-dependent SI signals. Importantly, they also demonstrated that these sPPases play a role in regulating *Papaver* pollen tube growth. Thus, the evidence suggests that modification of the Pr-p26.1a/b sPPase activity by Ca^{2+} and phosphorylation probably play a major role in SI-stimulated pollen tube inhibition. These findings provide an important part of the jigsaw puzzle of how SI operates.

11.4.3 Alterations to the Actin Cytoskeleton are Triggered by Papaver SI

11.4.3.1 Re-Organisation and Depolymerisation of F-Actin

Because the actin cytoskeleton is known to be a major target/effector of signals and $[Ca^{2+}]_i$ is known to play a key role in mediating these responses (Staiger 2000), as well as being well known to be crucial for pollen tube growth, its possible role in SI was investigated. Within 1 min of SI induction, dramatic alterations in F-actin organisation in incompatible, but not in compatible, pollen tubes were observed and further changes were seen over several hours (Geitmann et al. 2000). Subsequent studies established that early SI triggers massive depolymerisation of F-actin, which is sustained for at least 1 h (Snowman et al. 2002). Moreover, stimulating artificial increases of $[Ca^{2+}]_i$ in pollen tubes stimulated actin depolymerisation, suggesting that actin is a target of the SI-induced Ca^{2+} signals. Actin depolymerisation is known to result in rapid pollen tube inhibition (Gibbon et al. 1999), presumably because actin filaments are required to deliver secretory vesicles required for tip extension. Thus, SI signals, which alter actin dynamics, provide a further highly effective mechanism to inhibit incompatible pollen tube growth.

Many actin-binding proteins (ABPs) are Ca^{2+}-regulated (Staiger and Blanchoin 2006) for a recent review. Two ABPs that may be involved in mediating SI events have been identified. Profilin, which is highly abundant in pollen, was identified as a possible signalling intermediate that might be involved in the SI signalling network. Data suggested that it somehow regulates protein kinase or phosphatase activity, as it can stimulate the phosphorylation of pollen proteins *in vitro* (Clarke et al. 1998). However, data suggested that profilin activity alone was not sufficient to mediate the extensive depolymerisation stimulated by SI (Snowman et al. 2002). Another ABP, purified from *Papaver* pollen, PrABP80 (a putative gelsolin) has potent Ca^{2+} dependent severing activity (Huang et al. 2004). Thus, it is a good candidate for mediating the SI-mediated alterations. Analysis suggested that it could potentially act synergistically with profilin to mediate the large-scale depolymerisation stimulated by SI (Huang et al. 2004). This remains to be tested experimentally, but it begins to shed light on which ABPs may be involved in the SI-mediated actin depolymerisation.

Actin also has been shown to have another role, relating to signalling to programmed cell death (PCD); see Sect. 11.4.4.6 later.

11.4.3.2 SI also Mediates Apparent Depolymerisation of Microtubules

As cortical microtubules are also involved in signal-response coupling (Takemoto and Hardham 2004; Wasteneys 2003), it was recently investigated whether the pollen tube microtubule cytoskeleton are also a target for the SI signals. This established that SI stimulated very rapid apparent depolymerisation of cortical microtubules. As early as 1 min after SI, cortical microtubule bundles were virtually undetectable in incompatible pollen tubes (Poulter et al. 2008). However, the distinctive spindle-shaped microtubules around the generative cell remained relatively intact at this time point; at 30–60 min post-SI they were still evident, but were just starting to show signs of disintegration. Comparing the response of microtubules with that of the microfilaments, it is evident that the responses are quite distinct, as the microtubules do not reorganise later. Moreover, artificially depolymerising actin (using the drug, Latrunculin B; LatB) triggered microtubule depolymerisation, and stabilising actin with Jasplakinolide prior to SI prevented total microtubule depolymerisation. This suggested that the SI signals stimulate actin depolymerisation signal to trigger microtubule depolymerisation as a consequence (Poulter et al. 2008). Like actin, the cortical microtubules have also been found to have another role, relating to signalling to PCD; see Sect. 11.4.4.7.

11.4.4 SI Triggers Programmed Cell Death

Perhaps one of the most important findings relating to events triggered by SI in *Papaver* pollen was the demonstration of the involvement of PCD. This speculative investigation was prompted by the apparent superficial similarities between SI and the hypersensitive response (HR) in plants. Since the HR has long been known to involve PCD (Dangl et al. 1997; Greenberg et al. 1994; Lam et al. 2001), this seemed an obvious avenue to explore. These studies revealed a highly effective way to prevent fertilisation by incompatible pollen.

11.4.4.1 Features of Programmed Cell Death

Unwanted eukaryotic cells are often removed by apoptosis/PCD. There are a number of key features that are used to diagnose PCD, including cytochrome c leakage from the mitochondria into the cytosol, caspase activation and nuclear DNA fragmentation. Caspases are cysteine aspartate-specific proteases. Upon activation they cleave their substrates after an aspartic acid residue. They are key players in the initiation and execution of apoptosis; caspase-3 is a key executioner caspase (Riedl and Shi 2004). In animal cells, use of tetrapeptide substrates and inhibitors has established the substrate specificities of animal caspases. For example, caspase-3 has an optimal tetrapeptide recognition motif DEVD and is often called a DEVDase; its activity is inhibited by DEVD inhibitors; substrates that act as fluorogenic indicators for caspase activities allow a direct analysis of caspase-like activities. Through these approaches it has been established that PCD, involving a range of caspase-like activities, is activated in plant cells. However, the exact nature of plant caspases is currently a mystery, as there are no known caspase homologues (Lam and del Pozo 2000; Woltering 2004; Woltering et al. 2002).

11.4.4.2 PCD Involving a Caspase-3-like/DEVDase is Triggered by SI

An early study suggested that PCD might be involved in SI (Jordan et al. 2000), as S-specific nuclear DNA fragmentation was observed in incompatible pollen tubes. This was consolidated by other studies that provided several pieces of evidence, using a combination of key markers of PCD with caspase-inhibitors, to show that SI triggered this phenomenon in incompatible pollen. SI stimulated rapid leakage of cytochrome c into the cytosol, with significant increases detectable within 1–2 h (Thomas and Franklin-Tong, 2004). Moreover, SI-induced DNA fragmentation was inhibited by the caspase-3 inhibitor Ac-DEVD-CHO, but not the caspase-1 inhibitor Ac-YVAD-CHO (Thomas and Franklin-Tong 2004). This implicated a caspase-3-like/DEVDase activity involved in mediating the SI-induced DNA fragmentation. Data also implicated a role for Ca^{2+} signalling to PCD, suggesting that the SI-stimulated Ca^{2+} signals signal to PCD in *Papaver* pollen.

Since caspases are key mediators of PCD, the SI-stimulated DEVDase was further investigated. A key finding was that the DEVDase plays a key role in mediating SI. Pollen pre-incubated with the Ac-DEVD-CHO inhibitor was 'rescued' from inhibition, thereby implicating the DEVDase being functionally involved in mediating SI (Thomas and Franklin-Tong 2004).

Confirmation of the SI-induced DEVDase activity came from experiments showing that extracts from SI-induced incompatible pollen will cleave a classic caspase substrate, Poly (ADP-ribose) polymerase (PARP). This PARP-cleavage activity was inhibited by DEVD (Thomas and Franklin-Tong 2004). Use of a fluorescent caspase-3 substrate, Ac-DEVD-AMC, that acts as fluorogenic indicator for DEVDase activity has provided more direct evidence for the SI-induced DEVDase activity (Bosch and Franklin-Tong 2007; Li et al. 2007) and allowed detailed characterisation, in particular its temporal activation (Bosch and Franklin-Tong 2007). Imaging of the DEVDase activity in live cells has revealed that this caspase-like activity first appears in the cytosol around 1–2 h after SI induction and is later detected in the nuclei. The pollen generative cell exhibited high DEVDase activity, suggesting it is preferentially targeted.

11.4.4.3 Other Caspases Activated by SI in Incompatible Pollen

The use of AMC-based peptide caspase substrates has allowed further characterisation of the SI-induced caspase-like activities and has revealed the presence of VEIDase (caspase-6) and LEVDase (caspase-4) activities in addition to the DEVDase (caspase-3) activity previously identified in incompatible pollen tubes (Bosch and Franklin-Tong 2007). The DEVDase and VEIDase had similar temporal activation profiles, with activity detectable at 1–2 h after SI-induction, peaking relatively rapidly at 5 h post-SI, suggesting that they are involved in early PCD events. The LEVDase activity, in contrast, was activated rather slower and later, with activity still increasing 8 h, when the DEVDase and VEIDase activities had significantly decreased (Bosch and Franklin-Tong 2007). This hints that the LEVDase may be involved in later SI-mediated events.

11.4.4.4 Dramatic Cytosolic Acidification is Triggered by SI in Incompatible Pollen

Surprisingly, all three of the SI-activated caspase-like activities exhibit peak activity at pH 5. This hinted that a possible early SI-induced event was acidification of the cytosol. Investigation demonstrated that there was, indeed, a dramatic and rapid acidification of the pollen cytosol triggered in the first 1–2 h of SI (Bosch and Franklin-Tong 2007). What is involved in this phenomenon is not yet clear, but it points to major alterations to the nature of the cytosol being elicited by SI.

11.4.4.5 Involvement of MAPK Signalling in SI-Mediated PCD

MAPK cascades are often thought of as a universal signal transduction mechanism as they are involved in triggering numerous diverse signalling networks (Hamel et al. 2006). Dual phosphorylation of threonine and tyrosine residues in a TXY motif is responsible for their activation via a MAPKKK cascade. In plants, the involvement of MAPKs in biotic and abiotic stress responses has been particularly well studied; see Jonak et al. (2002), Nakagami et al. (2005) and Zhang and Klessig 2001 for reviews of MAPK signalling in plants. Investigations of signalling networks triggered by SI revealed the activation of a MAPK, p56 specifically in incompatible, but not compatible pollen (Li et al. 2007; Rudd et al. 2003). These data suggested that the p56-MAPK plays a role in integrating SI signals. However, as its activation peaked 10 min after SI induction, it could not be involved in the rapid arrest of pollen tube inhibition. p56 migrates to 56 kDa on SDS-PAGE, which hints that it may be a TDY-type MAPK, which are generally larger than other MAPKs.

Subsequent studies, led by knowledge that MAPKs are known to play a key role in mediating signalling to PCD in the plant-pathogen hypersensitive response, investigated the possible involvement of MAPKs in signalling to SI-mediated PCD. These studies, based primarily on use of the MAPK cascade inhibitor U0126, which blocks MAPK cascades in animal and plant cells (Lee et al. 2001) in combination with PCD markers, implicated a key role for MAPKs (most likely the p56-MAPK) in signalling to PCD in incompatible pollen (Li et al. 2007). U0126 was shown to prevent activation of the SI-stimulated p56-MAPK and concomitantly alleviate SI-induced DNA fragmentation, inhibit SI-induced caspase-3-like/DEVDase activity and SI-induced loss of viability (Li et al. 2007). This implicated involvement of a MAPK in signalling to PCD. As p56 is apparently the only MAPK activated by SI, this strongly implicated its involvement in participating in initiating PCD in incompatible pollen.

11.4.4.6 Actin Plays a Role in Mediating SI-Induced PCD

A surprising finding was that the actin depolymerisation participates in signalling to PCD in incompatible pollen. This possibility was investigated because the pattern of SI-induced actin depolymerisation was extensive and sustained, and this pattern of actin depolymerisation had been observed in apoptotic cells animal cells (Korichneva and Hammerling 1999; Morley et al. 2003). Studies in yeast also pointed to a role for alterations in actin dynamics playing a functional role in the initiation of apoptosis (Gourlay et al. 2004). Investigations examining the effect of the actin-depolymerising drug, latrunculin B (LatB), and the actin-stabilising/polymerising drug, jasplakinolide (Jasp), revealed that both treatments stimulated high levels of DNA fragmentation, which was mediated by a caspase-3 like/DEVDase activity (Thomas et al. 2006). Thus, it appears that disturbing actin polymer dynamics is likely to be responsible for triggering PCD in pollen. Interestingly, in yeast, actin stabilisation triggers PCD (Gourlay et al. 2004). Inducing relatively transient F-actin depolymerisation revealed that the threshold level for inducing PCD was very similar to that induced by 10 min SI (Snowman et al. 2002; Thomas et al., 2006), indicating that this is why the SI-induced actin depolymerisation is so extensive. Further experiments revealed that Jasp significantly reduced the extent of both LatB- and SI-induced DNA fragmentation (Thomas et al. 2006), indicating that preventing total actin depolymerisation alleviated PCD.

Together, these data implicated actin depolymerisation being functionally involved in the initiation of SI-induced PCD, and established a causal link between actin polymerisation status and initiation of PCD in plant cells. Thus, the rapid and substantial actin depolymerisation triggered by SI signalling not only results in the rapid inhibition of incompatible pollen tip growth, but it also activates a caspase-3-like/DEVDase activity, resulting in PCD. This suggests that the F-actin depolymerisation and PCD function together in a signalling network to prevent incompatible pollen from effecting fertilisation.

11.4.4.7 Microtubules Also Play a Role in Mediating SI-Induced PCD

As mentioned in Sect. 11.4.3.2, pollen cortical microtubules also are rapidly depolymerised in incompatible pollen tubes upon SI stimulus. Further evidence (see Sect. 11.4.4.6) showed that stabilising F-actin using jasplakinolide partially alleviated SI-induced PCD (Thomas et al. 2006), but because we found that F-actin triggers depolymerisation of microtubules, we wondered whether this might also implicate microtubules in signalling to PCD. We found that disruption of microtubule dynamics alone did not trigger PCD. However, taxol-induced stabilisation of microtubules alleviated the incidence of SI-induced PCD. If actin alone was involved, the SI-induced actin depolymerisation and PCD would progress normally in the presence of taxol. However, taxol-alleviation of PCD strongly suggest that microtubule depolymerisation is also involved in mediating PCD. Thus, SI signals target both the actin and the microtubule cytoskeleton and it appears that they work together, using signal integration between microfilaments and microtubules to implement PCD.

11.5 An Overall Model for Mechanisms Involved in Regulating SI in *Papaver*

From the studies described here, it is clear that there are several events triggered during interaction between incompatible pistil and pollen products. Figure 11.2 summarises what we know so far about the signals and targets for SI in incompatible *Papaver* pollen. Figure 11.2a shows a compatible situation, where the secreted pistil S-proteins cannot interact with the pollen *S*-receptor as they do not have the correct (matching *S*-haplotype) specificities. In this situation, the pollen germinates and grows normally and appears indistinguishable from unchallenged pollen tubes. Figure 11.2b shows an incompatible situation, where the pistil S-protein interacts with the pollen *S*-receptor as it has the correct specificity. This proposed receptor–ligand interaction causes Ca^{2+} influx in the shank of the pollen tube, rapid increases in $[Ca^{2+}]_i$ within seconds. This triggers a set (probably a network) of events in the incompatible pollen grain/tube. Studies indicate that these SI-specific events can be elicited in both pollen grains or pollen tubes, and that it does not matter at which stage pollen is affected, as long as polarity is established. Several SI-triggered events affect tip growth. The rapid actin depolymerisation will cause rapid inhibition of

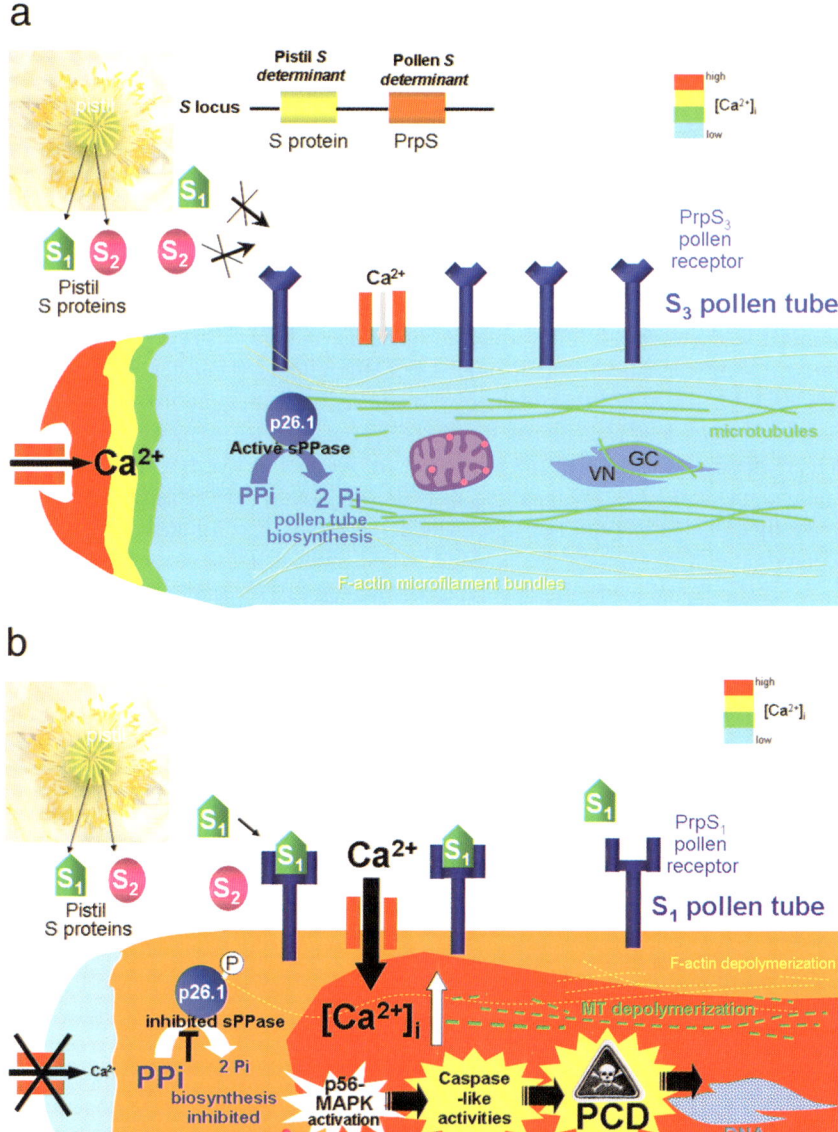

Fig. 11.2 Signals and targets involved in SI in Papaver. Comparisons between compatible and incompatible pollen tubes. The pollen tubes are shown in pseudo-colour to show $[Ca^{2+}]_i$; 'hot' colours indicate high $[Ca^{2+}]_i$ and 'cool' colours (*blue*) indicate low, basal levels of $[Ca^{2+}]_i$. The *S*-locus comprises linked pistil and pollen *S*-determinants. (**a**) A compatible situation, where pollen and pistil *S*-determinants do not match. Pistil S-proteins secreted by the stigma (S_1 and S_2 in this example) cannot interact with the pollen *S*-receptor, which has an S_3 specificity, as the haplotypes

growth. Inhibition of Pr-p26.1 sPPase activity by Ca^{2+} and phosphorylation will also cause inhibition of tip growth, but this is likely to be slower in affecting tip growth than the actin response. It probably also has a longer term effect on pollen tube metabolism, thereby preventing growth resuming. Loss of tip growth probably results in the loss of the apical $[Ca^{2+}]_i$ gradient and apical Ca^{2+} influx observed. Together these events comprise the *early* phase of the SI response, resulting in inhibition of tip growth (see Fig. 11.2b).

The second phase of SI is thought to involve a 'decision-making' phase, where the pollen tube enters PCD. The sustained actin and microtubule depolymerisation, which appear to be required in concert with actin, for PCD to proceed are upstream of activation of a caspase-like enzyme, resulting in execution of PCD (see Fig. 11.2b). However, how this is achieved is not yet known. It is possible that this integrates with the activation of the p56-MAPK, which is likely to signal to PCD. Identifying what is upstream of MAPK activation will be an important part of the jigsaw to establish. Nevertheless, the data indicate that it is not involved in tip growth inhibition as its activation occurs too late; data strongly implicate that it signals to PCD (see Fig. 11.2b). Cytochrome c release is an early event, occurring around 1 h post-SI, with implicated involvement in PCD, but how this relates to other events is not yet clear. Acidification of the cytosol is another event around 2 h post-SI, which is likely to constitute a crucial decision-making phase, allowing activation of the caspase-like activities, and probably marking the ultimate 'point of no return' for incompatible pollen tubes. Thus, a network of events ensure first that pollen tubes are inhibited and then that they do not achieve fertilisation, is set in train by the initial receptor–ligand interaction.

11.5.1 A Contrast to the S-RNase System and **Brassica** *SI Systems*

The poppy SI system appears to have evolved a number of mechanisms to exploit several 'weak points' of pollen–pistil interactions. This has provided a range of

Fig. 11.2 (continued) do not match. Thus, the pollen resembles an unchallenged pollen tube. Pollen germinates and grows, appearing indistinguishable from unchallenged pollen tubes. The pollen tube has a high apical $[Ca^{2+}]_i$ gradient and low, basal levels of $[Ca^{2+}]_i$ in the 'shank' of the pollen tube; normal levels of sPPase activity enables biosynthesis and actin microfilaments enable delivery of vesicles to the tip allowing growth. (**b**) An incompatible interaction, where pollen and pistil *S*-determinants match. One of the pistil S-proteins secreted by the stigma (S_1 in this example) interacts with the pollen *S*-receptor, which has an S_1 specificity. This results in rapid Ca^{2+} influx and high $[Ca^{2+}]_i$ in the shank (indicated by orange/red), while $[Ca^{2+}]_i$ at the tip decreases. F-actin and microtubules are rapidly depolymerised (indicated by dotted lines) and phosphorylation of sPPases occurs within minutes, resulting in inhibition of activity and biosynthesis. At 5–10 min, a MAPK is activated. Cytochrome c release occurs at ∼1 h. Cytosolic acidification of the cytosol occurs prior to activation of several caspase-like activities, resulting in PCD and ultimate DNA fragmentation and cellular dismantling

means to interfere with pollen tube growth, resulting in its inhibition. The regulation of pollen tube growth is complex and involves many components, and this is readily appreciated when one considers the mechanisms employed in this SI system. What is most striking is the contrast with the RNase SI system, which has made what appears to be a completely different choice with respect to means to inhibit pollen tube growth (see Chap. 10, this volume). Instead of employing cytotoxic RNase molecules, using ubiquitination and sequestration to regulate SI, *Papaver* appears to utilise a different set of existing cellular mechanisms that are important or crucial for pollen tube growth and function, using a cell signalling *trigger* of $[Ca^{2+}]_i$ to set in motion a network of events in incompatible pollen that results in its inhibition. This rapid inhibition is probably advantageous, as *Papaver* has no style, and so the slow response employed by the S-RNase system would be far less effective in this case. It is perhaps even more interesting that the *Papaver* SI system is also quite different to the Brassica SI system, which has evolved a rather different rapid response, but which is based on a signalling cascade triggered by a receptor-like kinase (RLK) system, with apparently rather different downstream events triggered (see Chap. 8, this volume). Thus, at face value, there appear to be great differences between the different SI systems. However, as few, if any, direct comparisons have been made between the components and mechanisms used for these different SI systems have been made to date, it is difficult to definitively state how different the mechanisms recruited for inhibition of incompatible pollen really are.

11.6 Future Perspectives

Although we have made great progress in our understanding of what is involved in the *Papaver* SI pollen incompatible SI response, many questions remain to be answered. For example, what other components are involved, and how do all the components identified interact and 'cross-talk' effectively to mount an integrated response?

Although much progress has been made with analysing the pollen response, the identification of the pollen components has been long-awaited. Hopefully this will soon be published. This will mark the starting point for an exciting phase of work exploring the exact nature of the pollen determinant and opens the way for receptor–ligand interaction studies.

From what we know so far, it appears that the *Papaver* SI system may be evolved from a pre-existing signalling system involved in programmed cell death. It would be interesting in the future to examine whether other parallels with the PCD system exist. For example, is SI really similar to the hypersensitive response (HR), or are these similarities only at a superficial level? Are there other signals used by HR, for example ROS and NO signalling, which are known to exist in pollen tubes (Prado et al., 2004), that may play a role in *Papaver* SI? Similarly, are other elements of PCD also involved in this SI system? Answers to these questions will not only help us in understanding this SI system, but will also elucidate how this SI system evolved.

It would also be extremely interesting to examine the other SI systems carefully to establish how different the mechanisms recruited for inhibition of incompatible pollen really are. Have the different systems arisen completely independently, recruiting different ways in which to inhibit pollen tube growth, or is there evidence of evolutionary links, suggesting that there may be some similar mechanisms utilised? Moreover, examining occurence of the *Papaver S*-determinants in other species has not been investigated to date. This type of study could provide important information about the distribution and phylogeny of these *S*-alleles, and could contribute to our knowledge towards understanding how SI evolved.

Acknowledgements Work in the lab of V.E.F-T. is currently funded by the Biotechnology and Biological Sciences Research Council (B.B.S.R.C.). Some of the work described was funded by a Royal Society China Fellowship. I acknowledge, in particular, my long-term collaborators Chris Staiger and Chris Franklin, as well as other collaborators and members of the SI community for useful and stimulating discussions over the years, and the many members of my lab who were involved in these studies.

References

Bosch M, Franklin-Tong VE (2007) Temporal and spatial activation of caspase-like enzymes induced by self-incompatibility in Papaver pollen. Proc Natl Acad Sci USA 104:18327–18332

Bredemeijer GMM, Blaas J (1981) *S*-Specific proteins in styles of self-incompatible *Nicotiana alata*. Theor Appl Genet 59:185–190

Campbell JM, Lawrence MJ (1981) The population genetics of the self-incompatibility polymorphism in *Papaver rhoeas*. I. The number and distribution of S-alleles in families from three localities. Heredity 46:69–79

Clarke SR, Staiger CJ, Gibbon BC, Franklin-Tong VE (1998) A potential signaling role for profilin in pollen of *Papaver rhoeas*. Plant Cell 10:967–979

Dangl JL, Morel J-B (1997) The hypersensitive response and the induction of cell death in plants. Cell Death Differ 4:671–683

de Graaf BHJ, Rudd JJ, Wheeler MJ, Perry RM, Bell EM, Osman K, Franklin FCH, Franklin-Tong VE (2006) Self-incompatibility in Papaver targets soluble inorganic pyrophosphatases in pollen. Nature 444:490–493

de Nettancourt D (1977) Incompatibility in angiosperms. Springer, Berlin Heidelberg New York

Dietrich A, Mayer JE, Hahlbrock K (1990) Fungal elicitor triggers rapid, transient, and specific protein-phosphorylation in parsley cell-suspension cultures. J Biol Chem 265:6360–6368

Foote HCC, Ride JP, Franklin-Tong VE, Walker EA, Lawrence MJ, Franklin FCH (1994) Cloning and expression of a distinctive class of self- incompatibility (S) gene from *Papaver rhoeas* L. Proc Natl Acad Sci USA 91:2265–2269

Franklin-Tong VE, Lawrence MJ, Franklin FCH (1988) An *in vitro* bioassay for the stigmatic product of the self-incompatibility gene in *Papaver rhoeas* L. New Phytol 110:109–118

Franklin-Tong VE, Ride JP, Read ND, Trewavas AJ, Franklin FCH (1993) The self-incompatibility response in *Papaver rhoeas* is mediated by cytosolic-free calcium. Plant J 4:163–177

Franklin-Tong VE, Ride JP, Franklin FCH (1995) Recombinant stigmatic self-incompatibility-(S-) protein elicits a Ca^{2+} transient in pollen of *Papaver rhoeas*. Plant J 8:299–307

Franklin-Tong VE, Hackett G, Hepler PK (1997) Ratio-imaging of $[Ca^{2+}]_i$ in the self-incompatibility response in pollen tubes of *Papaver rhoeas*. Plant J 12:1375–1386

Franklin-Tong VE, Holdaway-Clarke TL, Straatman KR, Kunkel JG, Hepler PK (2002) Involvement of extracellular calcium influx in the self-incompatibility response of *Papaver rhoeas*. Plant J 29:333–345

Geitmann A, Snowman BN, Emons AMC, Franklin-Tong VE (2000) Alterations in the actin cytoskeleton of pollen tubes are induced by the self-incompatibility reaction in *Papaver rhoeas*. Plant Cell 12:1239–1251

Geitmann A, Franklin-Tong VE, Emons AC (2004) The self-incompatibility response in *Papaver rhoeas* pollen causes early and striking alterations to organelles. Cell Death Differ 11:812–822

Gibbon BC, Kovar DR, Staiger CJ (1999) Latrunculin B has different effects on pollen germination and tube growth. Plant Cell 11:2349–2364

Gourlay CW, Carpp LN, Timpson P, Winder SJ, Ayscough KR (2004) A role for the actin cytoskeleton in cell death and aging in yeast. J Cell Biol 164:803–809

Greenberg JT, Guo A, Klessig DF, Ausubel FM (1994) Programmed cell death in plants: A pathogen-triggered response activated co-ordinately with multiple defence functions. Cell 77:551–563

Hamel L-P, Nicole M-C, Sritubtim S, Morency M-J, Ellis M, Ehlting J, Beaudoin N, Barbazuk B, Klessig D, Lee J, Martin G, Mundy J, Ohashi Y, Scheel D, Sheen J, Xing T, Zhang S, Seguin A, Ellis BE (2006) Ancient signals: Comparative genomics of plant MAPK and MAPKK gene families. Trends Plant Sci 11:192–198

Hearn MJ, Franklin, FCH, Ride JP (1996) Identification of a membrane glycoprotein in pollen of *Papaver rhoeas* which binds stigmatic self-incompatibility (S-) proteins. Plant J 9:467–475

Hetherington AM, Brownlee C (2004) The generation of Ca^{2+} signals in plants. Annu Rev Plant Biol 55:401–427

Hinata K, Nishio T, Kimura J (1982) Comparative studies on S-glycoproteins purified from different *S*-genotypes in self-incompatible *Brassica* species. II. Immunological specificities. Genetics 100:649–657

Huang S, Blanchoin L, Chaudhry F, Franklin-Tong VE, Staiger CJ (2004) A gelsolin-like protein from *Papaver rhoeas* pollen (PrABP80) stimulates calcium-regulated severing and depolymerization of actin filaments. J Biol Chem 279:23364–23375

Jonak C, Ökrész L, Bögre L, Hirt H (2002) Complexity, cross talk and integration of plant MAP kinase signalling. Curr Opin Plant Biol 5:415–424

Jones-Rhoades MW, Borevitz J, Preuss D (2007) Genome-wide expression profiling of the *Arabidopsis* female gametophyte identifies families of small, secreted proteins. PLoS Genet 3:1848–1861

Jordan ND, Kakeda K, Conner A, Ride JP, Franklin-Tong VE, Franklin FCH (1999) S-protein mutants indicate a functional role for SBP in the self-incompatibility reaction of *Papaver rhoeas*. Plant J 20:119–125

Jordan ND, Franklin FCH, Franklin-Tong VE (2000) Evidence for DNA fragmentation triggered in the self- incompatibility response in pollen of *Papaver rhoeas*. Plant J 23:471–479

Kakeda K, Jordan ND, Conner A, Ride JP, Franklin-Tong VE, Franklin FCH (1998) Identification of residues in a hydrophilic loop of the *Papaver rhoeas* S protein that play a crucial role in recognition of incompatible pollen. Plant Cell 10:1723–1731

Korichneva I, Hammerling U (1999) F-actin as a functional target for retro-retinoids: A potential role in anhydroretinol-triggered cell death. J Cell Sci 112:2521–2528

Kornberg A (1962) On the metabolic significance of phosphorolytic and pyrophosphorolytic reactions. Academic Press, New York, pp 251–264

Kurup S, Ride JP, Jordan N, Fletcher G, Franklin-Tong VE, Franklin FCH (1998) Identification and cloning of related self-incompatibility S- genes in *Papaver rhoeas* and *Papaver nudicaule*. Sex Plant Reprod 11:192–198

Lam E, del Pozo O (2000) Caspase-like protease involvement in the control of plant cell death. Plant Mol Biol 44:417–428

Lam E, Kato N, Lawton M (2001) Programmed cell death, mitochondria and the plant hypersensitive response. Nature 411:848–853

Lane MD, Lawrence MJ (1993) The population genetics of the self-incompatibility polymorphism in *Papaver Rhoeas*. VII. The number of S-alleles in the species. Heredity 71:596–602

Lawrence MJ (1975) The genetics of self-incompatibility in *Papaver rhoeas*. Proc R Soc Lond. Ser B 188:275–285

Lawrence MJ, O'Donnell S (1981) The population genetics of the. self-incompatibility polymorphism in. *Papaver rhoeas*. III. The number and frequency of S-alleles in two further natural populations (R102 and R104). Heredity 47:53–61

Lawrence MJ, Lane MD, O'Donnell S, Franklin-Tong VE (1993) The population-genetics of the self-incompatibility polymorphism in *Papaver Rhoeas*. V. Cross-classification of the S-alleles of samples from 3 natural-populations. Heredity 71:581–590

Lee J, Klessig DF, Nurnberger T (2001) A harpin binding site in tobacco plasma membranes mediates activation of the pathogenesis-related gene HIN1 independent of extracellular calcium but dependent on mitogen-activated protein kinase activity. Plant Cell 13:1079–1093

Li S, Samaj J, Franklin-Tong VE (2007) A mitogen-activated protein kinase signals to programmed cell death induced by self-incompatibility in *Papaver* pollen. Plant Physiol 145:236–245

Malho R, Read ND, Trewavas AJ, Pais MS (1995) Calcium channel activity during pollen tube growth and reorientation. Plant Cell 7:1173–1184

Mau SL, Raff J, Clarke AE (1982) Isolation and partial characterization of components of *Prunus-Avium* L styles, including an antigenic glycoprotein associated with a self-incompatibility genotype. Planta 156:505–516

Morley SC, Sun GP, Bierer BE (2003) Inhibition of actin polymerization enhances commitment to and execution of apoptosis induced by withdrawl of trophic support. J Cell Biochem 88:1066–1076

Nakagami H, Pitzschke A, Hirt H (2005) Emerging MAP kinase pathways in plant stress signalling. Trends Plant Sci 7:339–346

Poulter NS, Vatovec S, Franklin-Tong VE (2008) Microtubules are a target for self-incompatibility signaling in *Papaver* pollen. Plant Physiol 146:1358–1367

Prado AM, Porterfield DM, Feijo JA (2004) Nitric oxide is involved in growth regulation and re-orientation of pollen tubes. Development 131:2707–2714

Ride JP, Davies EM, Franklin FCH, Marshall DF (1999) Analysis of Arabidopsis genome sequence reveals a large new gene family in plants. Plant Mol Biol 39:927–932

Riedl SJ, Shi Y (2004) Molecular mechanisms of caspase regulation during apoptosis. Nat Rev Mol Cell Biol 5:897–907

Rudd JJ, Franklin-Tong VE (1999) Calcium signaling in plants. Cell Mol Life Sci 55:214–232

Rudd JJ, Franklin FCH, Lord JM, FranklinTong VE (1996) Increased phosphorylation of a 26-kDa pollen protein is induced by the self-incompatibility response in *Papaver rhoeas*. Plant Cell 8:713–724

Rudd JJ, Osman K., Franklin, FCH., Franklin-Tong VE (2003) Activation of a putative MAP kinase in pollen is stimulated by the self-incompatibility (SI) response. FEBS Lett 547:223–227

Snowman BN, Kovar DR, Shevchenko G, Franklin-Tong VE, Staiger CJ (2002) Signal-mediated depolymerization of actin in pollen during the self-incompatibility response. Plant Cell 14:2613–2626

Staiger CJ (2000) Signalling to the actin cytoskeleton in plants. Annu Rev Plant Physiol Plant Mol Biol 51:257–288

Staiger CJ, Blanchoin L (2006) Actin dynamics: Old friends with new stories. Cell Biol 9:554–562

Takayama S, Isogai A (2005) Self-incompatibility in plants. Annu Rev Plant Biol 56:467–489

Takemoto D, Hardham AR (2004) The cytoskeleton as a regulator and target of biotic interactions in plants. Plant Physiol 136:3864–3876

Thomas SG, Franklin-Tong VE (2004) Self-incompatibility triggers programmed cell death in *Papaver* pollen. Nature 429:305–309

Thomas SG, Huang S, Li S, Staiger CJ, Franklin-Tong VE (2006) Actin depolymerization is sufficient to induce programmed cell death in self-incompatible pollen. J Cell Biol 174:221–229

Walker EA, Ride JP, Kurup S, FranklinTong VE, Lawrence MJ, Franklin FCH (1996) Molecular analysis of two functional homologues of the S-3 allele of the *Papaver rhoeas* self-incompatibility gene isolated from different populations. Plant Mol Biol 30:983–994

Wasteneys GO (2003) Microtubules show their sensitive nature. Plant Cell Physiol 44:653–654

Woltering EJ (2004) Death proteases come alive. Trends Plant Sci 9:469–472

Woltering EJ, van der Bent A, Hoeberichts FA (2002) Do plant caspases exist? Plant Physiol 130:1764–1769

Wright S (1939) The distribution of self-sterility alleles in populations. Genetics 24:538–552

Zhang S, Klessig DF (2001) MAPK cascades in plant defense signaling. Trends Plant Sci 6:520–527

Chapter 12
Molecular Genetics of Sporophytic Self-Incompatibility in *Ipomoea*, a Member of the Convolvulaceae

Y. Kowyama, T. Tsuchiya, and K. Kakeda

Abstract Diploid *Ipomoea trifida* in the Convolvulaceae is a close relative of the cultivated hexaploid species, the sweet potato, and has sporophytic self-incompatibility controlled by a single multi-allelic *S*-locus. Genetic analyses of *I. trifida* plants collected from native populations in Central America have identified a number of different *S*-haplotypes, which show a linear dominance hierarchy with some co-dominance relationships. A linkage map of DNA markers showed that the *S*-locus is delimited to a 0.23 cM region and is located in the *S*-haplotype-specific divergent region (SDR) that has a physical size of 35–95 kb. Of the six genes located within the SDR, three stigma-specific novel genes, *SE1*, *SE2* and *SEA*, and an anther-specific gene, *AB2*, are candidates for encoding pistil and pollen determinants of self-incompatibility, respectively, suggesting that a unique recognition mechanism is involved in the self-incompatibility system of *Ipomoea*.

Abbreviations

AB1/2/3	Anther-specific genes expressed at the B-stage of development (1 week before anthesis); *AB2* is a candidate pollen *S*-determinant
AFLP	Amplified fragment length polymorphism
AMF	AFLP-based mRNA fingerprinting
BAC	Bacterial artificial chromosome
cM	Centimorgan
GSI	Gametophytic self-incompatibility

Y. Kowyama and K. Kakeda
Graduate School of Bioresources, Mie University, Tsu 514-8507, Japan, e-mail: kouyama@bio.mie-u.ac.jp

T. Tsuchiya
Life Science Research Center, Mie University, Tsu 514-8507, Japan

MX1	A spontaneous self-compatible mutant
SC	Self-compatible
SDR	*S*-haplotype-specific divergent region
SE1/2	Stigma-specific expression genes; candidate pistil *S*-determinants
SEA	Stigma-specific expression gene with alternative transcripts; a candidate pistil *S*-determinant
SFB/SLF	*S*-haplotype-specific F-box/*S*-locus F-box (the pollen *S*-determinant in many GSI systems)
SI	Self-incompatibility
SLG	*S*-locus glycoprotein
S-locus	Self-incompatibility locus
SP11/SCR	*S*-locus protein 11/*S*-locus cysteine-rich protein (the pollen *S*-determinant in *Brassica*)
SRK	*S*-locus receptor kinase (the pistil *S*-determinant in *Brassica*)
S-RNase	*S*-locus ribonuclease (the pistil *S*-determinant in many GSI systems)
SSI	Sporophytic self-incompatibility
SSP	*S*-locus-linked stigma protein

12.1 Introduction

Sporophytic self-incompatibility (SSI) occurs in several plant families, such as the Brassicaceae, Asteraceae, Malvaceae and Convolvulaceae (de Nettancourt 2001; see Chap. 4). In these families, self-incompatibility (SI) is genetically controlled by a single multi-allelic locus, the *S*-locus. In some of the plant species characterised to date, the *S*-locus has been shown to encode at least two distinct genes, the pollen and pistil determinant genes, responsible for the self-incompatibility reaction (Takayama and Isogai 2005; McClure and Franklin-Tong 2006). The set of genes coded in the *S*-locus, termed the *S*-haplotype (Nasrallah and Nasrallah 1993), is inherited as a single unit to maintain the self-incompatibility system. In SSI, the pollen phenotype is determined by the diploid *S*-haplotypes of the pollen-producing parent, whereas in gametophytic self-incompatibility (GSI), phenotype is determined by the haploid pollen genotype. The S-RNase-mediated system of GSI is widely distributed throughout angiosperms; this system is thought to be the ancestral form for other types of self-incompatibility (Steinbachs and Holsinger 2002; see Chap. 4). The most extensive molecular studies of SSI have been carried out in the Brassicaceae (see Chap. 7 for a review of the discovery of the *Brassica S*-locus determinants; Chap. 8 for a review of possible mechanisms involved in the SI response and Chap. 6 for a discussion of SRK and SP11/SCR polymorphism). To understand the evolution and molecular basis of self-incompatibility systems in higher plants, it will be essential to obtain information from a wide range of plant families. The present chapter is focused on the SSI system of *Ipomoea*, Convolvulaceae.

12.2 Sexual Reproduction in the Genus *Ipomoea*

About 150 years ago, Charles Darwin investigated outcrossing in higher plants and showed a diversity of sexual reproduction modes, such as dichogamy, monoecy, heterostyly and self-incompatibility. In his book, *The effects of cross- and self-fertilisation in the vegetable kingdom* (Darwin 1876), he provided a detail description of the phenomenon of 'inbreeding depression', based on the long-term experiments using the morning glory, *Ipomoea purpurea* (Convolvulaceae). The Convolvulaceae contains 55 genera and approximately 1,700 species (Austin 1997). The genus *Ipomoea*, the largest member of the Convolvulaceae, is taxonomically classified into several sections, based on their morphological and reproductive traits (Austin and Huáman 1996). For example, the section Pharbitis includes the self-incompatible species *I. serifera* and self-compatible species such as *I. purpurea*, *I. nil* and *I. hederacea*, that are known as horticultural varieties of the morning glory. Section Batatas consists of two groups of species, groups A and B, which are sexually isolated from each other (Nishiyama et al. 1975). Group A species, such as *I. lacunosa, I. triloba* and *I. tiliacea*, are self-compatible, whereas group B species such as *I. trifida*, *I. tabascana* and *I. batatas*, are self-incompatible. In group B, the diploid species *I. trifida* and the tetraploid species *I. tabascana* are most closely related to the hexaploid sweet potato *I. batatas* (Rajapakse et al. 2004). The sweet potato is one of the major staple food crops and is also used for the industrial production of starch and alcohol.

Since Terao (1934) first found cross-incompatibility groups in cultivars of the sweet potato, a large number of local varieties and lines have been classified into several incompatibility groups. These groups are defined by whether they can achieve successful cross-pollination, which is usually assessed by pollen-tube growth on the stigma and/or by seed set following reciprocal pollination with tester lines. Cross-incompatibility is presumed to result from pollination between parental plants with the same self-incompatibility phenotype. In the practical breeding of sweet potato, cross-incompatibility between parental plants is a barrier to production of hybrid progeny, because choice of parents for cross-pollination is sometimes limited to a small number of lines. *I. trifida*, a diploid species of the section Batatas, is a useful genetic resource in sweet potato breeding (Shiotani and Kawase 1987). Plant of *I. trifida* is herbaceous insect-pollinated weed native to Central America (Fig. 12.1A), and has a relatively small genome size (532 Mb per haploid) (Arumuganathan and Earle 1991).

12.3 Genetics of Self-Incompatibility in *Ipomoea*

In compatible cross-pollination, pollen germinates about 10–20 min after landing on the papilla cells of the stigma (Fig. 12.1B). In the case of self-pollination, however, complete arrest of pollen germination occurs even after several hours, resulting in failure of fertilisation and seed set. This indicates that in the self-incompatibility system of *Ipomoea*, recognition of self-pollen occurs rapidly after

Fig. 12.1 (**A**) Flowers of *Ipomoea trifida*, a diploid species of the section Batatas of the genus *Ipomoea*, family Convolvulaceae. (**B**) Pollen on the papilla cells of the stigma surface following cross-pollination. The scanning electron micrograph was taken by Dr. M. Iwano, NAIST

pollination, presumably following an interaction between *S*-gene products of the pollen and the stigma. Our earlier genetic studies in *Ipomoea* clearly showed that the self-incompatibility phenotypes segregated as expected for a single *S*-locus with multiple alleles in the progeny derived from crosses of different *S*-genotypes. Our investigations also showed that the pollen phenotype is sporophytically controlled with dominance relationships among *S*-haplotypes (Kowyama et al. 1980, 1994). The inhibition of pollen germination on the surface of the stigma and the rapid response to self-pollen recognition are consistent with the general features of SSI. Compared to self-incompatibility in Brassicaceae, *Ipomoea* shows a much more clear-cut and stronger incompatibility reaction. CO_2 treatments and bud pollination, which are known to induce a breakdown of self-incompatibility in *Brassica* (Hinata et al. 1994), have no effect on self-pollination in *Ipomoea*.

The number and frequencies of different *S*-haplotypes have been examined in *Brassica* (Nou et al. 1993) and in some GSI-type plants such as *Trifolium* (Williams and Williams 1947), *Lolium* (Fearon et al. 1994) and *Papaver* (Lane and Lawrence 1993). These studies identified many different *S*-haplotypes in their natural populations (Franklin et al. 1995). A genetic analysis of 224 plants in *I. trifida* collected from six natural populations in Central America identified 49 *S*-haplotypes (Kowyama et al. 1994). These *S*-haplotypes show a linear hierarchy of the dominant-recessive *S*-alleles that can be grouped into five classes of co-dominance (Fig. 12.2). The linear dominance relationship in *Ipomoea* might reflect the sequential generation of new *S*-haplotypes from recessive to more dominant alleles. Of 326 allelic pairs examined between 28 *S*-haplotypes, 20 pairs (6%) showed different interactions in the pollen and stigma. For example, S_{11} and S_{30} are co-dominant in the stigma, while S_{11} is dominant over S_{30} in the pollen. This variation in allelic interaction suggests that the expression of *S*-determinant genes is controlled in different ways in the pollen and stigma. Recent studies in *Brassica* demonstrate that the dominance relationship between the pollen *S*-alleles is regulated at the level of transcription, and that the expression of a recessive allele is specifically suppressed by the de novo DNA methylation of a promoter sequence of the gene (Shiba et al. 2002, 2006; see also Chap. 7).

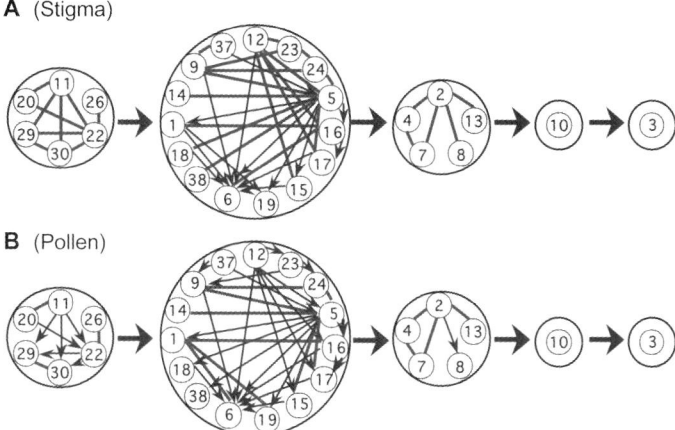

Fig. 12.2 Dominance relationships among S-alleles in the stigma (**A**) and pollen (**B**). For pairs of alleles connected by arrows, the allele indicated by the arrowhead is recessive. Alleles connected with a straight line are co-dominant

Analyses of self-compatible (SC) mutants have provided valuable information for elucidating the genetic features of self-incompatibility in *Lycopersicon* (Royo et al. 1994), *Brassica* (Watanabe et al. 1997) and *Pyrus* (Sassa et al. 1997). In *I. trifida*, only one SC plant (MX1) has been found as a spontaneous mutant from the hundreds of plants surveyed to date from a natural population of Central America (Kakeda et al. 2000). Genetic analysis of the F_1 progeny of crosses between MX1 and three different S homozygotes (S_1, S_3 and S_{22}) showed that self-compatibility in MX1 is due to a mutation in the S-locus: the mutated haplotype is designated S_c. The S_c haplotype is recessive to S_{22} and S_1, but is dominant to S_3, that is the dominance relationship is $S_{22} > S_1 > S_c > S_3$. Thus, the mutated S-haplotype is dominant over a functional S-haplotype, suggesting that the product of the S_c-haplotype confers a dominant negative effect on the functional product of the S_3-haplotype.

12.4 Stigma-Specific Proteins

In initial studies of the gene products associated with self-incompatibility, an S-locus glycoprotein (SLG) in *Brassica* (Nishio and Hinata 1977, Nasrallah et al. 1985) and an S-locus ribonuclease (S-RNase) in *Nicotiana* (Anderson et al. 1986) were identified as pistil proteins that co-segregate with S-haplotypes (see also Chaps. 7 and 9).

SLG and S-RNase are major proteins that are expressed abundantly in the pistil and are detectable by standard electrophoresis of tissue extracts. Currently, the functional role of SLG is somewhat controversial, since some functional S-haplotypes of *B. rapa* lack the *SLG* gene (Suzuki et al. 2003). Nevertheless, these pioneering studies stimulated further molecular studies on self-incompatibility in flowering plants. A similar approach was attempted with *Ipomoea*. Two-dimensional polyacrylamide gel electrophoresis (2D-PAGE) analysis of stigma proteins in *Ipomoea* detected highly polymorphic proteins of about 70 kDa in the acidic region (∼pH 4–6) of the gel. These proteins were found to be associated with specific *S*-haplotypes, and were therefore designated *S*-locus-linked stigma proteins (SSPs) (Kowyama et al. 2000). Analysis of cDNA clones showed that SSP belongs to the short-chain alcohol dehydrogenase (SCAD) family. SSPs are expressed predominantly in mature papilla cells of the stigma at the time of anthesis. The amino acid sequences of the SSPs are, however, more than 95% identical among different *S*-haplotypes. In addition, the *SSP* gene maps at 1.1 cM from the *S*-locus (Tomita et al. 2004a). These results suggest that SSP is unlikely to be involved in the self-pollen recognition of self-incompatibility in *Ipomoea*. Our surveys of 2D-PAGE profiles from stigma and pollen extracts have so far identified no *S*-haplotype-specific proteins other than SSP. This suggests that the *S*-locus gene products of *Ipomoea* are minor proteins that might be present at too low a level to be detectable by silver staining of the gel following standard 2D-PAGE.

12.5 Physical Size of the *S*-locus

Fine-scale mapping of DNA markers enables the identification of a gene of interest by a positional cloning strategy. AFLP (amplified fragment length polymorphism) analysis of genomic DNA is a useful approach for identifying molecular markers tightly linked to or co-segregating with a genetic trait (Agrama et al. 2002). The AFLP-based mRNA fingerprinting (AMF) method has also been successfully used to detect transcripts that are differentially expressed in various organs or at various stages of development, and employed to identify differences in gene expression among different genotypes (Simoes-Araujo et al. 2002). DNA markers linked to the *S*-locus of *Ipomoea* were identified by AFLP and AMF analyses using the combined genomic DNAs and cDNAs of 10–15 plants for each of four *S* genotypes in the F_1 progeny from a single cross, $S_1S_{22} \times S_{10}S_{29}$. Pooling the PCR templates for AFLP and AMF analyses is essential to achieve uniformity in the genetic background except for the gene of interest (Michelmore et al. 1991). For the AMF analysis, cDNA pools were separately prepared from the mature stigmas and anthers of each *S*-haplotype. The F_1 population, consisting of 873 plants, was further used for RFLP (restriction fragment length polymorphism) analysis to examine linkage of DNA markers.

From the AFLP and AMF analyses, eight DNA markers linked to the *S*-locus were found; three of these markers (SAM-23, AAM-68 and AF-41) mapped closely to the *S*-locus (Fig. 12.3A) (Tomita et al. 2004b). The marker SAM-23, derived from the stigma AMF, is a partial sequence of the *SSP* gene that was previously identified by 2D-PAGE of stigma extracts (see Sect. 12.4). The marker AAM-68, obtained from the anther AMF, is tightly linked to the *S*-locus and no recombinants

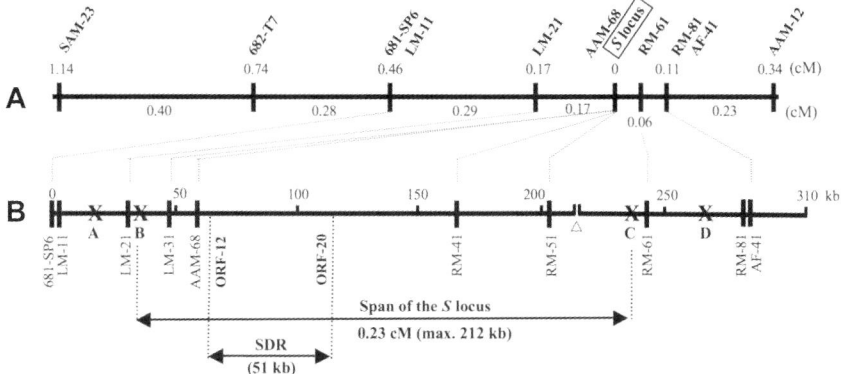

Fig. 12.3 Linkage map around the S-locus and the location of recombination breakpoints in the S_1-haplotype. (**A**) Linkage map of the DNA markers derived from AFLP and AMF analyses and from PCR amplified fragments. The numbers above and below the line indicate genetic distances in centimorgans (cM) from the S-locus. (**B**) Physical map of the DNA markers and positions of the recombination breakpoints. The numbers above the line indicate physical distance in kilobase pairs (kb). The breakpoints, A to D are marked with X's on the map. The physical span of the S-locus (~212 kb) was estimated from the distance between breakpoints B and C. The S-haplotype-specific divergent region (SDR) was estimated to be about 51 kb based on sequence comparison between the S_1- and S_{10}-haplotypes. The figure is redrawn from Rahman et al. (2007a) with permission from Springer

were detected in the 873 F_1 plants examined. The gene corresponding to the AAM-68 marker encodes a member of the glycosyltransferase family and is specifically expressed in the anther/pollen. However, AAM-68 cDNA clones isolated from different S-haplotypes exhibit a high similarity in their deduced amino acid sequences. This suggests that the AAM-68 gene is unlikely to be involved in the specificity of SI in *Ipomoea*. The AFLP marker AF-41 has sequence similarity to a histone deacetylase of *Arabidopsis*; it is located on the other side of the S-locus to SAM-23 and is 0.11 cM from the S-locus. Using these DNA markers as probes, BAC and cosmid clones covering the S-locus were screened from genomic libraries of the S_1-haplotype. Terminal end sequences of the BAC clones, 682-T7 and 681-SP6, map at 0.74 and 0.46 cM from the S-locus, respectively. From the mapping of these DNA markers, the S-locus has been delimited to a 0.57 cM region between the markers 681-SP6 and AF-41.

To date, no recombination events between the male and female determinants of self-incompatibility have been observed in any plant species examined, even among several hundred progeny of plants segregating for S-haplotypes (Casselman et al. 2000; Ikeda et al. 2005). Intergenic recombination would result in the breakdown of the self-incompatibility system. Therefore, the functional S-locus is thought to be located in a region in which recombination is suppressed, and the boundaries of the S-locus region can be defined by mapping recombination breakpoints in the flanking regions. Using recombinants detected between the S-locus and the closest DNA markers in *I. trifida*, recombination breakpoints were determined by sequencing genomic PCR fragments amplified from the neighbouring sequences of the markers (Rahman et al. 2007a). Four breakpoints have been identified and located on the linkage map (Fig. 12.3B). This analysis further delimited the S-locus of *Ipomoea* to a 0.23 cM region between the PCR markers LM-21 and RM-61. The physical size of the S-locus is estimated to be a maximum of 212 kb in the S_1-haplotype. In *Brassica campestris*, Casselman et al. (2000) showed that the physical limits of the S-locus in the S_8-haplotype lie within a 70 kb region. This region is encompassed by two recombination breakpoints that are 0.3 cM apart, and includes all the component genes involved in self-incompatibility of *Brassica,* that is S receptor kinase (*SRK*), *SLG* and S-locus protein 11 or S-locus cysteine-rich protein (*SP11/SCR*) (Schopfer et al. 1999; Takayama et al. 2000).

The relative amount of recombination in a chromosomal segment is shown by the ratio of DNA base pairs (kb) per recombination unit (cM). A chromosomal region that exhibits recombination suppression has a large kb/cM ratio compared to a segment that normally undergoes recombination. In the S_1-haplotype of *Ipomoea*, a large ratio, 920 kb/cM, was estimated for the S-locus region delimited between the markers LM-21 and RM-61, whereas the flanking region between the markers LM-21 and 681-SP6 had a ratio of 103 kb/cM. This demonstrates that genetic recombination is strongly suppressed in the S-locus region of *Ipomoea* to a level one-eighth to one-ninth of that in the neighbouring region. In the S-locus region of *Petunia inflata,* an extraordinarily high estimate of 17.6 Mb/cM is attributed to the centromeric location of the S-locus (Wang et al. 2003). Fluorescence in situ hybridisation (FISH) analysis of metaphase chromosomes in *I. trifida* showed that the S-locus region is located at the distal end of a chromosome, indicating that the remarkable suppression of recombination in the S-locus region of *Ipomoea* is not due to centromeric localisation (Suzuki et al. 2004). In *Brassica*, heterogeneity in the genomic sequences of the S-haplotypes is thought to be one of the genetic factors for the suppression of recombination. Genomic heteromorphisms in the S-locus have been established through chromosomal rearrangements, accumulation of transposons and insertion of haplotype-unique sequences (Cui et al. 1999; Suzuki et al. 1999). The suppression of recombination around the S-locus might contribute to the maintenance of the gene complex as a single genetic unit.

12.6 Genomic Organisation of the *S*-locus

The physical features of the S-locus have been revealed by genomic sequence analyses in several plant families. These analyses show that sequence polymorphism among S-haplotypes is a crucial factor for the male- and female-determinant genes in the S-locus that confer the specificity for self-pollen recognition in self-incompatibility (see Chap. 6 for a discussion of *Brassica* SRK and SP11/SCR polymorphisms). To analyse genomic organisation of the S-locus region in *Ipomoea*, sequence contigs of approximately 300 kb were constructed by map-based cloning of BAC and cosmid libraries from the S_1 homozygote (Tomita et al. 2004b). Sequence comparisons of the different S-haplotypes revealed highly variable regions, named S-haplotype-specific divergent regions (SDRs) (Fig. 12.4; Rahman et al.

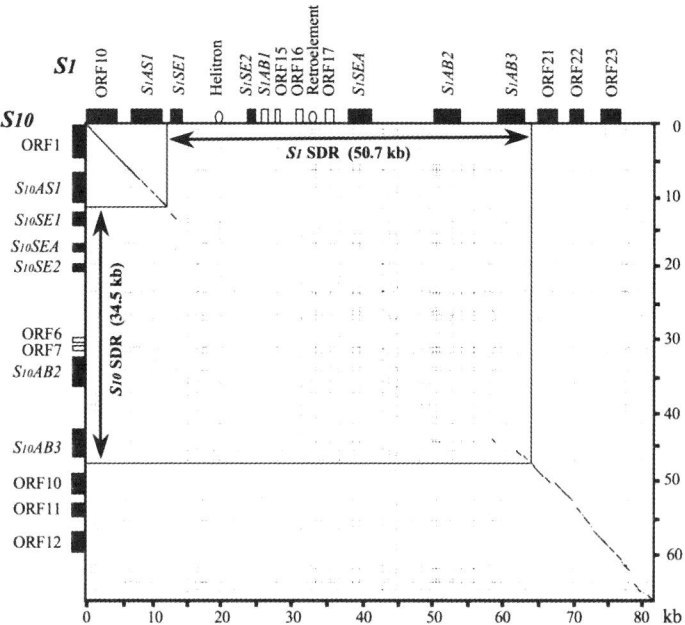

Fig. 12.4 Comparison of the nucleotide sequences spanning the S-loci of the S_1- and S_{10}-haplotypes by a Harr plot analysis. Highly polymorphic regions are shown as SDRs. The estimated sizes of the SDRs are 50.7 and 34.5 kb for the S_1- and S_{10}-haplotypes, respectively. The figure is from Rahman et al. (2007a) with permission from Springer

2007a). The SDR is approximately 50 kb in the S_1-haplotype, 35 kb in the S_{10}-haplotype and 95 kb in the S_{29}-haplotype. The regions flanking the SDRs have high sequence similarity between the S-haplotypes. Likewise, in some S-haplotypes of *Brassica rapa*, highly divergent regions of 30–56 kb are present in the S-locus regions, and are flanked by co-linear genomic regions that show high sequence similarities (Fukai et al. 2003; Shiba et al. 2003). In *Prunus* and *Malus* (Rosaceae), highly polymorphic regions are also found in the S-locus (Ushijima et al. 2001; Entani et al. 2003; Sassa et al. 2007). Overall, the sequence analyses in the Brassicaceae and Rosaceae, which have quite different systems of self-incompatibility, show that the male- and female-specific genes involved in the recognition specificity are located within the polymorphic divergent region of the S-locus. By analogy to these, the S-locus genes of *Ipomoea* are predicted to be located within the SDRs identified in the S-haplotypes. Differences in the sizes of the SDRs among S-haplotypes are due to the insertion of transposon-like sequences, retroelements and numerous interspersed repetitive sequences in the larger SDRs (Rahman et al. 2007a). An interesting correlation between allelic dominance relationship and SDR size has emerged from sequence comparisons of the S-locus regions: the more dominant the S-haplotype, the larger the SDR. This suggests that acquisition of sequence complexity by recessive S-haplotypes is responsible for the differentiation of more dominant S-haplotypes.

12.7 *S*-locus Genes in *Ipomoea*

Reproductive organ specific expression of *S*-locus genes is a prerequisite for the interaction of pollen and pistil in a SI system. In the S-RNase-based systems of GSI, S-RNases are highly abundant and are expressed in the transmitting tract of the style to inhibit the growth of the self-pollen tube (Lee et al. 1994). The pollen determinant gene *SFB/SLF* is expressed in developing pollen grains and pollen tubes (Ushijima et al. 2003; Entani et al. 2003; Sijacic et al. 2004; Sassa et al. 2007). In Papaveraceae, the female determinant of SI is a non-abundant extracellular signalling *S*-protein secreted on the stigma, which induces programmed cell death in the pollen tubes (Franklin-Tong and Franklin 2003; Thomas and Franklin-Tong 2004; see also Chap. 11). In the SSI system of *Brassica*, SLG and SRK are expressed in the mature papilla cells on the stigma surface (Nasrallah and Nasrallah 1993). The expression level of SRK is extremely low, compared to SLG. However, SRK plays a central role in self-pollen recognition in *Brassica*, and, in transgenic gain-of-function experiments, has been demonstrated to be the sole *S*-determinant on the pistil (Takasaki et al. 2000). The pollen determinant of *Brassica*, SP11/SCR, is expressed in the anther tapetal cells and microspores at late stages of pollen development (Schopfer et al. 1999; Takayama et al. 2000; Shiba et al. 2002). Thus, in all these examples, the *S*-determinants are expressed in a highly tissue-specific manner. Moreover, they show developmental expression coincident with the temporal acquisition of SI.

RNA-blot and RT-PCR analyses of the transcripts of genes located in the *S*-locus region of *Ipomoea* identified three stigma-specific genes (*SE1*, *SE2* and *SEA*) and three anther-specific genes (*AB1*, *AB2* and *AB3*) in the S_1-haplotype (Fig. 12.5; Rahman et al. 2007a, b). These six genes are located in the SDR and, with the exception of *AB1*, are present as single copies; two or three copies of *AB1* are present in the genome. *SE1*, *SE2* and *SEA* show similar patterns of gene expression with a high level of transcripts in the stigma of the young flower bud 1 week to 1 day before anthesis and with no detectable expression in vegetative tissues or reproductive organs other than the stigma. Thus, these genes exhibit tissue-specific and developmental expression expected of *S*-locus determinants.

The cDNA clones of the stigma-specific genes show a high level of allelic polymorphism among the *S*-haplotypes. The predicted amino acid sequences of SE1, SE2 and SEA are 53–76%, 61–67% and 57–65% identical, respectively, among the *S*-haplotypes. In GSI, S-RNases exhibit a high level of sequence polymorphism among *S*-haplotypes. The S-RNases of *Prunus* range in amino acid sequence identity from 55 to 82% among *S*-haplotypes (Entani et al. 2003; Ushijima et al. 2003). *Brassica* SRKs exhibit an amino acid identity of approximately 65% across the entire sequence: the majority of the sequence variation resides in the receptor domain that is responsible for *S* specificity (Nasrallah 2002; Watanabe et al. 2003). Thus, the allelic amino acid sequence diversity of the SE1, SE2 and SEA proteins of *Ipomoea* is comparable to those reported to date for the female *S*-determinants in both the GSI and SSI systems.

12 Molecular Genetics of Sporophytic Self-Incompatibility in *Ipomoea*

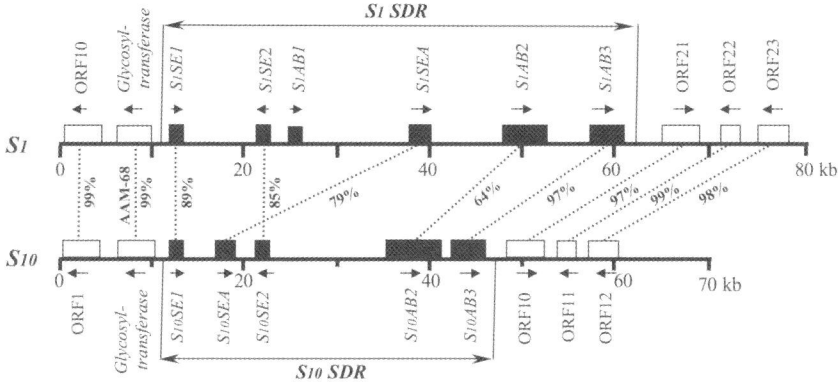

Fig. 12.5 Alignment of *S*-locus genes in the S_1- and S_{10}-haplotypes. The location of the SDRs in the S_1- and the S_{10}-haplotypes is shown. Black boxes indicate stigma- (*SE1, SE2, SEA*) or anther- (*AB1, AB2, AB3*) specific genes identified in the SDR. White boxes indicate genes or open reading frames (ORFs) outside the SDR. The arrows beside the boxes and the numbers below each line show the orientation of transcription and the physical distances in kb, respectively. Genes or ORFs in common between the two *S*-haplotypes are connected by dotted lines that display the percentage nucleotide identity. The figure is redrawn from Rahman et al. (2007a) with permission from Springer

Based on hydropathy plot analysis and secondary structure predictions, the SE1, SE2 and SEA proteins have several predicted membrane-spanning domains. In addition, the three proteins share partial sequence similarities, with 30–40% identity, in the S_1-haplotype. These findings suggest that the *SE1, SE2* and *SEA* genes are derived from a common ancestor and encode membrane-spanning proteins localised at the cell surface, where pollen-to-papilla cell interactions occur. Furthermore, in a database search, these three predicted proteins showed no significant homology to proteins with known function. Thus, they are novel. Together, these experimental data indicate that *SE1, SE2* and *SEA* are good candidates for genes encoding the stigma determinant in a novel SSI system in *Ipomoea*.

Sequence analyses of the three anther-specific genes *AB1, AB2* and *AB3* located in the SDR of the *Ipomoea* S_1-haplotype showed that *AB1* and *AB3* have more than 95% allelic similarity in their predicted amino acid sequences among some *S*-haplotypes. Furthermore, *AB1* is not found in the SDR of the S_{10}-haplotype, possibly indicating that it is located in the conserved region outside the SDR (Fig. 12.5). Thus, it is unlikely that *AB1* and *AB3* are involved in the determination of *S*-haplotype specificity in the anther/pollen of *Ipomoea*. The *AB2* gene, on the other hand, displays intriguing features in relation to anther/pollen specificity of self-incompatibility (Rahman et al. 2007b).

AB2 is expressed specifically in the early stages of anther development. In situ hybridisation analysis demonstrated that *AB2* is also expressed sporophytically in the tapetal cell layer at early stages of anther development. Since the tapetal cell layer is a diploid tissue that nourishes developing pollen grains and provides the components for the pollen coat, the expression pattern of *AB2* is consistent with a role for this gene in this particular SSI system. The predicted AB2 proteins from different S-haplotypes all have a putative signal peptide with a highly conserved sequence at the N-terminus, but show considerable sequence polymorphism in the C-terminal regions. Amino acid sequence identity in the different S-haplotypes ranges from 43 to 54%, indicating a high degree of polymorphism.

In addition, interestingly, the predicted AB2 proteins show homology to a class of cysteine-rich proteins termed plant defensins, which are members of the γ-thionin protein family. The defensins are small peptides with anti-microbial activity and are widely distributed throughout both flora and fauna (Boman 2003). Genes for defensin-like proteins include, for example, *PCP-A1* (Doughty et al. 1998) and *SP11/SCR* (Suzuki et al. 1999; Schopfer et al. 1999), which are also expressed in the developing anther of *Brassica*. The *Ipomoea* AB2 protein and the *Brassica* pollen *S*-determinant SP11/SCR share eight conserved cysteine residues (C1–C8) and a conserved glycine residue between C1 and C2. However, a conserved aromatic residue (tyrosine in SP11) located between C3 and C4, which is regarded as a fingerprint of SP11 proteins (Mishima et al. 2003), is not present in AB2. Alignment of the amino acid sequences of the eight conserved cysteine residues also indicates that AB2 is structurally different from PCP-A1 and SP11/SCR of *Brassica*. Thus the AB2 and SP11/SCR proteins show slight structural differences and probably also functional differences. Nevertheless, it is interesting to note that if AB2 is proved to be a pollen *S*-determinant, then the taxonomically distant species, *Brassica* and *Ipomoea*, employ a member of the γ-thionin protein family to confer the recognition specificity of SSI.

Taken together, the available data suggest that the *SE1*, *SE2* and *SEA* genes expressed in the stigma and the *AB2* gene expressed in the anther are strong candidates for the *S*-determinant genes in *Ipomoea*. Functional analyses of these genes by gain-of-function and loss-of-function experiments may provide definitive evidence that they are indeed involved in the SI system of *Ipomoea*. Further study will also be necessary to demonstrate that the AB2 protein is transferred to the pollen coat after its secretion from the tapetal cells and is translocated into the papilla cell walls of the stigma to interact with the female *S*-determinant.

12.8 Diversity of the SI Systems

The molecular mechanisms of SI have been elucidated in several plant families. In the GSI systems of Solanaceae, Plantaginaceae and Rosaceae, *S-RNase* and *SFB/SLF* genes have been identified as the pistil and pollen determinants of recognition specificity, respectively, suggesting the involvement of RNA and protein degradation during pollen-tube growth in the style (McClure and Franklin-Tong 2006; see also Chaps. 9 and 10, this volume). In the GSI system of the Papaveraceae, a small secreted *S*-protein has been identified as the pistil determinant that triggers a calcium-mediated signalling cascade to induce programmed cell death and pollen-tube growth inhibition (Franklin-Tong and Franklin 2003; Thomas and Franklin-Tong 2004; see also Chap. 11). In the SSI system of the Brassicaceae, SRK and SP11/SCR function as a receptor kinase anchored in the papilla cells and as its cognate ligand on the pollen coat, respectively, and are involved in the signal transduction pathway that induces self-pollen rejection through activation of downstream proteins in the stigma (Kachroo et al. 2002; Takayama and Isogai 2005; see also Chaps. 7 and 8). In other plant species with SSI, such as *Ipomoea* (Convolvulaceae) and *Senecio* (Asteraceae), the possible involvement of the *SRK* gene in recognition specificity has been postulated (Hiscock and McInnis 2003). In *I. trifida*, *SRK*-like genes are expressed in reproductive tissues and in vegetative tissues, and show no linkage to the *S*-locus (Kowyama et al. 1995, 1996). Similarly, *SRK*-like genes identified from *S. squalidus* are expressed both in the stigma and in vegetative tissues, and no allelic polymorphism has been found among *S*-haplotypes (Tabah et al. 2004). These observations suggest that the Convolvulaceae and Asteraceae do not use the SRK-mediated system of self-incompatibility that operates in the Brassicaceae.

Our studies described here propose that the SSI system of the Convolvulaceae involves molecular mechanisms different to those of the Brassicaceae, and supports the hypothesis that SI systems may have arisen independently on many occasions during the evolution of angiosperms (see Chap. 4). Likewise, quite distinct mechanisms operate in the S-RNase-based systems of the Solanaceae and Rosaceae and the Ca^{2+} signalling-mediated system of the Papaveraceae, although these are classified as the same type of GSI. The classification of self-incompatibility into two types, namely GSI and SSI, is based simply on differences in gene expression of pollen determinants, not on the molecular mechanisms that underlie the self-incompatibility systems in these various plant families. Investigation of a wide range of plant taxa will be required to understand the evolutionary lineages of the SI systems in flowering plants.

Acknowledgements We express our sincere thanks to Prof. Kokichi Hinata and Prof. Akira Isogai for their long-term support and encouragement. Our research was supported by a Grant-in-Aid for Scientific Research on Priority Areas (No. 11238203) and Grant-in-Aids for Scientific Research B (No. 16380005 and 19380005) from the Ministry of Education, Culture, Sports, Science and Technology, Japan to Y. Kowyama.

References

Agrama HA, Houssin SF, Tarek MA (2002) Cloning of AFLP markers linked to resistance to *Peronosclerospora sorghi* in maize. Mol Genet Genom 267:814–819

Anderson MA, Cornish EC, Mau S-L, Williams EG, Hoggart R, Atkinson A, Bonig I, Grego B, Simpson R, Roche PJ, Haley JD, Penschow JD, Niall HD, Tregear GW, Coughlan JP, Crawford RJ, Clarke AE (1986) Cloning of cDNA for a stylar glycoprotein associated with expression of self-incompatibility in *Nicotiana alata*. Nature 321:38–44

Arumuganathan K, Earle ED (1991) Nuclear DNA content of some important plant species. Plant Mol Biol Rep 9:208–218

Austin DF (1997) Convolvulaceae (Morning glory family). http://www.fau/divdept/biology/people/convolv.htm

Austin DF, Huáman Z (1996) A synopsis of *Ipomoea* (Convolvulaceae) in the Americas. Taxon 45:3–38

Boman HG (2003) Antibacterial peptides: Basic facts and emerging concepts. J Int Med 254:197–215

Casselman AL, Vrebalov J, Conner JA, Singhal A, Giovannoni J, Nasrallah ME, Nasrallah JB (2000) Determining the physical limits of *Brassica S* locus by recombinational analysis. Plant Cell 12:23–33

Cui Y, Brugiere N, Jackman L, Bi Y, Rothstein SJ (1999) Structural and transcriptional comparative analysis of the *S* locus regions in two self-incompatible *Brassica napus* lines. Plant Cell 11:2217–2231

Darwin CR (1876) The effects of cross- and self-fertilisation in the vegetable kingdom. John Murray, London

de Nettancourt D (2001) Incompatibility and incongruity in wild and cultivated plants. Springer, Berlin Heidelberg New York

Doughty J, Dixon S, Hiscock SJ, Willis AC, Parkin IAP, Dickinson HG (1998) PCP-A1, a defensin-like Brassica pollen coat protein that binds the *S* locus glycoprotein, is the product of gametophytic gene expression. Plant Cell 10:1333–1347

Entani T, Iwano M, Shiba H, Che FS, Isogai A, Takayama S (2003) Comparative analysis of the self-incompatibility (*S*-)locus region of *Prunus mume*: Identification of a pollen-expressed F-box gene with allelic diversity. Genes Cells 8:203–213

Fearon CH, Cornish MA, Hayward MD, Lawrence MJ (1994) Self-incompatibility in ryegrass. X. Number and frequency of alleles in a natural population of *Lolium perenne* L. Heredity 73:262–264

Franklin FCH, Lawrence MJ, Franklin-Tong VE (1995) Cell and molecular biology of self-incompatibility in flowering plants. Int Rev Cytol 158:1–64

Franklin-Tong VE, Franklin FCH (2003) Gametophytic self-incompatibility inhibits pollen tube growth using different mechanisms. Trends Plant Sci 8:598–605

Fukai E, Fujimoto R, Nishio T (2003) Genomic organization of the *S* core region and the *S* flanking regions of a class-II *S* haplotype in *Brassica rapa*. Mol Genet Genom 269:361–369

Hinata K, Isogai A, Isuzugawa K (1994) Manipulation of sporophytic self-incompatibility in plant breeding. In: Williams EG, Clarke AE, Knox RB (eds) Genetic control of self-incompatibility and reproductive development in flowering plants. Kluwer, Dordrecht, pp 102–115

Hiscock SJ, McInnis S (2003) Pollen recognition and rejection during the sporophytic self-incompatibility response: *Brassica* and beyond. Trends Plant Sci 8:606–613

Ikeda K, Ushijima K, Yamane H, Tao R, Hauck NR, Sebolt AM, Iezzoni AF (2005) Linkage and physical distances between the *S*-haplotype *S-RNase* and *SFB* genes in sweet cherry. Sex Plant Reprod 17:289–296

Kachroo A, Nasrallah ME, Nasrallah JB (2002) Self-incompatibility in the Brassicaceae: Receptor-ligand signaling and cell-to-cell communication. Plant Cell 14(Suppl):s227–s238

Kakeda K, Tsukada H, Kowyama Y (2000) A self-compatible mutant *S* allele conferring a dominant negative effect on the functional *S* allele in *Ipomoea trifida*. Sex Plant Reprod 13:119–125

Kowyama Y, Shimano N, Kawase T (1980) Genetic analysis of incompatibility in the diploid *Ipomoea* species closely related to the sweet potato. Theor Appl Genet 58:149–155

Kowyama Y, Takahashi H, Muraoka K, Tani T, Hara K, Shiotani I (1994) Number, frequency and dominance relationships of *S*-alleles in diploid *Ipomoea trifida*. Heredity 73:275–283

Kowyama Y, Kakeda K, Nakano R, Hattori T (1995) *SLG/SRK*-like genes are expressed in the reproductive tissues of *Ipomoea trifida*. Sex Plant Reprod 8:333–338

Kowyama Y, Kakeda K, Kondo K, Imada T, Hattori T (1996) A putative receptor protein kinase gene in *Ipomoea trifida*. Plant Cell Physiol 37:681–685

Kowyama Y, Tsuchiya T, Kakeda K (2000) Sporophytic self-incompatibility in *Ipomoea trifida*, a close relative of sweet potato. Ann Bot 85(Suppl A):191–196

Lane MD, Lawrence MJ (1993) The population genetics of the self-incompatibility polymorphism in *Papaver rhoeas*. VII. The number of *S*-alleles in the species. Heredity 71:596–602

Lee H-S, Huang S, Kao T-H (1994) *S* proteins control rejection of incompatible pollen in *Petunia inflata*. Nature 367:560–563

McClure BA, Franklin-Tong V (2006) Gametophytic self-incompatibility: Understanding the cellular mechanisms involved in "self" pollen tube inhibition. Planta 224:233–245

Michelmore RW, Paran I, Kesseli RV (1991) Identification of markers linked to disease-resistance genes by bulked segregant analysis: A rapid method to detect markers in specific genomic region by using segregating populations. Proc Natl Acad Sci USA 88:9828–9832

Mishima M, Takayama S, Sasaki K, Jee J, Kojima C, Isogai A, Shirakawa M (2003) Structure of the male determinant factor for *Brassica* self-incompatibility. J Biol Chem 278:36389–36395

Nasrallah JB (2002) Recognition and rejection of self in plant reproduction. Science 296:305–308

Nasrallah JB, Nasrallah ME (1993) Pollen–stigma signaling in the sporophytic self-incompatibility response. Plant Cell 5:1325–1335

Nasrallah JB, Kao T-H, Goldberg ML, Nasrallah ME (1985) A cDNA clone encoding an *S*-locus specific glycoprotein from *Brassica oleracea*. Nature 318:263–267

Nishio T, Hinata K (1977) Analysis of *S*-specific proteins in stigma of *Brassica oleracea* L. by isoelectric focusing. Heredity 38:391–396

Nishiyama I, Miyazaki T, Sakamoto S (1975) Evolutionary autoploidy in the sweet potato (*Ipomoea batatas* (L.) Lam.) and its progenitors. Euphytica 24:197–208

Nou IS, Watanabe M, Isuzugawa K, Isogai A, Hinata K (1993) Isolation of *S*-alleles from a wild population of *Brassica campestris* L. at Balcesme, Turkey and their characterization by *S*-glycoproteins. Sex Plant Reprod 6:71–78

Rahman MH, Tsuchiya T, Suwabe K, Kohori J, Tomita RN, Kagaya Y, Kobayashi I, Kakeda K, Kowyama Y (2007a) Physical size of the *S* locus region defined by genetic recombination and genome sequencing in *Ipomoea trifida*, Convolvulaceae. Sex Plant Reprod 20:63–72

Rahman MH, Uchiyama M, Kuno M, Hirashima N, Suwabe K, Tsuchiya T, Kagaya Y, Kobayashi I, Kakeda K, Kowyama Y (2007b) Expression of stigma- and anther-specific genes located in the *S* locus region of *Ipomoea trifida*. Sex Plant Reprod 20:73–85

Rajapakse S, Nilmalgoda SD, Molnar M, Ballard RE, Austin DF, Bohac JR (2004) Phylogenetic relationships of the sweet potato in *Ipomoea* series *Batatas* (Convolvulaceae) based on nuclear β-amylase gene sequences. Mol Phylogenet Evol 30:623–632

Royo J, Kunz C, Kowyama Y, Anderson MA, Clarke AE, Newbigin E (1994) Loss of histidine residue at the active site of *S*-ribonuclease leads to self-compatibility in *Lycopersicon peruvianum*. Proc Natl Acad of Sci USA 91:6511–6514

Sassa H, Hirano H, Nishio T, Koba T (1997) Style-specific self-compatible mutation caused by deletion of the S-RNase gene in Japanese pear (*Pyrus serotina*). Plant J 12:223–227

Sassa H, Kakui H, Miyamoto M, Suzuki Y, Hanada T, Ushijima K, Kusaba M, Hirano H, Koba T (2007) *S locus F-box brothers*; Multiple and pollen-specific F box genes with *S* haplotype-specific polymorphisms in apple and Japanese pear. Genetics 175:1869–1881

Schopfer CR, Nasrallah ME, Nasrallah JB (1999) The male determinant of self-incompatibility in *Brassica*. Science 286:1697–1700

Shiba H, Iwano M, Entani T, Ishimoto K, Shimosato H, Che F-S, Satta Y, Ito A, Takada Y, Watanabe M, Isogai A, Takayama S (2002) The dominance of alleles controlling self-incompatibility in *Brassica* pollen is regulated at the RNA level. Plant Cell 14:491–504

Shiba H, Kenmochi M, Sugihara M, Iwano M, Kawasaki S, Suzuki G, Watanabe M, Isogai A, Takayama S (2003) Genomic organization of the *S*-locus region of *Brassica*. Biosci Biotech Biochem 67:622–626

Shiba H, Kakizaki T, Iwano M, Tarutani Y, Watanabe M, Isogai A, Takayama S (2006) Dominance relationships between self-incompatibility alleles controlled by DNA methylation. Nat Genet 38:297–299

Shiotani I, Kawase T (1987) Synthetic hexaploids derived from wild species related to sweet potato. Jpn J Breed 37:357–376

Sijacic P, Wang X, Skirpan AL, Wang Y, Dowd PE, McCubbin AG, Huang S, Kao T-H (2004) Identification of the pollen determinant of S-RNase-mediated self-incompatibility. Nature 429:302–305

Simoes-Araujo JL, Rodrigues RL, de A Gerhardt LB, Mondego JM, Alves-Ferreira M, Rumjanek NG, Margis-Pinheiro (2002) Identification of differentially expressed genes by cDNA-AFLP technique during heat stress in cowpea nodules. FEBS Lett 515:44–50

Steinbachs JE, Holsinger KE (2002) S-RNase-mediated gametophytic self-incompatibility is ancestral in eudicots. Mol Biol Evol 19:825–829

Suzuki G, Kai K, Hirose T, Fukui K, Nishio T, Takayama S, Isogai A, Watanabe M, Hinata K (1999) Genomic organization of the S locus: Identification and characterization of genes in *SLG/SRK* region of S^9 haplotype of *Brassica campestris* (syn *rapa*). Genetics 153:391–400

Suzuki G, Kakizaki T, Takada Y, Shiba H, Takayama S, Isogai A, Watanabe M (2003) The S-haplotypes lacking *SLG* in the genome of *Brassica rapa*. Plant Cell Rep 21:911–915

Suzuki G, Tanaka S, Yamamoto M, Tomita RN, Kowyama Y, Mukai Y (2004) Visualization of the S-locus region in *Ipomoea trifida*: Toward positional cloning of self-incompatibility genes. Chromosome Res 12:475–481

Tabah DA, McInnis SM, Hiscock SJ (2004) Members of the S-receptor kinase multigene family in *Senecio squalidus* L. (Asteraceae), a species with sporophytic self-incompatibility. Sex Plant Reprod 17:131–140

Takasaki T, Hatakeyama K, Suzuki G, Watanabe M, Isogai A, Hinata K (2000) The S receptor kinase determines self-incompatibility in *Brassica* stigma. Nature 403:913–916

Takayama S, Isogai A (2005) Self-incompatibility in plants. Annu Rev Plant Biol 56:467–489

Takayama S, Shiba H, Iwano M, Shimosato H, Che FS, Kai N, Watanabe M, Suzuki G, Hinata K, Isogai A (2000) The pollen determinant of self-incompatibility in *Brassica campestris*. Proc Natl Acad Sci USA 97:1920–1925

Terao H (1934) Cross sterility groups in varieties of sweet potato. Agric Hort (Tokyo) 9:1163–1167

Thomas SG, Franklin-Tong VE (2004) Self-incompatibility triggers programmed cell death in *Papaver* pollen. Nature 429:305–309

Tomita RN, Fukami K, Takayama S, Kowyama Y (2004a) Genetic mapping of AFLP/AMF-derived DNA markers in the vicinity of the self incompatibility locus in *Ipomoea trifida*. Sex Plant Reprod 16:265–272

Tomita RN, Suzuki G, Yoshida K, Yano Y, Tsuchiya T, Kakeda K, Mukai Y, Kowyama Y (2004b) Molecular characterization of a 313-kb genomic region containing the self-incompatibility locus of *Ipomoea trifida*, a diploid relative of sweet potato. Breed Sci 54:165–175

Ushijima K, Sassa H, Tamura M, Kusaba M, Tao R, Gradziel TM, Dandekar AM, Hirano H (2001) Characterization of the S-locus region of almond (*Prunus dulcis*): Analysis of a somaclonal mutant and a cosmid contig for an S haplotype. Genetics 158:379–386

Ushijima K, Sassa H, Dandekar AM, Gradziel TM, Tao R, Hirano H (2003) Structural and transcriptional analysis of the self-incompatibility locus of almond: Identification of a pollen-expressed F-box gene with haplotype-specific polymorphism. Plant Cell 15:771–781

Wang Y, Wang X, McCubbin AG, Kao T-H (2003) Genetic mapping and molecular characterization of the self-incompatibility (S) locus in *Petunia inflata*. Plant Mol Biol 53:565–580

Watanabe M, Ono T, Hatakeyama K, Takayama S, Isogai A, Hinata K (1997) Molecular characterization of *SLG* and *S*-related genes in a self-compatible *Brassica campestris* L. var. yellow sarson. Sex Plant Reprod 10:332–340

Watanabe M, Takayama S, Isogai A, Hinata K (2003) Recent progresses on self-incompatibility research in *Brassica* species. Breed Sci 53:199–208

Williams RD, Williams W (1947) Genetics of red clover (*Trifolium pratense* L.) compatibility. III. The frequency of incompatibility S-alleles in two non-pedigree populations of red clover. J Genet 48:67–79

Chapter 13
Self-Incompatibility in the Grasses

P. Langridge and U. Baumann

Abstract The grasses are amongst the most important families of plants. The family includes major food crops and pastures and they dominate many natural ecosystems. Self-incompatibility (SI) is common in the grasses and is under the control of two unlinked loci, *S* and *Z*. Specification of SI in this family is gametophytic and is determined by the combination of *S* and *Z* alleles in the pollen grain. Available data suggests that the two-locus SI system is common to all self-incompatible grass species of the Pooideae and possibly all members of the Graminae. Genetic studies showed the presence of the *S-Z* system in the Triticeae, the Poeae and the Avenae. Linkage analyses have identified common markers linked to the *S*-gene in grass species from all three tribes. Molecular markers have confirmed the syntenous localisation of *S* and *Z* in several species. Although this SI system is complex relative to that in other families, the importance of this family and the detailed genetic and physical data emerging from many genomics programs makes this an important group for more detailed studies.

Abbreviations

BM2	A gene encoding thioredoxin h
cM	Centimorgan
GSI	Gametophytic SI
PCR	Polymerase chain reaction
SI	Self-incompatibility
SSI	Sporophytic SI
S, Z	*S*-locus and *Z*-locus that together determine GSI in the grasses
T	A third locus that also affects SI in the grasses

P. Langridge and U. Baumann
Australian Centre for Plant Functional Genomics, School of Agriculture Food and Wine, University of Adelaide, Urrbrae, SA 5064, Australia, e-mail: Peter.Langridge@adelaide.edu.au

V.E. Franklin-Tong (ed.) *Self-Incompatibility in Flowering Plants – Evolution, Diversity, and Mechanisms.*
© Springer-Verlag Berlin Heidelberg 2008

13.1 Introduction

The grass family Poaceae consists of around 700 genera and approximately 10,000 species (Watson and Dallwitz 1992). Amongst the flowering plants it is the fourth largest family and is the second largest family within the monocotyledons; only the Orchidaceae is larger (Watson 1990). Since the grasses provide all of our cereal crops, most of the world's sugar and a large proportion of the important feed and forage crops for grazing animals, the grasses are arguably the most important family of plants. The grasses also represent the most dominant family of plants on earth since they are found in almost every habitat available to flowering plants (Watson 1990).

Recent studies show that there were two major radiations in the grasses that led to the current six subfamilies. The first group includes the Pooideae (e.g. wheat, barley, ryegrass), Bambusoideae (the bamboos) and Oryzoideae (rice and its relatives), with the second group covering the Panicoideae (e.g. sugarcane, maize and sorghum), Chloridoideae (e.g. *Eragrostis*) and Arundinoideae (e.g. *Phragmites australis*) (Kellogg 1998).

Self-incompatibility (SI) has been known in the grasses for over a century. In 1877, Rimpau demonstrated the dominance of self-sterile individuals in populations of rye (Lundqvist 1954). Detailed analyses of SI in the Gramineae were also undertaken very early, for example Körnicke (1890), Troll (1930, 1931) and Beddows (1931). Connor (1979) provides a good overview of the early work and his review provides many specific references.

At least 16 genera of grasses show SI (Connor 1979), with self-incompatible and self-compatible species often found in the same genus. The perennial species tend to show a higher frequency of self-incompatibility relative to annuals (Körnicke 1890; Beddows 1931). Although the incompatible reaction is determined at the gametophytic level gametophytic SI (GSI), the grass system shows some features that are more typical of plants with sporophytic self-incompatibility (SSI; Heslop-Harrison 1982; see also Chap. 4, this volume); for example the grass pollen is tricellular and short-lived and the stigma is dry. Also in common with many of the sporophytic systems, the incompatible reaction usually occurs rapidly, although the strength of the reaction can vary between species and even genotypes (Shivanna et al. 1982). For example *Guardinia fragilis* (Pooideae) shows a block to pollen growth almost immediately upon landing on the stigma (Shivanna et al. 1982), whereas the incompatible reaction in *Cynodon dactylon* (Chloridoieae) does not stop pollen tube growth until it is within the style (Thomas and Murray 1975).

Self-incompatible, or species that show strong out-breeding, resulting in in-breeding depression under forced in-breeding, can be found in 5 of the 6 subfamilies of the Poaceae (summarised in Table 13.1). However, no SI species have yet been identified in the Bambusoideae (bamboos), Oryzidoideae (rice) or Centostecoideae. The occurrence of SI in the Oryzoideae is debatable. According to Nayar (1967), a population of *Oryza perennis* subsp. *barthii* from Sudan was completely self-sterile

Table 13.1 Summary of grass subfamilies and tribes in which self-incompatible or predominantly cross-fertilizing species have been identified

Subfamily	Tribe	Number of genera	Reference
Pooideae	Poeae	7	–
	Aveneae	9	Bush and Barrett 1993
	Agrogtideae	5	–
	Triticeae	4	–
	Bromeae	1	–
	Stipeae	1	–
Panicoideae	Paniceae	5	–
	Andropogoneae	8	McKone et al. 1998
Chloridoideae	Chlorideae	4	Daehler 1999
	Eragrostideae	1	–
Arundinoideae	Arundineae	1	–
	Danthonieae	1	–
Bambusoideae		0	–
Oryzidoideae		0	Oka and Morishima (1967)

The number of genera with SI species is presented. Note that no SI species have been identified amongst the bamboos and the occurrence of SI in the Oryzidoideae is controversial. (modified from Baumann et al. 2000). Except where indicated (reference), all data is from Hayman (1992) and Connor (1979)

in spite of its pollen stainability being over 80%. In contrast, Yeh and Henderson (1961) observed only partial sterility in their culture of *O. barthii*. Chu et al. (1969) examined *O. barthii* plants from West Africa and found seed fertility of around 15–43% upon selfing by hand.

SI has arisen many times in flowering plants. However, there have been no reports of differing self-incompatibility systems within a single family of plants; see Chap. 4 Hiscock, this volume. This seems to be the case in the dicotyledonous species analysed and also in the grasses.

13.2 Genetic Control of SI in the Grasses

The first studies of the genetics of SI in the Poaceae were conducted in the early 1950s in two systems; *Secale cereale* by Lundqvist (1954) and *Phalaris coerulescens* by Hayman (1956). Both studies indicated that SI in the grasses was under gametophytic control (i.e. utilising a GSI system) mediated by two unlinked and multi-allelic loci, named *S* and *Z*. An incompatible reaction occurs when both the *S* and *Z* alleles of the pollen are matched in the recipient pistil. Eight species have been shown to display the *S-Z* system and it is probable that a further eight also shows this mechanism for incompatibility (Table 13.2).

Table 13.2 Species where the two locus S-Z self-incompatibility system has been demonstrated (a) or suggested (b) (updated from Baumann et al. 2000)

Species	Tribe	Reference
S-Z system identified using parental genotypes of known relationship		
Secale cereale	Triticae	Lundqvist (1954)
Festuca pratensis	Poeae	Lundqvist (1955)
Phalaris coerulescens	Avenae	Hayman (1956)
Hordeum bulbosum	Triticae	Lundqvist et al. (1962a)
Dactylis aschersoniana	Poeae	Lundqvist (1965)
Briza media	Poeae	Murray (1974)
Lolium perenne	Poeae	Cornish et al. (1979)
Lolium multiflorum	Poeae	Fearon et al. (1983)
Gametophytic system identified and differences between reciprocal crosses		
Alopecurus myosuroides	Avenae	Leach and Hayman (1987)
Cynosurus cristatus	Poeae	Weimarck (1968)
Holcus lanatus	Avenae	Weimarck (1968)
Alopecurus pratensis	Avenae	Weimarck (1968)
Arrhenatherum elatius	Avenae	Weimarck (1968)
Festuca rubra	Poeae	Weimarck (1968)
Deschampia flexuosa	Avenae	Weimarck (1968)
Phalaris arundinacea	Avenae	Weimarck (1968)
Panicum virgatum	Panicodae	Martinez-Reyna and Vogel (2002)

13.2.1 Features of the S-Z System

The two gene nature of SI in the grasses results in several aspects that distinguish this system from the single locus SI systems. The most significant differences are the varying outcome of reciprocal crosses and the extent of compatibility (namely, the percentage of compatible pollen). Compatibility can be 0, 50, 75 or 100%, depending on the genotypes.

The key feature of the grass SI system can be summarised as follows:

1. There is complementary interaction between *S* and *Z* alleles. A pollen grain is incompatible when both its *S* and *Z* alleles are matched in the recipient pistil. A pollen grain is compatible and fully functional with only an *S* or *Z* gene in common with the stigma (Hayman 1992).
2. There are reciprocal differences in the compatibility between parents (Leach 1988). The $S_{11}Z_{12} \times S_{12}Z_{12}$ cross is 50% compatible, while the reciprocal cross $S_{12}Z_{12} \times S_{11}Z_{12}$ is fully incompatible. The compatible degree can be 100% compatible ($S_{12}Z_{12} \times S_{34}Z_{34}$), 75% compatible ($S_{12}Z_{12} \times S_{23}Z_{23}$), 50% compatible ($S_{12}Z_{12} \times S_{13}Z_{12}$) and 0% compatible ($S_{12}Z_{12} \times S_{11}Z_{12}$).
3. Polyploid grasses are as self-incompatible as their diploid relatives and parents. Only one S-Z pair of alleles in the diploid pollen of tetraploid grasses needs to be matched in recipient pistil for the self-incompatibility reactions to occur. This

effect has been shown in *Secale cereale*, *Festuca pratense* and *Dactylis glomerata* (Lundqvist 1957, 1962b, 1969) and in *Lolium perenne* (Fearon et al. 1984a, b).

4. In contrast to single locus systems there are no dominant or competitive interactions between alleles of the diploid pollen or between the alleles of the tetraploid style. In an autotetraploid grass, a pollen grain will be incompatible if any one of the possible *S-Z* allele combinations is present in the recipient pistil (Lewis 1947).

There are also genetic and environmental factors affecting self-compatibility; see also Chap. 2, this volume. It is a common observation that a small number of seeds are obtained in many self-incompatible grass species after self-pollination (Hayman 1992). The partial breakdown of self-incompatibility is referred to as pseudo-compatibility. Pseudo-compatibility has been investigated in *S. cereale* and *F. pratensis* (Lundqvist 1958, 1968). The amount of seed set decreases with increased homozygosity at *S* or *Z*. Also the individual genes or gene pairs can be ranked in their capacity to self pollinate. Consistent results were observed, in that S_1 was more pseudo-compatible than S_2, and Z_3 more than Z_4. Apart from genetic factors, environmental factors, such as high temperature and relative humidity, also cause pseudo-compatibility (Wilkins and Thorogood 1992; Wricke 1978). Normal seed set in the field of about 1% could be increased to 25% if temperatures reached as high as 30–35°C during flowering. The highest seed set was obtained if plants were exposed to high temperatures together with 60–80% relative humidity for 2–4 days before anthesis. The relevance of the environmental interaction is unclear but may be due to direct inhibition of the SI mechanism.

13.2.2 The Third Locus 'T'

In addition to the *S*- and *Z*-loci that determine the specificity of the incompatible reaction, a third locus, named *T*, also affects SI in the grasses. Lundqvist (1968) studied self-fertile mutants of rye and suggested that self-fertility resulted from mutations at a locus other than *S*- and *Z*-. Interestingly all mutants were pollen-only, namely the pollen had lost SI specificity while the stigma remained its ability to differentiate pollen genotypes. This suggestion was confirmed in *L. perenne* (Thorogood and Hayward 1991; Thorogood et al. 2005), *P. coerulescens* (Hayman and Richter 1992) and through the study of a further mutant in *S. cereale* (Voylokov et al. 1993). *T* was shown to be linked to the Esterase 5–7 gene complex from chromosome 5R in rye (Fuong et al. 1993). Hayman (1992) argued that *T* was either active or inactive and showed no allelic variation as seen at *S*-loci and *Z*-loci. He suggested two hypotheses for the function of *T*. It might have formerly shown allelic variability and contributed to the specification of the pollen and style, but one allele became fixed. Alternatively, *T* may not be connected with specific recognition, but might be required for signal transduction in the SI reaction.

13.2.3 Mechanism of Action of S- and Z-Gene Products

Neither the *S*- nor the *Z*-gene has been cloned. Therefore, attempts to understand the mechanism of action of the products of these genes are based largely on physiological and biochemical information.

Heslop-Harrison (1982) put forward a general hypothesis on the SI reaction of grasses. He suggested that the pistil incompatibility determinants are proteins (possibly glycoproteins), with lectin-like properties. These could be present in the stigma surface secretions and in the transmitting tracts. The binding specificity of these factors could be such that they are complementary to sugar sequences or arrays displayed by wall carbohydrate in the growth zone of incompatible pollen tubes, but not complementary to those present in compatible tubes. Binding of the stigma proteins at the tip of an incompatible pollen tube could lead to a disruption of apical growth by preventing the dissociation of the polysaccharide content of the wall precursor bodies and interfering with the extension of polysaccharide micro-fibres in the sub-apical zone (Heslop-Harrison 1982).

Wehling et al. (1994a) took a different route, suggesting that pollen protein phosphorylation might be part of the signal transduction cascade of the SI response in rye. In both *Brassica* SSI and *Papaver* GSI there is evidence for kinase-based signalling cascades operating to control SI (see Chaps. 8 and 11, this volume). A higher level of protein phosphorylation was detected in germinated pollen treated with self-incompatible stigmatic extracts than in those treated with self-compatible stigmatic extracts (Wehling et al. 1994a). Separation of phosphorylated pollen proteins by SDS-PAGE revealed four major proteins in the MW range of 43–82 kDa that were differentially phosphorylated in self-incompatible and self-compatible genotypes. These results led Wehling et al. (1994a) to propose a model for grass SI whereby the stigma SI determinants act as ligands, interacting with their specific pollen receptors, to stimulate a signals transduction cascade which could lead to the inhibition of incompatible pollen. This is rather similar to the model proposed for the single locus GSI system found in *Papaver* (see Chap. 11). In contrast, when protein phosphorylation of pollen from self-incompatible *Phalaris coerulescense* was studied in the presence of compatible and incompatible stigma extracts, no evidence was found for increased phosphorylation activity in an incompatible reaction. Furthermore, the analysis of pollen protein phosphorylation of self-fertile mutants (*S*-pollen only, *Z*-pollen only and *T* mutants) did not show altered phosphorylation of any of the pollen proteins (Baumann 1995). Thus, whether kinase-based signalling generally may play a role in GSI in the grasses, or whether it is just found only in a few species, is currently unclear.

13.3 Approaches and Progress in Cloning *S* and *Z*

Both forward and reverse genetic approaches have been taken in attempts to identify the genes at the *S* and *Z* loci.

13.3.1 Reverse Genetics

Attempts to clone *S*- or *Z*-loci from the grasses using a candidate gene approach have been largely unsuccessful. Grass pollen is tri-nucleate, the stigma is dry and the incompatibility reaction is rapid, all of which are similar to the characteristics found also in *Brassica* (see Chap. 4). These similarities led Wehling et al. (1994b) to investigate sequences related to the *Brassica SLG* gene in rye. Rye inbred lines segregating at the *S*-locus and homozygous at the *Z*-locus were used as templates for PCR, amplified with primers derived from the *Brassica* SLG sequences, in an attempt to amplify homologous sequences. After denaturing gradient gel electrophoresis, although a 280 bp PCR-fragment displayed a polymorphism correlated to the *S*-genotypes, there have been no further publications extending this work. As noted earlier (Sect. 13.2.3), Wehling et al. (1994a) also examined protein phosphorylation patterns in rye, in the hope that this might provide insights into identifying components involved in regulating SI in rye, but again this did not result in the identification of clear candidates for *S*- or *Z*-loci.

Differential screening of proteins or mRNA from different genotypes has provided the basis for reverse genetic approaches. Tan and Jackson (1988) examined the proteins of *P. coerulescens* pollen but failed to identify a protein related to *S* or *Z*. A later study by Li et al. (1994) used a differential screen of a cDNA library from mature *P. coerulescens* pollen. While this revealed a strong candidate for the *S*-determinant in this species (a thioredoxin; Li et al. 1994), no candidate for the *Z*-determinant was found. Moreover, later evidence revealed that the *S*-determinant candidate represented a gene which, although it was closely linked to the *S*-locus, it did not represent the *S*-locus determinant (Baumann et al. 2002; Langridge et al. 1999).

Kakeda and colleagues (personal communication) have used differential AFLP analysis of cDNA to identify candidates for both *S*- and *Z*-determinants from *H. bulbosum* (see Chap. 12, for details of their AFLP approach to characterise the *S*-locus in *Ipomoea*, which has a single *S*-locus SSI system). Their screens have resulted in promising leads for candidates for both pollen and stigma *S*- and *Z*-determinants but further analysis will be needed before this approach can be validated.

13.3.2 Forward Genetics

We are aware of four research groups currently taking the forward genetics approach in three grass species: *P. coerulescens* (Baumann and Langridge, Australia), *L. perenne* (Forster, Australia and King, UK), *H. bulbosum* (Kakeda, Japan) and *S. cereale* (Wehling and Hackauf, Germany). The high conservation of gene order within the grasses has the advantage that maps and markers identified in one species are usually suitable in others. This has been particularly true for the four species used to study SI in the grasses since they are all quite closely related.

No physical maps have been made of the *S*-locus region in any grass. However, the genome sizes of *Lolium* (4 pg; Rees and Durrant 1986) and *Phalaris* (3pg; Watson 1990) are quite small suggesting that, on average, the genetic to physical distance will be amenable to positional cloning. High quality BAC libraries have also been produced for *Lolium* (Farrar et al. 2007; Shinozuka et al. 2007), *Phalaris* (Shi et al. unpublished) and rye (Shi et al. unpublished). Positional cloning will be substantially more difficult in rye and *H. bulbosum*, where the estimates of genome size are considerably larger than for *Phalaris* or *Lolium*, at around 15 pg (Vilhar et al. 2001; Jakob et al. 2004).

Cornish et al. (1980) found linkage between *S*-locus and the isoenzyme, phosphoglycoisomerase (PGI-2) in *L. perenne*. This linkage relationship was also found in *P. coerulescens, F. pratensis, Alopecurus mysorides, Holcus lantanus* and *S. cereale* (Leach and Hayman 1987). The *S*-locus was located on chromosome 6 in *Lolium* and 1R in rye (Lewis et al. 1980; Wricke and Wehling 1985). Linkage between *S*- and restriction fragment length polymorphism (RFLP) markers *Xiag249* (2.7 cM) and *Xpsr544* (4.5 cM) were described in rye by Senft and Wricke (1996). These two markers are located on chromosome 1R supporting the earlier work with isozymes. For Z, linkage was found with the β-glucosidase locus on chromosome 2R, with a recombination value of 14.4% (Gertz and Wricke 1989). Voylokov et al. (1998) reported that S was closely linked to *Xpsr634* on 1R and Z to *Xbcd266* on 2R in *S. cereale*. Similar results were obtained in *H. bulbosum* (Kakeda, Personal Communication 2000). These linkage data therefore show that the genetic organisation of the *S*- and *Z*-loci is similar in several grass genera. Thus, information from one species can be used in another.

Because of gametic selection, markers or genes that are closely linked to either *S*- and *Z*-loci will show disturbed segregation in crosses where the pollen is only partially compatible with the stigma. Analysis of such crosses allows the identification of markers or genes linked to *S* or *Z* and the locations of SI loci may be inferred from the linked markers or genes. Leach (1988) described the statistical methods that can be used to analyse such disturbed segregation populations. This will aid the analysis of any candidate *S*- or *Z*-determinants in the grasses when they are identified.

13.3.2.1 Mapping the *S*-locus

High resolution mapping of *S*-locus has been conducted in *Lolium* (Shinozuka et al. 2007) and *Phalaris* (Bian et al. 2004). The resolution of these maps is approaching 0.1 cM. However, the region still spans about 2 Mb when the closest flanking markers are mapped to rice genome sequence (see Fig. 13.1). A large number of markers can be identified in this region but they all co-segregate with *S*- in populations of around 1,000 lines for both *Lolium* and *Phalaris*. This strong clustering of markers in the vicinity of S can be seen in Fig. 13.1. Interestingly, the gene *BM2* (thioredoxin h; Li et al. 1994) that was initially identified as a candidate for *S* maintains co-segregation with *S* in these large populations.

13 Self-Incompatibility in the Grasses

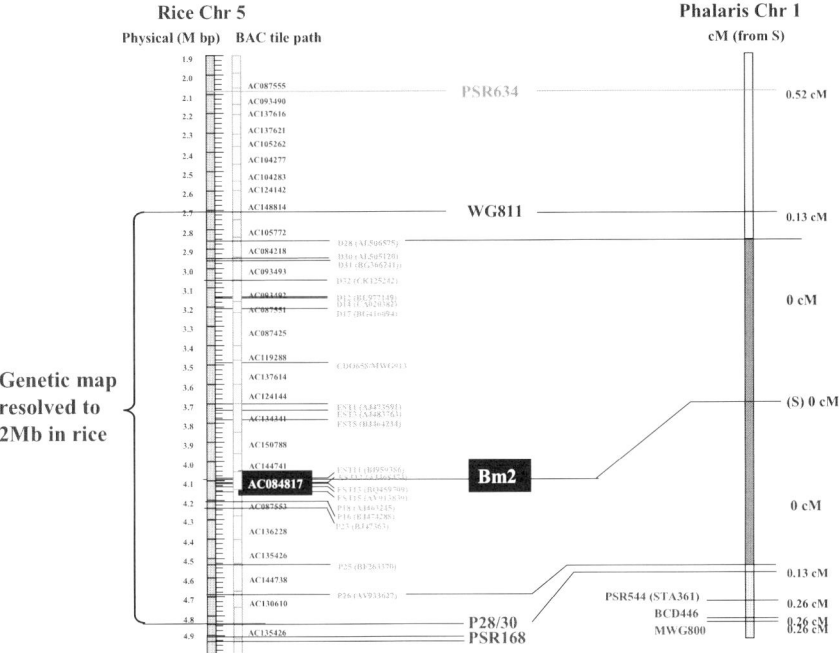

Fig. 13.1 Comparison of the genetic map of the *S*-locus on *Phalaris* chromosome 1 with the syntenous region in rice. In *light-grey font*: RFLP probes generated from EST sequences homologous to rice genes. In addition to Bm2 another 18 probes co-segregate with S in *Phalaris*. For details on the Z-mapping population, see Bian et al. (1994)

Overall, these results imply that there is strong suppression of recombination in the vicinity of *S*-locus, as has been observed in other species (see also Chaps. 4, 6, 12, and 14). This in turn suggests that *S*-determinant will be very difficult to clone via a positional cloning approach.

13.3.2.2 Mapping the Z-Locus

The prospects for positional cloning of the *Z*-locus are substantially better than for the *S*-locus. High resolution maps have been made for the region surrounding the *Z*-locus in *Phalaris* (Bian et al. 2004), *Lolium* (Shinozuka et al. 2007) and rye (Hackauf and Wehling 2005). Figure 13.2 shows the comparison between genetic maps of *Phalaris* and rye with the physical map of rice. The genetic resolution is not as high as for the *S*-locus, at around 1.5 cM, but this distance corresponds to only 380 kb in the rice genome. Using rice as an approximation of physical distances, this means that recombination in the vicinity of *S*- is suppressed over 1,000-fold relative to *Z*. Consequently, the *Z*-locus is now viewed as the best target for positional cloning.

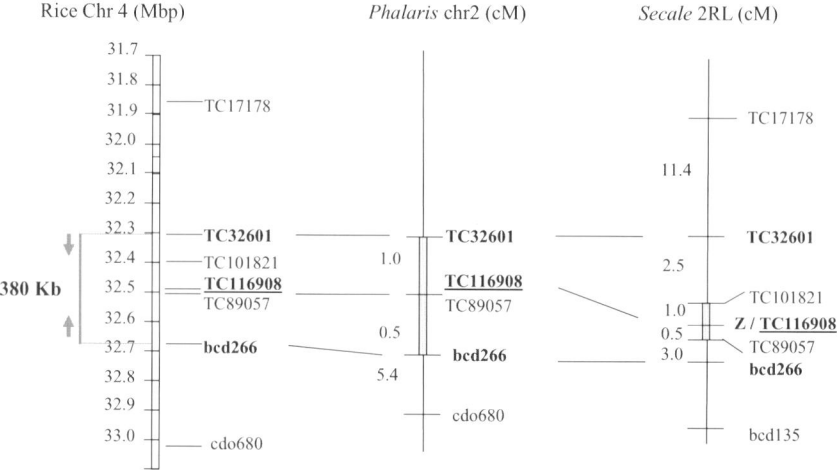

Fig. 13.2 The Z region of *Phalaris* and rye compared to the equivalent region on the rice physical map. The distances on the rice map are in Mbp, while for *Phalaris* and rye the distances are given in cM. The 1.5 cM region of the SI grasses is equivalent to around 380 Kbp in rice. The rye data is from Hackauf and Wehling (2005)

13.4 Conclusions

Progress in the isolation of the genes controlling SI in the grasses has been slow. Initial attempts to use differential screening approaches have yielded several candidates but confirmation of their role in the SI process has proved difficult. Indeed one of the early candidates, the thioredoxin *h* clone BM2 (Li et al. 1994), led researchers down a false path. This emphasises the dangers of following candidate gene approaches in species where the background genetic and molecular information is generally poor and where there is likely to be high allelic variation due to the strong out-breeding nature of these species.

The positional cloning approach has been slow and is proving problematic for the *S*-locus. However, good progress has been made in zeroing in on the *Z*-locus and we are optimistic that the *Z*-locus determinant will soon be isolated from *Lolium*, *Secale* or *Phalaris*. To our knowledge, the analysis of SI in the grasses is not the major focus of any research groups at present. The groups working on cloning *S*- and *Z*-locus determinants in the grasses are doing this either as part of larger program on genetic analysis or as a lingering hobby and abiding fascination for this tantalising SI system. However, the scene may be changing rapidly since the expanding research effort in biofuels has revived interest in SI in grasses as a means for controlling fertility in the potential cellulosic crops such as *Miscanthus* and switchgrass (Martinez-Reyna and Vogel 2002). This may result in a renaissance and renewed emphasis on directed research on establishing the molecular basis of SI studies in the grasses. If this occurs, hopefully we can soon look forward to identification of the *S*- and *Z*-locus determinants, which will pave way for a better understanding of the basis of how SI is controlled in this very important group of plants.

References

Baumann U (1995) Pollen mRNAs of *Phalaris coerulescens* and their possible role in self-incompatibility. PhD thesis. University of Adelaide

Baumann U, Juttner J, Bian XY, Langridge P (2000) Self-incompatibility in the grasses. Ann Bot 85(Suppl A):203–209

Bian XY, Friedrich A, Bai JR, Baumann U, Hayman DL, Barker SJ, Langridge P (2004) High-resolution mapping of the S and Z loci of *Phalaris coerulescens*. Genome 47:918–930

Beddows AR (1931) Seed setting and flowering in various grasses. University College of Wales Series H 12:5–99

Bush EJ, Barrett SCH (1993) Genetics of mine invasions by *Deschampsia cespitosa* (Poaceae). Can J Bot 71:1336–1348

Chu YE, Morishima H, Oka HI (1969) Partial self-incompatibility found in *Oryza perennis subsp. barthi*i. Jpn J Genet 44:225–229

Conner JA, Conner P, Nasrallah ME, Nasrallah JB (1998) Comparative mapping of the *Brassica* S-locus region and its homeolog in *Arabidopsis*: Implications for the evolution of mating systems in the Brassicaceae. Plant Cell 10:801–812

Connor HE (1979) Breeding systems in the grasses: A survey. N Z J Bot 17:547–574

Cornish MA, Hayward MD, Lawrence MJ (1979) Self-incompatibility in diploid *Lolium perenne* L. Heredity 43:95–106

Cornish MA, Hayward MD, Lawrence MJ (1980) Self-incompatibility in ryegrass. III. The joint segregation of S and PGI-2 in *Lolium perenne* L. Heredity 44:55–62

Daehler CC (1999) Inbreeding depression in smooth cordgrass (*Spartina alterniflora*, Poaceae) invading San Francisco Bay. Am J Bot 86:131–139

Farrar K, Asp T, Lubberstedt T, Xu ML, Thomas AM, Christiansen C, Humphreys MO, Donnison IS (2007) Construction of two *Lolium perenne* BAC libraries and identification of BACs containing candidate genes for disease resistance and forage quality. Mol Breed 19:15–23

Fearon CH, Hayward MD, Lawrence MJ (1983) Self-incompatibility in rye grass. V. Genetic control, linkage and seed set in diploid *Lolium multiflorum* Lam. Heredity 50:35–46

Fearon CH, Hayward MD, Lawrence MJ (1984a) Self-incompatibility in rye grass. VII. The determination of incompatibility genotypes in autotetraploid families of *Lolium perenne* L. Heredity 53:403–413

Fearon CH, Hayward MD, Lawrence MJ (1984b) Self-incompatibility in rye grass. VIII. The mode of action in the pollen of autotetraploids of *Lolium perenne* L. Heredity 53:415–422

Fuong FT, Vovlokov AV, Smirnov VG (1993) Genetic studies of self-fertility in rye (*Secale cereale* L.) 2. The search for isozyme marker genes linked to self-incompatibility loci. Theor Appl Genet 83:619–623

Gertz A, Wricke G (1989) Linkage between the incompatibility locus Z and a beta-glucosidase locus in rye. Plant Breed 102:255–259

Hackauf B, Wehling P (2005) Approaching the self-incompatibility locus Z in rye (*Secale cereale* L.) via comparative genetics. Theor Appl Genet 110:832–845

Hayman DL (1956) The genetic control of incompatibility in *Phalaris coerulescens* DESF. Austr J Biol Sci 9:321–331

Hayman DL (1992) The S-Z incompatibility system. In: Chapman GP, (ed) Grass evolution and domestication. Cambridge University Press, Cambridge, pp 117–137

Hayman DL, Richter J (1992) Mutations affecting self-incompatibility in *Phalaris coerulescens*. Heredity 68:495–503

Heslop-Harrison J (1982) Pollen-stigma interaction and cross-compatibility in the grasses. Science 215:1358–1364

Jakob SS, Meister A, Blattner FR (2004) The considerable genome size variation of *Hordeum* species (Poaceae) in linked to phylogeny, life form, ecology and speciation rates. Mol Biol Evol 21:860–869

Kellogg EA (1998) Relationships of cereal crops and other grasses. Proc Natl Acad Sci USA 95:2005–2010

Körnicke F (1890) Über die autogenetische und heterogenetische Befrüchtung bei den Pflanzen. Verhandlungen des naturhistorischen Vereines des preussischen Rheinlandes Band 5 47:84–99

Langridge P, Baumann U, Juttner J (1999) Revisiting and revising the self-incompatibility genetics of *Phalaris coerulescens*. Plant Cell 11:1826

Leach CR (1988) Detection and estimation of linkage for a co-dominant structural gene locus linked to a gametophytic self-incompatibility locus. Theor Appl Genet 75:882–888

Leach CR, Hayman DL (1987) The incompatibility loci as conserved linkage groups in the Poaceae. Heredity 58:303–305

Lewis D (1947) Competition and dominance of incompatibility alleles in diploid pollen. Heredity 1:85–108

Lewis EJ, Humphreys MW, Caton MF (1980) Chromosomal location of two isozyme loci using primary trisomics. Theor Appl Genet 57:237–239

Li X, Nield J, Hayman D, Langridge P (1994) Cloning a putative self-incompatibility gene from the pollen of the grass *Phalaris coerulescens*. Plant Cell 6:1923–1924

Lundqvist A (1954) Studies on self-sterility in rye, *Secale cereale* L. Hereditas 40:278–294

Lundqvist A (1955) Genetics of incompatibility in *Festuca pratensis* Huds. Hereditas 47:542–562

Lundqvist A (1956) Self-incompatibility in rye. I. Genetic control in the diploid. Hereditas 42:293–348

Lundqvist A (1957) Self-incompatibility in rye. II. Genetic control in the autotetraploid. Heriditas 43:467–511

Lundqvist A (1958) Self-incompatibility in rye. IV. Factors relating to self-seeding. Hereditas 44:193–256

Lundqvist A (1962a) Self-incompatibility in diploid *Hordeum bulbosum* L. Hereditas 48:138–152

Lundqvist A (1962b) The nature of the two-loci incompatibility system in grasses. I. Hypothesis of a duplicative origin. Heriditas 48:153–168

Lundqvist A (1965) Self-incompatibility in *Dactylis aschersoniana* Graebn. Hereditas 54:70–87

Lundqvist A (1968) The mode of origin of self-fertility in grasses. Hereditas 59:413–426

Lundqvist A (1969) Self-incompatibility in *Dactylis glomerata* L. Hereditas 61:353–360

Martinez-Reyna JM, Vogel KP (2002) Incompatibility system in switchgrass. Crop Sci 42:1800–1805

McKone MJ, Lund CP, O'Brien JM (1998) Reproductive biology of two dominant prairie grasses (*Andropogon gerardii* and *Sorghastrum nutans*, Poaceae) – male-biased sex allocation in wind pollinated plants. Am J Bot 85:776–783

Murray BG (1974) Breeding systems and floral biology in the genus *Briza*. Heredity 33:285–292

Nayar NM (1967) Prevalence of self-incompatibility in *Oryza baythii* Cheval.: Its bearing on the evolution of rice and related taxa. Genetica 38:521–527

Oka H-I, Morishima H (1967) Variations in the breeding systems of wild rice, *Oryza perenni*. Evolution 21:249–258

Rees H, Durrant A (1986) Recombination and genome size. Theor Appl Genet 73:72–76

Senft P, Wricke G (1996) An extended genetic map of rye (*Secale cereale* L.). Plant Breed 115:508–510

Shinozuka H, Cogan N, Smith K, Forster J (2007) Fine-structure genetic and physical mapping of the perennial ryegrass [*Lolium perenne* L.] self-incompatibility loci. Molecular breeding of forage 2007, Sapporo, Japan, p 98

Shivanna KR, Heslop-Harrison Y, Heslop-Harrison J (1982) The pollen–stigma interaction in the grasses. III. Features of the self-incompatibility response. Acta Bot Neerl 31:307–319

Tan LW, Jackson JF (1988) Stigma proteins of the two loci self-incompatible grass *Phalaris coerulescens*. Sex Plant Reprod 1:25–27

Thomas SM, Murray BG (1975) A new site for the self-incompatibility reaction in the Gramineae. Incompat Newslett 6:22–23

Thorogood D, Hayward MD (1991) The genetic control of self-compatibility in an inbred line of *Lolium perenne* L. Heredity 67:175–181

Thorogood D, Armstead IP, Turner LB, Humphreys MO, Hayward MD (2005) Identification and mode of action of self-incompatibility loci in *Lolium perenne* L. Heredity 94:356–363

Troll HJ (1930) Die Bedeutung der Blüh- und Befruchtungsverhältnisse von Gräsern für ihre Züchtung. Der Züchter Band II:330–336

Troll HJ (1931). Untersuchungen über die Selbststerilität und Selbstfertilität bei Gräsern. Zeitschrift für Züchtung. A. Pflanzenzüchtung XVI:105–136

Vilhar B, Greilhuber J, Koce JD, Temsch EM, Dermastia M (2001) Plant genome size measurement with DNA image cytometry. Ann Bot 87:719–728

Vovlokov AV, Fuong FT, Smirnov VG (1993) Genetic studies of self-fertility in rye (*Secale cereale* L.) 1. The identification of genotypes of self-fertile lines for the *Sf* alleles of self-incompatibility genes. Theor Appl Genet 83:616–618

Watson L (1990) The grass family, Poaceae. In: Chapman GP (ed) Reproductive versatility in the grasses. Cambridge University Press, Cambridge, pp 1–31

Watson L, Dallwitz MJ (1992) The grass genera of the world. CAB International, Wallingford

Wehling P, Hackauf B, Wricke G (1994a) Phosphorylation of pollen proteins in relation to self-incompatibility in rye (*Secale cereale* L.) Sex Plant Reprod 7:67–75

Wehling P, Hackauf B, Wricke G (1994b) Identification of *S*-locus linked PCR fragments in rye (*Secale cereale* L.) by denaturing gradient gel electrophoresis. Plant J 5:891–893

Weimarck A (1968) Self-incompatibility in the Gramineae. Hereditas 60:157–166

Wilkins PW, Thorogood D (1992) Breakdown of self-incompatibility in perennial ryegrass at high temperature and its uses in breeding. Euphytica 64:65–69

Wricke G (1978) Pseudo-self-compatibility in rye and its utilization in breeding. Zeitschrift für Pflanzenzüchtung 81:140–148

Wricke A, Wehling P (1985) Linkage between an incompatibility locus and a peroxidase isozyme locus (*Prx-7*) in rye. Theor Appl Genet 71:289–291

Yeh B, Henderson MT (1961) Cytogenetic relationship between cultivated rice, *O. sativa* L., and five wild diploid forms of Oryza. Crop Sci 1:445–450

Chapter 14
Heteromorphic Self-Incompatibility in *Primula*: Twenty-First Century Tools Promise to Unravel a Classic Nineteenth Century Model System

A. McCubbin

Abstract Research on heteromorphic self-incompatibility systems has a long and prestigious history particularly in the *Primula* genus. This being said, the rise of attractive alternative model systems, coincident with the development of molecular techniques, led to a significant hiatus in this field. As a result, our comprehension of these systems at the molecular level lags dramatically behind that of several other types of self-incompatibility. The body of research on heteromorphic SI in *Primula* has not recently been reviewed in detail. In this article I reassess historical data regarding heteromorphic SI in *Primula*, attempting to update interpretations where appropriate to account for developments in our understanding of self-incompatibility systems and plant biology in general. In addition, I review recent developments, particularly at the molecular level, which, though still in their early stages, are generating tools that promise to dramatically improve our understanding of the operation of this system in the near future.

Abbreviations

A	Male (androecium) characteristics
C	The cross sectional area of stylar conducting tissue
G	Female (gynaecium) characteristics
Gm	A second stylar length factor
IP	Pollen incompatibility
IS	Style compatibility
L	Thrum homozygote lethality

A. McCubbin
School of Biological Sciences and Center for Reproductive Biology, Washington State University, Pullman, WA 99164, USA, e-mail: amccubbin@wsu.edu

Mpm	Pollen mating type dominance
Mpp	Pollen size dominance
P	Pollen size
P*	Abnormal pin morph
Pm	Pollen compatibility type
Pp	Pollen size
PvSLL1	cDNA from *Primula* inked to the *S*-locus; encodes a small putative transmembrane protein of unknown function
PvSLL2	cDNA from *Primula* loosely linked to the *S*-locus; has homology to the *CONSTANS-LIKE* gene
S	Stigmatic papillar length
s	The pin allele
S	The thrum allele
SCR/SP11	*S*-locus cysteine rich – the *Brassica* pollen S gene
SI	Self-incompatibility
ss	The pin form genotype, a homozygous recessive
Ss	The thrum form genotype, a heterozygote

14.1 Introduction

The vast majority of flowering plants are not only hermaphrodite, but in most cases bear flowers in which male and female organs are positioned in close proximity. As is clear throughout this book, one consequence of this has been the evolution of a diverse array of strategies to prevent inbreeding. In line with both the volume of research carried out, as well as frequency of occurrence in the plant kingdom, the bulk of this text focuses on homomorphic self-incompatibility (SI) systems. In these systems there are no morphological differences between mating types, and reproductive specificity is controlled by a large multi-allelic series of one or more self-incompatibility (*S*)-loci. Another category, heteromorphic SI systems, though less pervasive is at least superficially more analogous to the single sex breeding systems found widely in the animal kingdom, in that they are associated with only two or three mating types with visibly different physical characteristics (Fisher and Mather 1943; Lewis 1949; Gibbs 1986; Barrett 1988; de Nettancourt 2001). Heteromorphic SI is a breeding system that combines genetically controlled floral polymorphism with diallelic, sporophytically controlled biochemical SI. Though remarkably consistent in its phenomenology, heteromorphic SI is polyphyletic and has been reported in 28 plant families scattered throughout the Angiosperms, leading to the belief that it is has arisen independently in a number of Angiosperm lineages (Ganders 1979; Barrett et al. 2000). See Chap. 1, this volume, for a detailed account of recent progress in our understanding of the biology, ecology and genetics of heterostyly.

14.2 Floral Characteristics of the Mating Types of *Primula*

Of the 426 species within the genus *Primula*, 91% possess a dimorphic heteromorphic SI system (Wedderburn and Richards 1990). In all species that possess dimorphic heteromorphic SI, the two mating types differ grossly, in that one of the two morphs bears flowers with long styles and short stamens and is descriptively termed the 'pin' form (the style and stigma resemble a pin). The other bears flowers with reciprocally positioned short styles and long stamens – the 'thrum' form. This is descriptively (if somewhat archaically) named; the Oxford English Dictionary defines thrum as 'an unwoven end of a warp thread, or a fringe of such ends, left in the loom when the finished cloth is cut away' which accurately describes the appearance of the anthers in the mouth of corolla in this morph (see Fig. 14.1). Though these are the most obvious differences, closer examinations have revealed additional characters that differ between morphs (see Fig. 14.1), though these characters do not necessarily vary in all *Primula* species. The following floral characteristics have been reported to be dimorphic: style length, stigma size and shape, corolla mouth size, anther positioning, pollen size and amount.

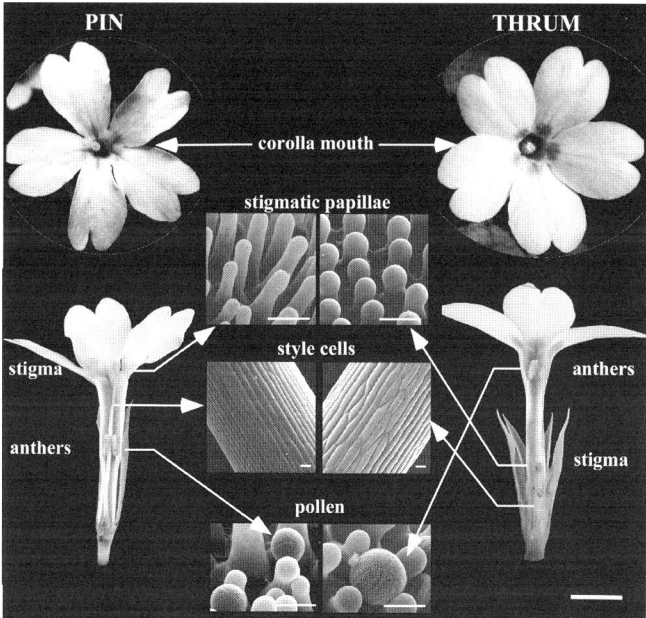

Fig. 14.1 Heteromorphic characters in pin and thrum *Primula* floral morphs. Scale bar for images of overall floral form = 5 mm. The central panel are scanning electron microscope images of cell types as labelled, scale bars = 25 μm. Images courtesy of Amy Hetrick

14.2.1 Style

In pin flowers, stigmas are positioned at the corolla mouth, whereas in thrum the stigmas are positioned approximately two-thirds of the way down the corolla tube (Darwin 1877; see Fig. 14.1). To bring about this positioning, pin style length is approximately double that of thrum (Webster 2006). Studies at the cellular level have shown that this polymorphism is a result of pin style cells being both longer and thinner than thrum (Darwin 1877), and thus the difference in style length is largely a result of differential cell elongation (Heslop-Harrison et al. 1981; Webster and Gilmartin 2006). As a consequence of the difference in cell shapes, the cross-sectional area of pin styles is also smaller in pin than thrum flowers (Dowrick 1956). These three traits are inherently interconnected and most likely regulated by a single genetic factor.

14.2.2 Stigma

The stigmatic papillae are longer in the pin than in the thrum morph, mirroring the difference in length of style cells (see Fig. 14.1). Though the processes underlying style and stigma cell development are not fully understood, it seems likely that this trait is a result of the same genetic polymorphism that leads to style dimorphism (Richards 1997). Stigma shape is also dimorphic in the majority of, but not all *Primula* species, such that pin stigmas are hemispherical, whereas thrum stigmas are somewhat flattened with a central depression (Darwin 1877; Vuilleumier 1967; Dulberger 1974). In contrast to papilla length, there is evidence that this trait is independent of cell shape in the style and stigma and governed by a separate gene (Kurian and Richards 1997).

14.2.3 Corolla Mouth Size

A novel dimorphism has been identified surprisingly recently. Though visually no polymorphism had previously been noted in the corolla tube, a particularly thorough analysis revealed that the corolla mouth is wider in thrum than in pin flowers (Webster and Gilmartin 2006). The corolla of *Primula* is divided into two developmentally distinct domains, the region below the point of anther attachment being designated the 'lower corolla' and that above the 'upper corolla' (Sporne 1974; Erbar 1991). The dimorphism in corolla mouth size results from differences in cell shape in the upper corolla, the cells in the thrum morph being wider than those in the pin, whereas there is no detectable difference between morphs in the cells of the lower corolla (Webster and Gilmartin 2006).

14.2.4 Anthers

The anthers are positioned reciprocally in relation to stigmas between morphs, such that they are in the mouth of the corolla in the thrum morph and approximately two-thirds of the way down the corolla tube in the pin (Darwin 1877; see Fig. 14.1). Anther filaments are fused to the corolla in *Primula* species, causing anther positioning to be inherently related to lower corolla development. There is only a very slight difference in lower corolla cell length between morphs, and consequently dimorphism in anther positioning is likely to be a result of increased cell division in the lower corolla in the thrum morph (Webster and Gilmartin 2006).

14.2.5 Pollen

The pollen is dimorphic in both size and amount produced. Thrum pollen grains are slightly more than twofold larger in volume than pin (Darwin 1862; Dowrick 1956). Interestingly and perhaps associated with this size difference, there is a reciprocal difference in pollen production such that slightly more than two fold more pollen is produced by pin than thrum flowers (Ornduff 1979; Piper et al. 1986). It is tempting to suggest (though not proven) that the overall investment of resources in pollen is approximately equal between the two morphs.

14.2.6 Self-Incompatibility Specificity

Even after manual pollination to fully circumvent potential barriers caused by floral morphology, pollen is only able to fertilise stigmas of the alternate morph (Bateson and Gregory 1905). This is a result of a biochemical SI system in which distinct molecular specificities on or in the pin and thrum stigmas are matched with those present on or in the pollen. Though this incompatibility may be somewhat 'leaky', fully compatible pollen tube growth is facilitated only when the pollen and pistil are from different morphs. Genetic evidence suggests that the pollen and pistil factors that provide the mating type specificity are encoded by separate genes (Dowrick 1956; Kurian and Richards 1997).

14.3 Functions of Heteromorphic Characters

How the various dimorphic morphological characters contribute to heteromorphic SI has been a contentious issue. The fact that the phenomenology of this breeding system is remarkably consistent across the diverse genera in which it is found, despite having multiple evolutionary origins, strongly suggests that many of these characters are important for its operation. The point of contention concerns whether morphological polymorphisms act purely to maximise legitimate (between morph) and minimise illegitimate (within morph) pollen transfer between mating types (Darwin 1877; Ganders 1974; Ornduff 1979), or whether these characters contribute directly to active pollen rejection (Lewis 1942; Mather and de Winton 1941; Dulberger 1975). As there are multiple characters, an intermediate viewpoint is also possible, where some characters might participate in each function (Ganders 1974; Yeo 1975; Richards 1986; see also Chap. 1, this volume).

It makes intuitive sense that the reciprocal stigma and anther positioning would promote legitimate (and minimise illegitimate) pollen transfer between pin and thrum morphs. This hypothesis was first proposed by Darwin (1877) and was supported by his observation that pin and thrum pollen was largely restricted to different body positions on bumblebees and moths visiting *P. veris*. It is surprising, therefore, that the majority of data from studies of pollen loads upon stigmas are not consistent with this hypothesis, and in many cases stigmas have been observed to hold more intra- than extra-morph pollen (Ornduff 1979; Ganders 1979). A possible explanation proposed by Ganders (1979) and supported by Richards (1986) is that most studies have ignored intra-flower pollination. Intra-flower pollination is considerable in all bisexual flowers and can easily obscure analysis of cross-pollination. As a result, to monitor pollen transfer between flowers it is necessary to emasculate flowers to be used as stigma sources before anthesis. Such an experiment has not been reported for *Primula*, but has upheld Darwin's hypothesis in the dimorphic species *Jepsonia heterandra* (Ganders 1974).

The morphological characters most frequently proposed to be involved in SI are pollen size and stigmatic papilla length. It has been hypothesised that the presence of greater energy reserves in the larger thrum pollen is essential for successful growth of pollen tubes to the base of the long pin style (Darwin 1877). Direct empirical data refuting this hypothesis comes from a mutant in which dominance of pollen size had been lost such that it produced both pin- and thrum-sized pollen. Both pollen sizes possessed thrum pollen specificity when used to pollinate pin stigmas (Kurian and Richards 1997), hence pollen size apparently does not play role in the SI reaction.

Stigmatic papilla length has also been postulated to be key to the SI reaction (Dulberger 1975). In some species (but not *Primula*), pollen adheres to stigma only in inter-morph pollinations. However, this is more likely to be a result of biochemical than mechanical interaction (Richards 1986). Again, empirical evidence against the involvement of papilla length in SI comes from the mutant mentioned above. In plants possessing this mutant allele, both style and stigmatic papilla length were intermediate between pin and thrum, yet controlled pollinations showed that these plants had normal pin stigma specificity in SI reactions (Kurian and Richards 1997).

In my view, the data discussed above effectively disprove the hypothesis that morphological characters are directly involved in the SI response. Nonetheless, the consistency with which these characteristics are found in heteromorphic SI systems suggests that they must have important functions. The alternative hypothesis that morphological polymorphisms act purely to maximise legitimate and minimise illegitimate pollen transfer between mating types (Darwin 1877; Ganders 1974;

Ornduff 1979) is plausible, but raises the question as to why these features are important to heteromorphic SI systems when they are not found in homomorphic SI systems.

A plausible solution to this conundrum was suggested by Ganders 1979, and is based simply on the number of *S*-alleles (or *S*-haplotypes) in each class of SI system. In homomorphic systems a large number of *S*-alleles exist (see, e.g. Chaps. 5 and 9, this volume), hence statistically speaking a large percentage of pollen brought by pollinators to an individual stigma is likely to be compatible. Thus, even though a considerable amount of self pollen is likely to be deposited on a stigma, there should be adequate compatible pollination for good seed set. In contrast, in heteromorphic SI, which only has two *S*-alleles, in the absence of disassortative pollination 50% of incoming pollen will be self incompatible in addition to the intra-flower pollen present on the stigma. It is conceivable that this might result in reduced legitimate pollination to a level below that required for adequate seed set. As a result, there may be a distinct selective advantage to genotypic disassortative pollination in heteromorphic SI that is not applicable to multi-allelic homomorphic systems.

Whilst it cannot be said with certainty that morphological characters do not participate in a subtle way to SI per se, they are clearly not required for self-recognition and rejection. Thus, it seems likely that all the dimorphic physical characteristics act in a combinatorial manner to promote inter-morph pollination, reduce intra-morph pollination and/or to counteract the differences in ease of pollination between morphs generated by stigma and anther positioning.

14.4 Physiological Nature of SI

14.4.1 Site of Operation

The cessation of pollen germination or tube growth after intramorph pollination occurs at multiple sites in *Primula* (Shivanna et al. 1981; Wedderburn and Richards 1990). Pollen may fail to germinate on, or penetrate the stigma surface and/or cease tube growth within the stigma or stylar transmitting tract (Wedderburn and Richards 1990). The vast majority of pollen ceases growth in or on the stigma surface, as expected for a sporophytically controlled SI system (Shivanna et al. 1981). However, there are consistent differences between morphs in regard to the extent of growth of self pollen tubes that succeed in penetrating the stigma. In within-thrum pollinations, self incompatible pollen tubes rarely reach the style, whereas in within-pin pollinations, $\sim$5% of pollen tubes reach the transmitting tract before ceasing growth (Shivanna et al. 1981). Whether these differences reflect the speed with which a single SI reaction operates, or the presence of multiple inhibitory mechanisms is a point of contention. Based upon the cessation of pollen/pollen tube growth at multiple sites in the pistil, and bioassays of the effects of crudely fractionated pistil extracts upon pollen tube growth (Golynskaya et al. 1976), it has been suggested that as many as six gene products might be involved in several different physiological SI mechanisms (Wedderburn and Richards 1990). This view was challenged by Barrett and Cruzan (1994) and defended by Richards (1997). In assessing these contrasting viewpoints, one important consideration is that in the original reports there has been an unfortunate, and not entirely correct, use of terminology.

The term *multiple sites of inhibition*, coined by Shivanna et al. (1981) is misleading. The results documented in this report show that there are multiple sites at which pollen or pollen tubes cease growth. However, it does not necessarily follow that there are multiple sites of inhibition. A plausible alternative explanation is that inhibition invariably starts on the stigma surface, but varies in effectiveness, such that the point at which pollen tube growth ceases is not uniform. A second consideration is that, with the notable exception of the gametophytic SI system of *Papaver rhoeas* (Franklin-Tong et al. 1988; see Chap. 11, this volume), it is prudent to view data from in vitro bioassays of SI systems with caution, especially those employing relatively crude extracts (Jackson and Linskens 1990). Genetic background can significantly impact the results of these studies in unpredictable ways, for example inclusion of PEG 4000 in the germination medium causes S_2-specific inhibition of pollen tube growth in *Nicotiana alata* (Jahnen et al. 1989). Overall, these factors weaken the evidence used to infer that SI in *Primula* involves multiple mechanisms, but there are also empirical data and theoretical considerations inconsistent with this hypothesis.

Experimental support for the stigma being the primary and perhaps only site of inhibition, at least in the thrum morph, comes from the demonstration that removal of the stigma of either morph severely reduces intra-morph inhibition of pollen tube growth (Stevens and Murray 1982). In this study, pollinations were scored as the percentage of tubes at the top of the style which grew to the base, pin pistils scored 33% vs. 21% in legitimate vs. illegitimate pollinations and thrum 66% vs. 72%. Overall, this suggests that a single female recognition factor is predominantly located in the stigma in both morphs. However, it does suggest that the style might be a secondary region of inhibition in the pin morph. Initiation of the inhibition of pollen on the stigma surface is consistent with the sporophytic nature of heteromorphic SI, and it is difficult to envisage a mechanism by which sporophytic gene products are retained after extended pollen tube growth. Nonetheless, acceptance of stylar recognition and inhibition in the pin morph is not necessarily inconsistent with a single pair recognition molecules, and might indicate a novel feature of this system.

Though heteromorphic SI is and has always been classified as sporophytic, this is inferred from, and only necessarily the case, in the thrum genotype. Thrum plants are genetically heterozygous, but thrum pollen displays a uniform dominant *S* allele phenotype and is classically sporophytic. In contrast, the homozygous pin pollen is genotypically and phenotypically uniform. Hence, from a genetic standpoint one cannot determine whether pin pollen phenotype is sporophytically and/or gametophytically controlled. Lewis and Jones (1993) hypothesised that cryptic gametophytic SI (see Chap. 4, this volume) might operate in *Primula*, and in the pin morph it is at least theoretically feasible that a single mechanism could operate both sporophytically and gametophytically. This could be achieved by simply broadening the spatial and temporal expression of *s* gene products. In pollen, coat proteins synthesised by the (sporophytic) tapetal tissue could potentially mediate recognition on the stigma surface, and gametophytic expression of pollen s by the haploid pollen/pollen tubes could provide a recognition function after more extended growth. Expanding the expression of female s specificity from the stigma to include the style could potentially position all the components required to provide an SI mechanism that is both gametophytic and sporophytic at the molecular level. However, as the genes encoding specificity are yet to be identified for either pollen or pistil, no molecular evidence from *Primula* exist to support this scenario. Nonetheless, as the *Brassica* pollen *S* gene (SCR/SP11; see Chaps. 6 and 7, this volume) exhibits precisely the expression pattern required for the pollen component of such a system (Takayama et al. 2000), it is certainly within the realms of possibility. Consequently, perhaps the most parsimonious explanation of the data available is that in thrum pollinations pollen rejection is entirely sporophytic and takes place on or in the stigma, but in pin pollinations, though the majority of pollination rejection occurs in the same manner, additional stylar inhibition may occur as a result of gametophytic expression of pollen s specificity and stylar expression of pistil s specificity. It is not necessary to invoke more than one pair of *S*-locus gene products to operate such a system.

14.4.2 Candidate Molecules in the Operation of Heteromorphic SI

Clear differences in water economy between pin and thrum pollen have been reported, but appear unlikely to be involved in SI per se (Shivanna et al. 1983). In terms of candidate proteins that might be involved in the operation of heteromorphic SI in *Primula*, the available published data proposes that phytohemagglutinins (Golynskaya et al. 1976) and stylar apoplastic peroxidase (Carraro et al. 1996) are candidates. Though these reports are the most recent available on this topic, they stem from an era when peroxidases and lectins were being also proposed to be involved in the operation of homomorphic SI systems. Similar reports can be found linking peroxidases and lectin interactions to the operation of gametophytic SI (e.g. Bredemeijer and Blass 1981; Sharma et al. 1985), and whilst these were certainly not invalid at the time of publication, in view of current knowledge they are largely overlooked. In *Primula*, more recent studies are not available, but it is prudent to view these propositions cautiously.

14.5 The *Primula* S-locus

14.5.1 Genetic Structure

Studies of the inheritance of heteromorphic SI predate the rediscovery of Mendel's work (Hildebrand 1863; Darwin 1877). As early as 1905, Bateson and Gregory demonstrated that the short styled thrum morph is dominant and the long styled pin morph, recessive (Bateson and Gregory 1905). One of the most intriguing aspects of heteromorphic SI is that this breeding system, though for the most part segregating as a single genetic unit, is clearly comprised of a number of dimorphic characters in multiple floral organs. The first proposition that these characters might be controlled by more than one linked gene has been attributed to Pellow in an Annual Report of the John Innes Horticultural Institute (Pellow 1928, cited in Lewis and Jones 1993). This report hypothesised that two genes might reside at this genetic locus, a concept later expanded to *a series of closely linked genes* (Haldane 1933).

The locus that controls the various morphological characteristics and the SI mating type that combines to make the heteromorphic SI system is designated, as are all SI loci, the *S*-locus. This use of terminology is somewhat misleading for though all *S*-loci control SI interactions, as is clear throughout this volume, the various types of SI system are biochemically distinct and controlled by very different genes. The collective term *supergene complex* has been applied to the multigene linkage group governing heteromorphic SI (Darlington and Mather 1949; Lewis and Jones 1993) and provides a useful term that highlights the apparently complex nature of this genetic locus; see also Chap. 1, this volume.

Like the *S-locus* of *Brassica*, heteromorphic SI *S*-loci function sporophytically (i.e. pollen phenotype = parental genotype), but unlike *Brassica* and other homomorphic SI systems, the *Primula* heteromorphic SI *S*-locus possesses only two functional alleles. One of the two alleles shows complete dominance for all traits. The allele determining thrum form is dominant in *Primula* (Bateson and Gregory 1905; Mather 1950), the thrum being genotypically a heterozygote (*Ss*) and the pin homozygous recessive (*ss*). Though the genes controlling biochemical SI and associated floral polymorphisms usually segregate as a single unit, crossover events do occur rarely within the locus, leading to breakdown of heteromorphic SI and phenotypes known as *homostyles*. Analyses of such recombinants have provided insights into the structure of the supergene complex and empirical evidence for its multi-gene composition.

Homostyles exhibit floral phenotypes with juxtaposed anthers and stigmas, in either a 'long' or 'short' configuration. Such plants have been found (occasionally stabilised) in wild populations and in large segregation analyses. Consistent with these phenotypes resulting from recombination within the *S*-linkage group, characteristics usually specific to one or other morph are found combined within a single flower. Homomorphic variants were reported as early as 1877 by Darwin, but the majority of our understanding of the structure of the *Primula S-locus* is attributable to a wealth of genetic data collected over more than 30 years by Ernst, which is briefly describe here.

Using *P. hortensis* and *P. viscosa*, Ernst and authors who later reanalysed his data (Ernst 1955; Dowrick 1956; Lewis and Jones 1993) provided evidence for at least three recombinable sub-units within the *S* linkage group: *G* (female gynaecium characteristics), *P* (pollen size) and *A* (male androecium characteristics) (Ernst 1955; Dowrick 1956). Lewis and Jones (1993) provided a generally accepted argument that Ernst's data was best reconciled by the sub-units being in the order *GPA*. In their reappraisal of data generated by Ernst for *P. hortensis*, Lewis and Jones (1993) calculated that recombination between *GP* and *A* occurred at ~0.37% and *G* and *PA* at ~0.19%, giving a total locus length of ~0.56% of a chromosome.

There have been several propositions to expand (further sub-divide) the genetic components assigned to the *Primula* S-locus beyond *G*, *P* and *A*. Dowrick (1956) suggested the addition of *C*, *S*, I^S and I^P, where *C* is the cross-sectional area of stylar conducting tissue, *S* is stigmatic papillar length, I^S is style compatibility and I^P is pollen incompatibility. There is weak or no empirical support for *C*, *G* and *S* being separate factors; they have not been demonstrated by recombinants and, more significantly, these characters appear to result from a common developmental process and hence are likely to be encoded by the same gene (Webster and Gilmartin 2006). In contrast, there is substantial support for style compatibility specificity (I^S) and pollen incompatibility specificity (I^P) being encoded by different genes (Dowrick 1956), as well as suggest the presence of additional components within the locus (Kurian and Richards 1997).

The most recent report on this subject resulted from the identification of abnormal pin morph plants (designated P*) of *Primula X tommasinii* (Kurian and Richards 1997). These plants were not true homostyles, as their style length was intermediate between that of pin and thrum. These mutants were compatible when selfed or crossed with pin mothers, but incompatible with thrum mothers, that is the pollen possessed thrum specificity for biochemical SI, but stigmas possessed pin specificity. Surprisingly, P* plants possessed dimorphic pollen, such that the pollen was 50% pin and 50% thrum in size. Style and stigma cell sizes of plants possessing this mutant allele were intermediate relative to pin and thrum morphs, but stigma shape remained pin. Lastly, analyses of progeny resulting from controlled pollinations showed that the mutant allele was heritable in combination only with the recessive pin allele (*s*), that is the mutant allele was lethal when homozygous or combined with the thrum allele (*S*). A number of inferences can be drawn from these results with respect to the components of the *S*-locus complex (Kurian and Richards 1997).

The phenotype of plants possessing the P* allele corroborates the hypothesis that male and female SI are genetically distinct (Dowrick 1956), as the traits were recombined within the mutant allele (Kurian and Richards 1997). Further P* appears to harbour the lethal gene hypothesised to reside in the thrum allele, as P* was never found to be homozygous or in combination with *S*, providing recombinational evidence for the existence of this gene. A particularly intriguing and novel inference is that pollen size is regulated by at least two genetic components, one acting gametophytically and the other sporophytically, such that thrum pollen is uniformly large. In P* plants the sporophytic 'over-ride' was absent and pollen carrying the dominant thrum allele remained large, but pollen carrying the recessive allele was small. To account for the intermediate style length of P* plants, the authors further suggested that there may be two loci within the *S*-locus that act in an additive fashion to control style and stigma cell size, but not stigma morphology (which is encoded by a separate gene) (Kurian and Richards 1997). Richards (1997) proposed an expansion of the *G*, *P* and *A* model of the *S-locus* of *P. X tommasinii* based on this data, adding *Mpp* (pollen size dominance), *Pp* (pollen size), *Pm* (pollen compatibility type), *Mpm* (pollen mating type dominance), *L* (thrum homozygote lethality) and *Gm* (a second stylar length factor). There is solid experimental support for most, but not all of these components.

One genetic component for which there is no recombinational evidence is pollen mating type dominance (*Mpm*), but a strong theoretical argument can be made for its existence. The sporophytic behaviour of the pollen SI phenotype is generally thought to be a result of secretion of pollen *S* recognition molecules by diploid tapetal tissue of the anther, providing a uniform sporophytic coating on the pollen grains irrespective of haploid genotype (Shivanna et al. 1983). This scenario has been confirmed in the homomorphic sporophytic system of *Brassica* (Schopfer et al. 1999; see Chap. 8, this volume). In heteromorphic SI, whilst being sporophytically controlled, the thrum pollen phenotype is not diallelic as *S* is fully dominant and phenotypically the pollen behaves as *S*. As a result, it is necessary to invoke silencing of the recessive allele of the gene controlling pollen specificity. A summary of the sub-components of the *Primula S-locus*, the existence of which is well supported through recombinants, developmental and/or strong theoretical justification, along with references to the relevant reports is provided in Table 14.1.

Table 14.1 Substantiated sub-components of the *Primula S-locus*

Locus (if assigned)	Character	Evidence	Reference(s)
G/g	Style length A (stigmatic papilla length, style cell size)	Recombination	Ernst 1955, Dowrick 1956
IS/is	Pistil biochemical specificity	Recombination	Dowrick 1956, Kurian and Richards 1997
A/a	Anther height	Recombination	Ernst 1955, Dowrick 1956
IP/ip	Pollen biochemical specificity	Recombination	Dowrick 1956, Kurian and Richards 1997
Pp/pp	Pollen size	Recombination	Kurian and Richards 1997
Mpp/mpp	Pollen size dominance	Recombination	Kurian and Richards 1997
Mpm/mpm	Pollen biochemical specificity dominance	Theoretical	Kurian and Richards 1997
l/L	Lethal	Recombination	Kurian and Richards 1997
Gm/gm	Style length B	Recombination	Kurian and Richards 1997
–	Corolla mouth size	Developmentally distinct	Webster and Gilmartin 2006
–	Stigma shape	Recombination	Kurian and Richards 1997

14.5.2 Location and Size of the S-locus

The nature of the *S-locus* of *Primula* as a supergene complex raises the issue of how recombination might be suppressed in a presumably substantial chromosomal region. A likely answer is that suppression of recombination is an incidental result of chromosomal location (Darlington 1929). Prior to the advent of modern cytogenetic techniques and genome sequencing, the classical genetic technique of 'double reduction' in autotetraploids was used to demonstrate linkage to a centromere (Darlington 1929). In this technique a 'triplex' plant (tetraploid with alleles in a 3:1 ratio) with the genotype *SSSs* is employed. In such a plant, as centromeres segregate during the first meiotic division and products of duplication separate at the second, each of the four diploid gametes produced at meiosis cannot possess both halves of the same centromere. Consequently, in the triplex plant double recessive (*ss*) gametes cannot occur without recombination between the *S*-locus and the centromere of the chromosome on which it resides. The frequency with which double recessives are recovered is thus an indicator of genetic distance from the centromere. This technique has been used to show that in *P. sinensis S* is located within the region of recombination suppression of a centromere, i.e. complete linkage was observed, and in *P. obconica S* is close to, but just outside, this region (Darlington 1929).

Given the sub-centromeric location of the *S*-locus in *Primula*, it is perhaps not surprising that a considerable number of traits are linked to the *S*-locus. These include *blue flower* (De Winton and Haldane 1931), *green stigma* (De Winton and Haldane 1931), *Hose in Hose* (Ernst 1931), *sepaloid* (Webster and Gilmartin 2003), *staminoid carpels* (Webster and Gilmartin 2003), *Oak leaf* (Webster 2005) and the floral colours *magenta* and *maroon* (Kurian 1986, cited in Richards 2001). None of these traits show complete linkage to *S*, and they appear to flank one or other side of the locus. The sub-centromeric location of the *Primula S*-locus and the fact that many of the traits both within, and linked to it, are reproductive characteristics is reminiscent of another class of *S* loci. Solanaceous species possess S-RNase based, gametophytic SI (see Chaps. 9 and 10), and the *S loci* regulating this system are also sub-centromeric (Entani et al. 1999) and also habour a considerable number of reproductive traits (Bernacchi and Tanksley 1997). S-RNase based SI is believed to be ancestral in the angiosperms, whereas heteromorphic SI is derived (see Chap. 4). Though we do not have sufficient information to determine whether there is synteny between the *S* loci of the two systems, it is tempting to speculate that heteromorphic SI might have recruited the sub-centromeric linkage group once used for S-RNase based SI.

14.5.3 Allelic Dominance

The dominance of *S* over *s* is a fascinating aspect of diallelic heteromorphic SI. Dominance is complete in both the pollen and pistil, and prevails even in tetraploid plants with the genotype *Ssss* (Dowrick 1956). Though we currently know nothing about how dominance is achieved in heteromorphic SI, considerable progress has recently been made in the field of gene silencing in plants. An exciting recent report on how this phenomenon operates in SI in *Brassica* has opened a tantalising possibility for the mode of operation of dominance in heteromorphic SI (Shiba 2007). In *Brassica*, which possesses sporophytic SI, though pollen can be diallelic with respect to *S* phenotype, there is a dominance hierarchy between *S*-alleles. In heterozygous combinations where alleles low in the dominance series are combined with those that are high, complete dominance occurs and the pollen exhibits a monoallelic phenotype (analogous to thrum pollen in *Primula*). This dominance correlates with changes in the expression of alleles of the pollen *S* gene, *SP11/SCR*, pollen *S* phenotype being determined by reversible suppression of recessive by dominant alleles (Shiba et al. 2002; Kakizaki et al. 2003). This suppression involves transcriptional silencing by DNA methylation. Cytosine residues in the 5′ promoter sequences of recessive alleles of *SCR/SP11* are specifically methylated only when these alleles are heterozygous with a dominant allele (Shiba et al. 2007). This methylation is developmentally regulated, tissue specific and acts to block transcription of recessive *SCR/SP11* alleles. Such developmentally regulated epigenetic gene silencing is well documented in animals, but this is the first report of its importance in plants (Shiba et al. 2007; Shiba and Takayama 2007; see also Chap. 7, this volume). The identity of the *S-locus* component(s) of dominant alleles that causes this methylation is not yet clear. One possibility has emerged from the characterisation of the sporophytically silenced floral repressor *FWA*. Methylation and silencing of *FWA* is mediated by transcription of *SINE*-related direct repeat sequences, which are targeted around the transcription start site of this gene (Kinoshita et al. 2007). The intergenic region around *SCR/SP11* is both highly polymorphic between alleles and contains short repetitive sequences with similarity to the 5′ promoter region of *SCR/SP11*. As a result, small RNAs derived from the intergenic region have been hypothesised to be responsible for regulating allelic silencing by targeting cytosine methylation of the *SCR/SP11* promoter (Shiba and Takayama 2007).

While there is no direct evidence that developmentally regulated, reversible DNA methylation is involved in allelic dominance in heteromorphic SI, it is an attractive possibility. Silencing of the recessive alleles of genes encoding pollen and pistil biochemical SI must occur in heteromorphic SI, and this silencing must be reversible, as these genes are subsequently functional in pin progeny of thrum plants. Though only limited sequence data are available for the *Primula S*-locus, there is already ample evidence that this region possesses the repetitive structure required for encoding the small RNAs necessary (Manfield et al. 2005).

14.6 Floral Development

Recent investigations into the biology of heteromorphy in *P. vulgaris* have employed scanning electron microscopy to characterise divergence in floral morphology from a developmental viewpoint. These studies have provided novel insights concerning whether or not traits are developmentally correlated, and hence whether they might be encoded by a common gene or independently. In early stages of floral development, the two morphs have been found to be indistinguishable, only after all the floral organs start to differentiate do differences become apparent (Webster and Gilmartin 2003). The first visible divergence (in 5 mm buds) is the elevation of the stigma above the anthers in the pin morph, whereas in the thrum, the reciprocal situation occurs (Webster and Gilmartin 2006). As morphological divergence appears to start with style elongation, it has been suggested that part of the *G* component of the *S-locus* is the first to be implemented, the function of the dominant *G* allele being to inhibit style elongation from this point in development of thrum flowers (Webster and Gilmartin 2006).

Within the corolla, the upper corolla tube (above the point of anther attachment) develops before the lower in both morphs, but elongation of the lower corolla tube initiates earlier in the thrum than the pin morph. Divergence in anther position, driven by this lower corolla expansion (genetically an *A* function trait), is not visible until quite late in development when the buds are approximately half (11 mm) of their mature size (20 mm). In contrast to the inhibitory role of thrum *G* function, thrum *A* function plays a dominant role in promoting anther elevation by increasing growth of the lower corolla tube (to which the anther filament is fused). Hence the *A* and *G* components of the *S-locus* differ in spatial, temporal and physiological functions and, most likely, act through distinct mechanisms (Webster and Gilmartin 2006).

14.7 Molecular Genetic Characterisation of the *Primula S-locus*: Current Status

Given the wealth of traditional genetic data on heteromorphic SI, it is somewhat surprising that few molecular analyses of any heteromorphic SI system are to be found in the literature. Such studies have begun only recently and are underway in *Turnera* sp. (Anthanasiou and Shore 1997; Shore et al. 2006), Buckwheat (*Fagopyrum homotropicum*) (Jotaro et al. 1999; Matsui et al. 2004) as well as *Primula* (Manfield et al. 2005; McCubbin et al. 2006; Li et al. 2007; see also Chap. 1, this volume).

The first molecular data relating to *Primula* heteromorphic SI was provided by Manfield et al. (2005). In this study, a genomic DNA marker linked to the dominant thrum *S* allele was identified by random amplification of polymorphic DNA (RAPD)-PCR. The genomic fragment encoding this marker was converted to a sequence characterised amplified (SCAR) marker for the *S* allele. Segregation analysis showed complete linkage of the marker to the dominant allele in a population of 191, and using long and short homostyles, the marker was mapped to the *A* side of the *S-locus*. An 8.8 kb genomic DNA clone encoding this SCAR fragment has been sequenced, revealing a highly repetitive structure and a large number of retrotransposon-like sequences. This structure is reminiscent of *S*-linkage groups of species with homomorphic SI systems and has been suggested to reflect accumulation of structural elements that may help to suppress recombination within these linkage groups. Unfortunately, no genes were identified within this genomic fragment.

cDNAs specific to floral organs of one or other floral morph have been identified by subtractive suppressive hybridisation (McCubbin et al. 2006). Subtracting combined cDNAs from mature and developing floral organs of pin and thrum morphs, led to the identification of 11 classes of cDNA that exhibited differential expression between developing pin and thrum flowers in at least one organ (McCubbin et al. 2006). None of these cDNAs produced an *S* allele specific RFLP on blots of *Eco*R I digested genomic DNA, suggesting that they are probably not linked to the *S*-locus and more likely represent components of developmental pathways regulated by genes at the *S-locus*. In support of this hypothesis, a number of the genes identified exhibited homology to gene families known to be involved in developmental processes, including rapid alkalinization factors and *CHX* ion transporters (implicated in cell expansion), and *DExH* box RNA helicases and *SKS* multicopper oxidases (implicated in the regulation of cell size). Though conclusive evidence for these genes playing a functional role in heteromorphic SI is lacking, it seems likely that at least some are involved processes that regulate floral morphology.

The most recent report on this subject also used a cDNA-based approach to identify genes that differ in expression levels or sequence polymorphism between pin and thrum morphs. Using differential display, Li et al. (2007) also identified a novel cDNA that is differentially expressed between morphs, but more significantly, also identified two cDNA fragments linked to the *S*-locus. The genes encoding these cDNAs have been termed *PvSLL1* and *PvSLL2*. *PvSLL2* has homology to the *CONSTANS-LIKE* gene family from *Arabidopsis* and is loosely linked to *S*, as two recombinants were identified in 146 plants. This gene did, however, map to the *G* side of the *S*-locus using recombinant homostyles and represents the first molecular marker identified for this region (Li et al. 2007). *PvSLL1*, in contrast, showed complete linkage to *S* and mapped to the *A* side of the *S*-locus. Current data suggests that *PvSLL1* is a single copy gene that resides within the *S*-locus, and it is the first gene reported to do so. *PvSLL1* encodes a small putative trans-membrane protein with similarity to *Arabidopsis* At1g72020, a gene of unknown function (Li et al. 2007).

Whether this gene might play a role in heteromorphic SI is unknown, but sequence analysis of *PvSLL1* alleles uncovered a novel finding, which necessitates that we modify our view of allelic diversity in this SI system. Three alleles of *PvSLL1* were identified, rather than two as predicted by our traditional view of a diallelic locus. Two sequence variants of *PvSLL1* ($PvSLL1^{P1}$ and $PvSLL1^{P2}$) were associated with recessive *s* allele and a third with the dominant *S* allele ($PvSLL1^{T}$). This suggests that while heteromorphic SI in *Primula* is necessarily regulated by only two alleles, this is true only in a functional sense. At the DNA level, polymorphisms may accumulate within functional alleles, such that their sequences diverge, generating functionally identical alleles that are polymorphic at the molecular genetic level. Nucleotide alignments of the cDNA sequences of the three alleles revealed eight sites of nucleotide substitution between them, $PvSLL1^{P1}$ and $PvSLL1^{P2}$ differing at seven of these positions, $PvSLL1^{P1}$ and $PvSLL1^{T}$ at four and $PvSLL1^{P2}$ and $PvSLL1^T$ at four. In addition, $PvSLL1^{P1}$ has a longer intron resulting from tandem sequence duplication. The fact that $PvSLL1^{P1}$ and $PvSLL1^T$ exhibit higher pairwise sequence similarity than $PvSLL1^{P1}$ and $PvSLL1^{P2}$ suggest that $PvSLL1^{P1}$ and $PvSLL1^{P2}$ diverged before $PvSLL1^{P1}$ and $PvSLL1^T$, which casts doubt on this gene being involved in the operation of heteromorphic SI.

To say that our current knowledge, at the molecular level, of the components of the *S*-locus and operation of heteromorphic SI in *Primula* is incomplete is a considerable understatement. However, the studies outlined above provide useful data concerning which approaches may be most effective. Importantly, they have generated molecular markers for the *S-locus*. In the homomorphic SI systems characterised to date, combining physical mapping and large insert DNA libraries has proven to be highly effective for identifying genes involved in the operation of these systems (Suzuki et al. 1999; Schopfer et al. 1999; Sijacic et al. 2004; see also Chap. 6, this volume). Similar efforts are underway in *Primula*. Although not yet published, we are aware of the construction of at least two bacterial artificial chromosome (BAC) libraries of lines of *P. vulgaris*, and the successful identification of *S* linked BAC clones from these libraries (cited in Li et al. 2007, McCubbin unpublished results).

14.8 Conclusions and Future Prospects

As is clear throughout this article (and see also Chap. 1), though we have considerable knowledge of the physical aspects of heteromorphic SI, our understanding of the specific roles of the various traits of which these systems are comprised and the genes that encode them is meagre. Elucidation of these factors promises to provide considerable insight into, not just another SI system, but several fundamental aspects of plant biology. The genetic linkage of the genes that regulate the various components of heteromorphic SI provides an excellent starting point from which to study the molecular mechanisms behind subtle aspects of floral biology, such as organ positioning, pollen size and pollen number. These are critical to plant/pollinator interactions and not readily tractable in other model systems. In addition, the genetics of heteromorphic SI affords an excellent opportunity to study the regulation of allelic dominance and reversible gene silencing in plant development in relation to several different floral characteristics.

Although recent progress in the molecular characterisation of heteromorphic SI has not yet provided answers to any of the key questions, it has generated useful information and invaluable tools that promise to lead to dramatic advances in this field in the near future. Recent improvements in the speed and cost of large scale DNA sequencing will greatly expedite this progress. One of the most significant hurdles to overcome to take full advantage of this data will be the development of a transformation system for *Primula*. Though recombinant homostyles may provide some basis for ascribing gene function, confirmation of function in plant transformation experiments remains the acid test, and is an important goal.

Acknowledgements The author gratefully acknowledges funding from the National Science Foundation (USA) through grant No. IOB-0543922.

References

Athanasiou A, Shore JS (1997) Morph-specific proteins in pollen and styles of distylous *Turnera* (Turneraceae). Genetics 146:669–679

Barrett SCH (1988) The evolution, maintenance, and loss of self-incompatibility systems. In: Lovett Doust J, Lovett Doust L (eds) Plant reproductive ecology: Patterns and strategies. Oxford University Press, New York, pp 98–124

Barrett SCH, Cruzan MB (1994) Incompatibility in heterostylous plants. In: Williams EG (ed) Genetic control of self-incompatibility and reproductive development in flowering plants. Kluwer, Dordecht, pp 189–229

Barrett SCH, Jesson LK, Baker AM (2000) The evolution and function of stylar polymorphisms in flowering plants. Ann Bot 85:253–265

Bateson W, Gregory RP (1905) On the inheritance of heterostylism in *Primula*. Proc R Soc Lond B 76:581–586

Bernacchi D, Tanksley SD (1997) An interspecific backcross of *Lycopersicon esculentum* x *L. hirsutum*: Linkage analysis and a QTL study of sexual compatibility factors and floral traits. Genetics 147:861–877

Bredemeijer GMM, Blass J (1981) S-specific proteins in styles of self-incompatible *Nicotiana alata*. Theor Appl Genet 57:429–434

Carraro L, Lombardo G, Gerola P (1996) Style peroxidase and heteromorphic incompatibility reactions in *Primula acaulis* Hil ("thrum morph") Caryologica 49:101–112

Darlington CD (1929) The significance of chromosome behaviour in polyploids for the theory of meiosis. In: Conference on polyploidy. John Innes Horticultural Institute, Merton, p 42

Darlington CD, Mather K (1949) The elements of genetics. Allen and Unwin, London

Darwin C (1862) On the two forms or dimorphic conditions in the species of *Primula* and on their remarkable sexual relations. Proc Linn Soc Bot 6:77–96

Darwin C (1877) The different forms of flowers on plants of the same species. John Murray, London

de Nettancourt D (2001) Incompatibility in angiosperms, 2nd edn. Springer, Berlin Heidelberg New York

Dowrick VPJ (1956) Heterosyly and homostyly in *Primula obonica*. Heredity 10:219–236

Dulberger R (1974) Structural dimorphism of stigmatic papillae in distylous *Linum* species. Am J Bot 61:238–243

Dulberger R (1975) S-gene action and the significance of characters in the heterostylous syndrome. Heredity 35:407–415

Entani T, Iwano M, Shiba H, Takayama S, Fukui K, Isogai A (1999) Centromeric localization of an *S-RNase* gene in *Petunia hybrida* Vilm. Theor Appl Genet 99:391–397

Erbar CS (1991) Sympetaly-a systematic character? Botanishce Jahrbuecher fuer Systematik Pflanzengesiche and Pflanzengeographie 112:417–451

Ernst A (1931) Weitere studuen über die verevbung der calycanthemie bei *Primula*. Arch Juilius Klaus Stift Verebungsforsch 6:277–375

Ernst A (1955) Self-fertility in monomorphic *Primulas*. Genetica 27:391–448

Fisher RA, Mather K (1943) Inheritance of style length in *Lythrum salicaria*. Ann Eugen 12:1–12

Franklin-Tong VE, Lawrence MJ, Franklin FCH (1988) An in vitro bioassay for the stigmatic product of the self-incompatibility gene in *Papaver rhoeas* L. New Phytol 110:109–118

Ganders FR (1974) Disassortive pollination in the disytlous plant *Jepsonia heterandra*. Can J Bot 52:2401–2406

Ganders FR (1979) The biology of heterostyly. N Z J Bot 17:607–635

Gibbs PE (1986) Do homomorphic and heteromorphic self- incompatibility systems have the same sporophytic mechanism? Plant Syst Evol 154:285–323

Golynskaya EL, Bashrikova NV Tonchuk NN (1976) Phytohaemagglutinins of the pistil of *Primula* as possible proteins of generative incompatibility. Sov Plant Pysiol 23:169–176

Haldane (1933) Two new allelomorphs for heterostylism in *Primula*. Am Nat 67:559–560

Heslop-Harrison Y, Heslop-Harrison J, Shivanna KR (1981) Heterostyly in *Primula*. 1. Fine structural and cytochemical features of the stigma and style of *Primula vulgaris* Huds. Protoplasma 107:171–187

Hildebrand F (1863) De la variation des animeaux et des plantes a l'etat domestique. C Reinwald, Paris

Jackson JF, Linskens HF (1990) Bioassays for incompatibility. Sex Plant Reprod 3:207–212

Jahnen W, Lush WM, Clarke AE (1989) Inhibition of in vitro pollen tube growth by isolated S-glycoproteins of *Nicotiana alata*. Plant Cell 1:501–510

Jotaro A, Nagano M, Woo SH, Campbell C (1999) Development of SCAR markers linked to the *Sh* gene in buckwheat. Fagopyrum 16:19–22

Kakizaki T, Takada Y, Ito A, Suzuki G, Shiba H, Takayama S, Isogai A, Watanabe M (2003) Linear dominance relationship among four class-II *S*-haplotypes in pollen is determined by the expression of SP11 in *Brassica* self-incompatibility. Plant Cell Physiol 44:70–75

Kinoshita Y, Saze H, Kinoshita T, Miura A, Soppe WJ, Koornneef M, Kakutani T (2007) Control of *FWA* gene silencing in *Arabidopsis thaliana* by *SINE*-related direct repeats. Plant J 49:38–45

Kurian V, Richards AJ (1997) A new recombinant in the '*S*' supergene in *Primula*. Heredity 78:383–390

Lewis D (1942) The physiology of incompatibility in plants 1. The effect of temperature. Proc R Soc Lond B 131:13–26

Lewis D (1949) Incompatibility in flowering plants. Biol Res 24:472–496

Lewis D, Jones DA (1993) The genetics of heterostyly in evolution and function of heterostyly. Springer, Berlin Heidelberg New York, pp 129–150

Li J, Webster M, Furuya M, Gilmartin PM (2007) Identification and characterization of pin and thrum alleles of two genes that co-segregate with the *Primula S*-locus. Plant J 51:18–31

Manfield IW, Pavlov VK, Li J, Cook HE, Hummel F, Gilmartin PM (2005) Molecular characterization of DNA sequences from the *Primula vulgaris S-locus*. J Exp Bot 56:1177–1188

Mather K (1950) The genetical architecture of heterostyly in *Primula sinensis*. Evolution 4:340–352

Mather F, De Winton D (1941) Adaptation and counter-adaptation of the breeding system in *Primula*. Ann Bot 11:297–311

Matsui K, Nishio T, Tetsuka T (2004) Genes outside the *S* supergene suppress *S* functions in buckwheat (*Fagopyrum ecsculentum*) Ann Bot 94:805–809

McCubbin AG, Lee C, Hetrick A (2006) Identification of genes showing differential expression between morphs in developing flowers of *Primula vulgaris*. Sex Plant Reprod 19:63–72

Ornduff R (1979) Pollen flow in a population of *Primula vulgaris* Huds. Bot J Linn Soc 78:1–10

Pellow C (1928) Annual Report of John Innes Horticulture Institute. Innes Institute, Norwich, England, p 13

Piper JG, Charlesworth B, Charlesworth D (1986) Breeding system evolution in *Primula vulgaris* and the role of reproductive assurance. Heredity 56:207–217

Richards AJ (1986) Plant breeding systems, 1st edn. Allen and Unwin, London

Richards AJ (1997) Plant breeding systems, 2nd edn. Chapman and Hall, London

Schopfer CR, Nasrallah ME, Nasrallah JB (1999) The male determinant of self-incompatibility in *Brassica*. Science 286:1697–1700

Sharma N, Bajaj M, Shivanna KR (1985) Over-coming self-incompatibility through the use of lectins and sugars in *Petunia* and *Eruca*. Ann Bot 55:139–141

Shiba H, Iwano M, Entani T, Ishimoto K, Shimosato H, Che FS, Satta Y, Ito A, Takada Y, Watanabe M, Isogai A, Takayama S (2002) The dominance of alleles controlling self-incompatibility in *Brassica* pollen is regulated at the RNA level. Plant Cell 14:491–504

Shiba H, Kakizaki T, Iwano M, Tarutani Y, Watanabe M, Isogai A, Takayama S (2007) Dominance relationships between self-incompatibility alleles controlled by DNA methylation. Nat Genet 38:297–299

Shiba H, Takayama S (2007) RNA silencing systems and their relevance to allele-specific DNA methylation in plants. Biosci Biotechnol Biochem 71:2632–2646

Shivanna KR, Heslop-Harrison J, Heslop-Harrison Y (1981) Heterostyly in *Primula* 2. Sites of pollen inhibition, and effects of pistil constituents on compatible and incompatible pollen tube growth. Protoplasma 107:319–337

Shivanna KR, Heslop-Harrison J, Heslop-Harrison Y (1983) Heterostyly in *Primula* 3. Pollen water economy – a factor in the intramorph-incompatibility response. Protoplasma 117:175–184

Shore JS, Arbo MM, Fernandez A (2006) Breeding system variation, genetics and evolution in the Turneraceae. New Phytol 171:539–551

Sijacic P, Wang X, Skirpan AL, Wang Y, Dowd PE, McCubbin AG, Huang S, Kao T-H (2004) Identification of the pollen determinant of S-RNase-mediated self-incompatibility. Nature 429:302–305

Sporne KR (1974) The morphology of angiosperms. Hutchinson University Library, London

Stevens VAM, Murray BG (1982) Studies on heteromorphic self-incompatibility systems: Physiological aspects of the incompatibility system of *Primula obconica*. Theor Appl Genet 61:245–256

Suzuki G, Kai N, Hirose T, Fukui K, Nishio T, Takayama S, Isogai A, Watanabe M, Hinata K (1999) Genomic organization of the *S*-locus: Identification and characterization of genes in *SLG/SRK* region of S^9 haplotype of *Brassica campestris* (syn. *rapa*). Genetics 153:391–400

Takayama S, Shiba H, Iwano M, Shimosato H, Che F-S, Kai N, Watanabe M, Suzuki G, Hinata K, Isogai A (2000) The pollen determinant of self-incompatibility in *Brassica campestris*. Proc Natl Acad Sci USA 97:1920–1925

Vuilleumier BS (1967) The origin and evolutionary development of heterostyly in the Angiosperms. Evolution 21:210–226

Webster MA (2005) Floral morphogenesis in *Primula*: Inheritance of mutant phenotypes, heteromorphy and linkage analysis. Dissertation, University of Leeds, UK

Webster MA, Gilmartin PM (2003) A comparison of early floral ontogeny in wild-type and floral homeotic mutant phenotypes of *Primula*. Planta 216:903–917

Webster MA, Gilmartin PM (2006) Analysis of late stage flower development in *Primula vulgaris* reveals novel differences in cell morphology and temporal aspects of floral heteromorphy. New Phytol 171:591–603

Wedderburn F, Richards AJ (1990) Variation in within-morph incompatibility inhibition sites in heteromorphic *Primula* L. New Phytol 116:149–162

Yeo PF (1975) Some aspects of heterostyly. New Phytol 75:147–153

Index

120K, 38, 204, 220, 222, 224, 225, 232
    uptake, 225

actin cytoskeleton, 246
actin depolymerization, 246, 250
actin-binding proteins (ABPs), 246
AFLP, 23, 264, 265, 281
Aliciella, 6
allelic dominance, 267, 301, 302, 305
Amsinckia, 7, 10
ancestral, 7–9, 46, 55, 65, 73, 77, 78, 81–83, 85–87, 91, 93, 105, 108–111, 114, 117, 127, 260, 301
Anchusa, 14
ANITA grade, 78, 79, 83, 94
anther-height dimorphism, 21
anther-specific genes, 268, 269
Antirrhinum, 115, 116, 195, 198, 200, 201, 221, 226, 229
Antisense, 157, 182, 183, 196, 223–225, 232, 242, 246
apoptosis, 248, 250
Arabidopsis, 24, 93, 127, 137, 176, 178, 179, 184, 201, 226, 242, 265, 304
    A. lyrata, 36, 56, 126, 128, 129, 131, 133, 136–142, 164, 180
    A. thaliana, 55, 115, 125, 180
ARC1, 152, 162, 181, 182, 184
Armeria, 14
assortative mating, 12
Asteraceae, 36, 38, 46, 55, 82, 87, 88, 90, 91, 110, 154, 260
autogamous self-fertilization, 44
Averrhoa carambola, 22

bacterial artificial chromosome (BAC) libraries, 305

Balancing selection, 24, 62, 104, 111, 130, 221
Basal angiosperms, 74, 76, 78, 79, 81, 83, 85, 87, 88, 93, 94
biased morph ratios, 12
BnExo70A1, 184
Boraginaceae, 25
bottleneck, 13, 106, 108, 112, 136
Brassica, 38, 77, 93, 126–128, 131, 133, 137, 140, 153, 155–163, 174–176, 178–180, 182, 184–185, 239, 254, 262, 263, 267, 268, 270, 281, 297, 302
Brassicaceae, 36, 38, 40, 55, 82, 88, 105, 112, 114, 124, 125, 127, 135, 153, 165, 174, 179, 260
breakdown of heterostyly, 25
breeding system, 34, 39, 55, 79, 94, 105, 108, 126, 152, 153, 290, 293, 298

Ca^{2+} imaging, 244, 245
Ca^{2+} influx, 244, 247, 248, 251
Ca^{2+}-based GSI, 86, 87
Ca^{2+}-signalling, 93, 165, 244–246, 271
Capsella, 93, 127–129, 133, 137, 142, 154, 180
caspase-like activities, 237, 248, 249, 253
Caspases, 248
chimera, 134, 219, 220
Chloranthaceae, 79
chromosome walking, 24, 133, 159
clonal propagation, 12, 57, 67
clonality, 54, 57, 62–67
    SI-clonality paradox, 66
CO_2 treatment, 153, 262
co-evolve, 130, 134, 142, 174
co-suppression, 157
Commelinaceae, 81
Compatible interaction (Brassica), 175

Convolvulaceae, 88, 261
corolla mouth dimorphism, 292
cross-incompatibility, 261
cryptic SI, 40
Cytosolic free Ca^{2+} ($[Ca^{2+}]_i$), 178, 244–246, 254

Darwin, 4, 10, 11, 14, 34, 76, 124, 261, 292–294, 297–298
Decodon verticillatus, 15
defensin-like protein, 127, 134, 270
DEVDase, 248–251
diallelic incompatibility, 9–11, 14, 21, 25, 26
differential display, 23, 24, 151, 152, 160, 200, 304
Differential screening, 281, 284
dimorphism, 6, 8–9, 11, 15–16, 21, 23, 292, 293
distyly, 4–9, 14–16, 18–21, 76
disassortative mating, 3, 11
Diversification, 110
diversification, 65, 67, 74, 77, 83, 85–87, 91, 93, 94, 103, 106, 108, 110–112, 117, 125, 130, 135
diversifying selection, 131, 132
DNA fragmentation, 248, 250, 258
DNA methylation, 134, 164, 262, 302
Dodecatheon, 9
Dollo's law, 109
dominance, 14, 16–18, 35, 36, 75, 90, 130, 133, 136, 140, 154, 160, 163, 164, 262, 263, 267, 276, 294, 298–301, 302, 305

EGF-like domain, 131, 161
Eichhornia, 10, 15, 18
enantiostyly, 6
endomembrane, 38, 185, 204, 220, 230–233
evolution of SI, 46, 53, 54, 57, 74, 78, 81, 82, 91–94, 164
extinction rate, 110

F_1-hybrid, 152, 153
F-actin organization, 246, 251
F-box, 93, 103, 115–117, 199–205, 210, 219, 220–222, 226, 228, 229, 233
Fagopyrum, 7, 16, 22, 23, 25
floral polymorphism, 10, 290, 298
forward genetics, 142, 281
frequency-dependent selection, 12, 53, 60, 104, 105, 108, 111, 117, 130

G gene, 90
gametophytic SI (GSI), 35–38, 43, 55, 57, 67, 75–77, 80, 86–90, 92–94, 105, 109, 194, 195, 201, 210, 239, 262, 268, 270, 276, 280
γ-thionin protein family, 270
genetic drift, 13, 42, 58, 59, 130
genotype by environment interactions, 41, 42
glycosyltransferase family, 265
grasses, 75, 76, 80, 88, 239, 275–284
gymnosperm, 77

herkogamy, 8–10, 20–21
heteromorphic SI, 4, 12, 13, 19, 40–76, 83, 89, 289–305
heterostyly, 4–11, 14, 19–20, 22, 25, 26, 80, 86, 261, 290
homomorphic SI, 7, 12, 14, 21, 24, 35, 67, 76, 295, 297, 298, 303, 305
homostyles, 10, 16, 18–19, 23, 24, 298, 299, 303, 304
Houstonia, 7
HT-B, 38, 203, 204, 208, 222–225, 230, 231, 233
    degradation, 223, 232
    gene family, 223
    structure, 223
Hugonia, 6
Hypericum, 7, 14
Hypersensitive Response (HR), 247, 254
hypervariable region, 123, 126, 132, 198, 219, 220

in vitro bioassay, 158, 241, 242, 296
inbreeding depression, 5, 35, 42, 46, 76, 79, 80, 84, 124, 135, 136, 261, 276
inhibitor model, 206–208, 227, 228
inversostyly, 6
Iochrominae, 106
Ipomoea, 261

joint gametophytic/sporophytic, 76, 85, 90

Late-acting ovarian SI, 11, 73, 76–82, 84–86, 89, 90, 92–95
lectin-like domains, 131, 132, 280
LEVDase, 238, 249
ligand-independent SRK self-association, 131
Liliaceae, 80
Limonium, 14
Linum, 7
Lithodora, 9
long homostyle, 10, 16, 19, 23, 24

Lycium, 108
Lythraceae, 7, 14
Lythrum, 15, 18, 22

Magnoliids, 79
Maloideae, 112
Malus, 112, 115, 195, 203, 221, 267
MAPK, 250
mapping, 7, 20, 21, 23, 24, 127, 131, 156, 264, 265, 282, 283, 305
mating system evolution, 86, 117, 124, 142
Menyanthaceae, 25
microtubules, 247, 251
mixed mating, 10, 35, 43, 66, 136, 140
MLPK (M-Locus Protein Kinase), 162, 173, 177, 180, 181, 183, 186
modifier genes, 38, 203, 222
  pistil modifier genes, 203, 222
modifier loci, 75, 136, 137, 141
molecular markers, 23, 24, 218, 264, 275, 305
monocotyledons, 73, 78, 80–83, 86–88, 94, 276
mutation, 15–21, 35–38, 41–45, 56, 58–61, 63, 64, 66, 67, 69, 109, 111, 116, 134–137, 139–141, 158, 178, 180, 196, 198, 206, 218, 221, 223, 243, 263, 279

Narcissus, 7, 8, 11, 12, 14
Nicotiana, 38, 77, 105, 109, 195, 196, 221–223, 225, 263
  N. alata, 107, 196, 199, 203, 218, 219, 220, 224, 225, 232, 296
  N. plumbaginifolia, 203
Nymphoides, 12, 13

Orchidaceae, 80
out-crossing, 5, 10, 43, 46, 74, 78, 91, 105, 110, 111, 135, 136, 138, 261
overdominance, 35
Oxalidaceae, 7, 14, 25
Oxalis, 15, 18

P1-derived artificial chromosome, 152, 159
PAN_APPLE domain, 126, 131, 132, 161
Papaver, 76, 86, 88, 93, 158, 239, 241–254, 262, 280, 296
Papaver Self-Incompatibility-Like PSIL genes, 242
Papaveraceae, 82, 87, 93, 105, 155, 239, 268, 270–271
PARP-cleavage, 248
parthenocarpy, 79
pathogen defence, 76, 77
PCPs, 158, 159, 176, 177, 270

Petunia, 200–202, 204, 205, 208
phosphorylation, 93, 157, 161, 162, 173, 177, 181, 183, 186, 237, 245, 246, 247, 249, 252, 280, 281
phylogenetic analysis of SI, 57, 81, 86, 93, 94, 106, 107, 110, 112, 124, 197
Physalis, 65, 106
physical size of the S-locus, 159, 259
pistil S, 105, 114, 160, 174, 193–196, 203, 221, 224, 227, 238, 241–244, 251, 252, 297
PLANT U-BOX 8, 140
Plasticity in Self-Fertility, 40
Plumbaginaceae, 25
Poaceae, 36, 38, 75, 80, 82, 83, 87, 91, 92, 276, 277
Polemoniaceae, 92
pollen capture, 175, 179
pollen S-, 114–117, 135, 136, 158–160, 163, 164, 174, 180, 194–196, 198–208, 210, 218, 221, 227–229, 231, 232, 238, 243–245, 251, 252, 262, 269, 270, 297, 300, 302
pollen coat proteins (PCPs), 152, 158, 176, 177
pollen discounting, 35, 42
$PrpS_1$, 243
pollen germination, 22, 39, 79, 154, 166, 178, 184, 186, 239, 261, 262, 295
  hydration, 159, 173, 176–179, 184, 189
pollen limitation, 12, 33, 43–46, 66
pollen transfer, 5–7, 10–12, 21, 66, 294
pollen-dominant alleles, 133
pollen size, 10, 15–16, 18–20, 291, 294, 298, 299, 300, 305
polyphyletic origin, 77, 78, 93, 94, 290
Pontederia, 15
Pontederiaceae, 7, 9
positional cloning of the S-locus, 20, 21, 23, 24, 156, 180, 264, 282–284
PPi levels, 246
Pr-p26.1, 245
PrABP80, 247
Primulaceae, 82, 92
Primula, 6, 7, 9, 13–17, 19, 22–25, 291–305
Programmed Cell Death (PCD), 247–254
Prunus, 36, 114–116, 195, 200–202, 206, 221, 227, 229, 231, 267, 268
Pseudo-compatibility, 279
pseudo-self-fertility, 39
Pseudogenes, 138
PvSLL1/2, 4, 290, 304
Pyrus, 203

random amplification of polymorphic DNA-PCR, 303

Raphanus sativus, 127, 164
  ARC1, 152
recombination, 9, 15–18, 23, 24, 42, 56, 75, 90, 114, 116, 128–130, 139–141, 200, 265, 266, 282, 283, 298–301
  breakpoint, 265, 266
  suppression, 24, 75, 266
reproductive assurance, 10, 34, 35, 44, 46, 66, 135, 136, 140
RFLP, 264, 282, 283, 304
ribonuclease (non-S)
  Aspergillus oryzae, 196
Rubiaceae, 7, 10, 25

S protein Binding Protein (SBP), 243
S protein homologues (*SPHs*), 242
S Receptor Kinase see SRK
S-alleles, 15, 19, 23, 33, 35–37, 41–46, 53, 54, 56, 69, 90, 105–109, 111–114, 158, 196, 198, 199, 206, 243, 255, 295, 302
  genealogies, 12, 54, 62, 63, 67, 106, 114–115, 219
  most recent common ancestor, 62–64
S-haplotype, 56, 90, 126–130, 133–136, 139–141, 157, 158, 160–164, 174, 194, 195–200, 203, 206–209, 220, 227, 229, 239, 240, 243, 251, 260, 262, 271, 295
  linear hierarchy, 262
S-linked load, 42, 44–46
S-locus, 14–17, 19–25, 35–38, 41–45, 54, 56, 60, 62, 65–69, 75, 88, 90, 92, 104–109, 111, 115, 125, 126–130, 132, 133, 136–140, 153–157, 159, 160, 165, 174, 176, 179, 194–196, 200, 202–203, 205, 210, 218, 219–222, 229, 239, 241–243, 260, 262–269, 271, 281–284, 297–305
  mutations, 35, 36
S-RNase, 36, 38, 41, 57, 77, 86, 87, 91–93, 104, 105, 112, 114–117, 165, 195–210, 218–233, 239, 253, 254, 260, 263, 268, 270, 271, 301
  binding proteins, 224
  compartmentalization, 230–232
  cytotoxicity, 198, 208, 227
  degradation model, 208, 209, 228–231
  evolution, 112, 114, 220
  inhibitor model, 206
  SCF^{SLF} complex, 205
  uptake, 204, 220, 232
S-Z system, 275, 277, 278
Salvia, 6, 9
SBP, 243
SBP1, 204, 205, 226, 229, 230
SCF complex, 204–205, 208–209, 226

SE1, SE2, SEA, 268
seed discounting, 35, 42–44
self-compatible (SC), 9, 15–17, 19, 36, 43, 82, 84, 87, 89, 111, 114, 125, 134, 137, 157, 158, 162, 199, 202, 203, 218, 261, 263, 276, 280
SI
  breakdown, 199, 202, 218
  Brassicaceae-type, 88, 123–142, 152–165, 174–186
  modifier, 136, 137, 141
  Papaveraceae-type, 86, 87, 238–254,
  partial expression, 56
  pseudo-, 55
  Solanaceae-type, 86, 103–117, 194–208, 218–231
self-sterility, 46, 73–86, 89, 92–95
shared ancestral polymorphism, 105, 108–111, 114, 117
sheltered load, 42–45
site-directed mutagenesis, 132, 243
SLF binding partners, 226
SLF/SFB, 93, 116, 200–206, 208–210, 219, 221, 226, 227, 229–230, 232–233, 270
SLG, 127–129, 131, 132, 137, 155–161, 164, 176, 263, 266, 268, 281
SLR1, 156, 176, 177
Solanaceae, 55, 65
Solanum, 56, 57, 65, 106
soluble inorganic pyrophosphatases (sPPases), 245–246, 252
Sorbus, 112
SP11/SCR, 88, 114, 115, 125, 127, 128, 130, 132–134, 137, 138, 159–165, 174, 177, 179–181, 183, 186, 260, 266, 268, 270, 271, 297, 302
speciation rate, 110
sporophytic self-incompatibility(SSI), 73, 90, 110, 125, 130, 136, 163, 302
sPPases, 245, 246, 252
eSRK, 126–128, 131–134, 161
SRK, 55, 56, 88, 93, 105, 114, 125–141, 156–164, 174, 177, 179–184, 266, 268, 271
SRK-like gene, 138, 271
stigma height dimorphism, 6, 8–9, 11
stigmatic SI, 79, 83, 85, 88, 93, 94, 180, 239, 240
structural heteromorphism, 128–130
Supergene, 15, 17, 18, 298, 301
suppression of recombination, 130, 266, 283, 301
sweet potato, 261

Index

T-locus, 279
Tapetum, 133, 159, 164
Thioredoxin *h*, 162, 177, 181, 182, 225, 281, 282, 284
thrum, 76, 89, 291–304
transcriptional silencing, 302
transient SI, 136, 140
transition (mating type), 7, 9–10, 25, 43, 46, 54–57, 91, 105, 109–111, 140, 165
transmission advantage, 34
tristylous populations, 4
tristyly, 5, 6, 8, 14, 17, 18, 22, 89

Turnera, 7, 10, 16, 19, 22, 24
Tylosema, 6
ubiquitination, 93, 173, 177, 183, 184, 204–206, 208

vacuolar network, 177, 185, 186, 220, 225, 232
VEIDase, 238, 249

Witheringia, 106

ZmPK1, 156

Printing: Krips bv, Meppel, The Netherlands
Binding: Stürtz, Würzburg, Germany